Discovering the Chesapeake

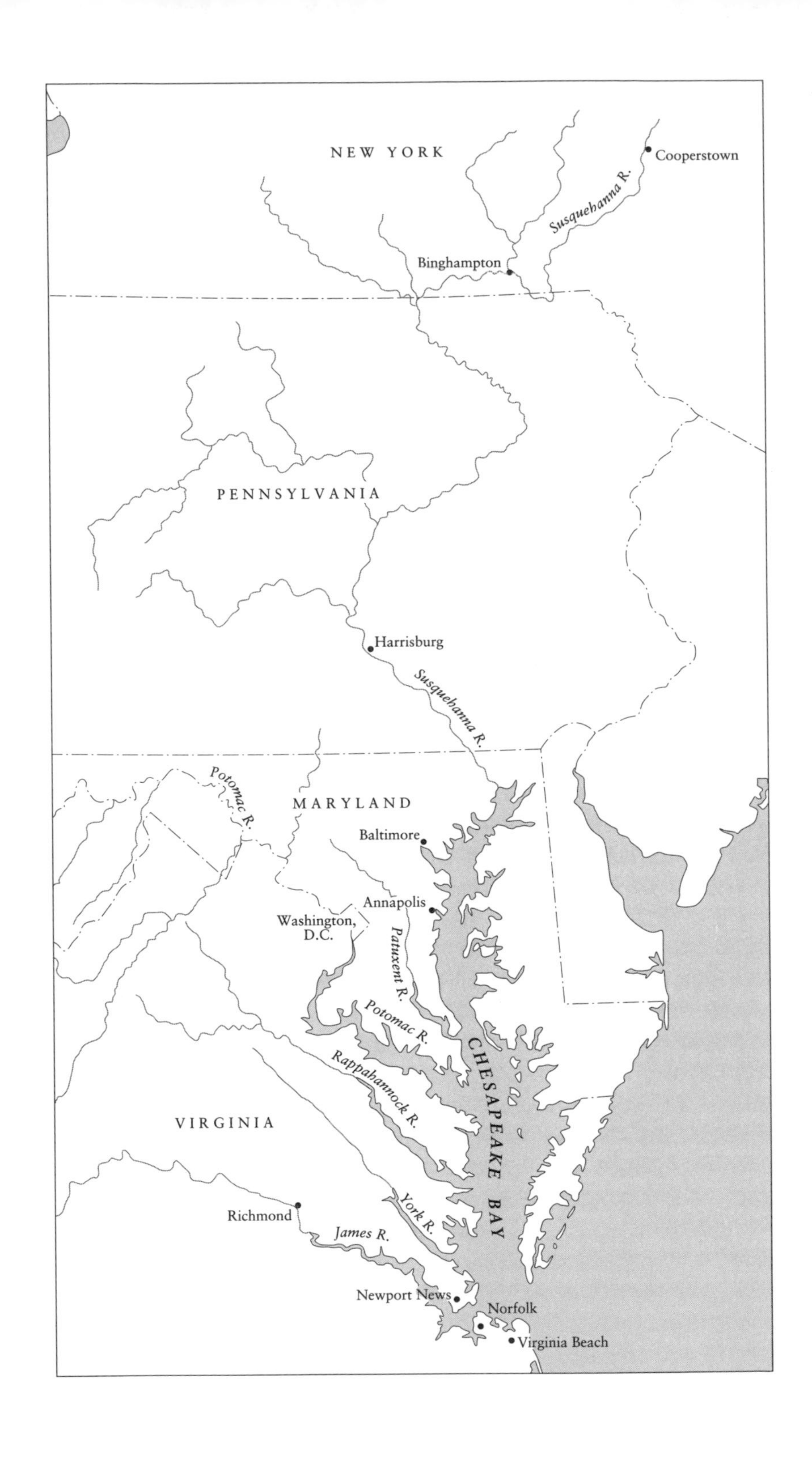

NEW YORK
Cooperstown
Susquehanna R.
Binghampton
PENNSYLVANIA
Harrisburg
Susquehanna R.
Potomac R.
MARYLAND
Baltimore
Annapolis
Washington, D.C.
Patuxent R.
Potomac R.
Rappahannock R.
CHESAPEAKE BAY
VIRGINIA
York R.
Richmond
James R.
Newport News
Norfolk
Virginia Beach

Discovering the Chesapeake

The History of an Ecosystem

Edited by

PHILIP D. CURTIN, GRACE S. BRUSH,

and GEORGE W. FISHER

THE JOHNS HOPKINS UNIVERSITY PRESS
Baltimore & London

This book has been brought to publication with the generous assistance of the Morton K. Blaustein Department of Earth and Planetary Sciences, The Johns Hopkins University.

Printed in the United States of America on acid-free paper
9 8 7 6 5 4 3 2 1

The Johns Hopkins University Press
2715 North Charles Street
Baltimore, Maryland 21218-4363
www.press.jhu.edu

Library of Congress Cataloging-in-Publication Data will be found at the end of this book.
A catalog record for this book is available from the British Library.

ISBN 0-8018-6468-2 (pbk.)

Frontispiece: William L. Nelson

To the memory of

James F. Lynch and Donald W. Pritchard

Contents

Acknowledgments *xi*
List of Contributors *xiii*
Introduction *xv*
PHILIP D. CURTIN

ONE
The Chesapeake Ecosystem
Its Geologic Heritage 1
GEORGE W. FISHER AND JERRY R. SCHUBEL

TWO
Climate and Climate History in the Chesapeake Bay Region 15
JOHN E. KUTZBACH AND THOMPSON WEBB III

THREE
Forests before and after the Colonial Encounter 40
GRACE S. BRUSH

FOUR
Human Influences on the Physical Characteristics of the Chesapeake Bay 60
DONALD W. PRITCHARD AND JERRY R. SCHUBEL

FIVE
A Long-Term History of Terrestrial Birds and Mammals in the Chesapeake-Susquehanna Watershed 83
DAVID W. STEADMAN

SIX

Living along the "Great Shellfish Bay"
The Relationship between Prehistoric Peoples and the Chesapeake 109
HENRY M. MILLER

SEVEN

Human Biology of Populations in the Chesapeake Watershed 127
DOUGLAS H. UBELAKER AND PHILIP D. CURTIN

EIGHT

A Useful Arcadia
European Colonists as Biotic Factors in Chesapeake Forests 149
TIMOTHY SILVER

NINE

Reconstructing the Colonial Environment of the Upper Chesapeake Watershed 167
ROBERT D. MITCHELL, WARREN R. HOFSTRA, AND EDWARD F. CONNOR

TEN

Human Influences on Aquatic Resources in the Chesapeake Bay Watershed 191
VICTOR S. KENNEDY AND KENT MOUNTFORD

ELEVEN

Land Use, Settlement Patterns, and the Impact of European Agriculture, 1620–1820 220
LORENA S. WALSH

TWELVE

Chesapeake Gardens and Botanical Frontiers 249
ANNE E. YENTSCH AND JAMES L. REVEAL

THIRTEEN
Genteel Erosion
The Ecological Consequences of Agrarian Reform in the Chesapeake, 1730–1840 279
CARVILLE EARLE AND RONALD HOFFMAN

FOURTEEN
Farming, Disease, and Change in the Chesapeake Ecosystem 304
G. TERRY SHARRER

FIFTEEN
Bird Populations of the Chesapeake Bay Region
350 Years of Change 322
JAMES F. LYNCH

COMMENTARY
Reading the Palimpsest 355
WILLIAM CRONON

Index 375

Acknowledgments

The editors wish first of all to acknowledge the National Science Foundation for its generous support of the conferences that led to the organization of this volume. We are grateful to Anne K. Schwartz and Anne G. Curtin for their editorial assistance at several stages in the preparation of the manuscript. We would also like to express our gratitude to Edward C. Papenfuse, environmentalist and Archivist of the State of Maryland, for his advice and assistance at many points along the way.

The editors would like to acknowledge the role of Cynthia H. Requardt, the Kurrelmeyer Curator of Special Collections at the Eisenhower Library, Johns Hopkins University. Requardt was director of a parallel project, a library exhibit and lecture series on colonial encounters in the Chesapeake, which involved many of the contributors to this volume. Participation in that project first called our attention to the importance of looking at the Chesapeake ecosystem from an interdisciplinary and historical point of view.

More than anything, we are grateful for the pains that the contributors have taken to step away from their professional roles in specific disciplines, in which they typically talk only to other specialists, to tell each other and the broader public what we can learn from them about the Chesapeake watershed.

Contributors

GRACE S. BRUSH is a professor in the Department of Geography and Environmental Engineering, the Johns Hopkins University, Baltimore, Maryland.

EDWARD F. CONNOR is a professor of biology at San Francisco State University, San Francisco, California.

WILLIAM CRONON is the Frederick Jackson Turner Professor of History, Geography, and Environmental Studies at the University of Wisconsin–Madison.

PHILIP D. CURTIN is a professor emeritus in the Department of History, the Johns Hopkins University, Baltimore, Maryland.

CARVILLE EARLE is the Carl O. Sauer Professor of Geography, Department of Geography and Anthropology, Louisiana State University, Baton Rouge.

GEORGE W. FISHER is a professor in the Department of Earth and Planetary Sciences, the Johns Hopkins University, Baltimore, Maryland.

RONALD HOFFMAN is director of the Omohundro Institute of Early American History and Culture and professor of history at the College of William and Mary in Williamsburg, Virginia.

WARREN R. HOFSTRA is a professor of history at Shenandoah University, Winchester, Virginia.

VICTOR S. KENNEDY is a professor at the Horn Point Laboratory, University of Maryland Center for Environmental Science, Cambridge, Maryland.

JOHN E. KUTZBACH is a professor at the Institute for Environmental Studies and the Department of Atmosphere and Ocean Sciences, University of Wisconsin–Madison.

JAMES F. LYNCH was a zoologist at the Smithsonian Environmental Research Center, Edgewater, Maryland, before his death in 1998.

HENRY M. MILLER is director of research at Maryland's state museum, Historic St. Mary's City, St. Mary's City, Maryland.

ROBERT D. MITCHELL is a professor of geography, retired, in the Department of Geography, University of Maryland, College Park.

KENT MOUNTFORD is a senior scientist at the U.S. Environmental Protection Agency, Chesapeake Bay Program Office, Annapolis, Maryland.

DONALD W. PRITCHARD was a professor and associate dean at the Marine Sciences Research Center, State University of New York at Stony Brook until 1988, when he retired. He died in 1999.

JAMES L. REVEAL is a professor in the Department of Botany, University of Maryland, College Park.

JERRY R. SCHUBEL is president of the New England Aquarium, Boston, Massachusetts.

G. TERRY SHARRER is curator of health sciences at the Museum of American History, Smithsonian Institution, Washington, D.C.

TIMOTHY SILVER is a professor of history at Appalachian State University, Boone, North Carolina.

DAVID W. STEADMAN is associate curator of ornithology in the Department of Natural Sciences, Florida Museum of Natural History, University of Florida, Gainesville.

DOUGLAS H. UBELAKER is curator of physical anthropology at the National Museum of Natural History, Smithsonian Institution, Washington, D.C.

LORENA S. WALSH is a historian at the Colonial Williamsburg Foundation, Williamsburg, Virginia.

THOMPSON WEBB III is a professor in the Department of Geological Sciences, Brown University, Providence, Rhode Island.

ANNE E. YENTSCH is an associate professor in the History Department, Armstrong Atlantic State University, Savannah, Georgia.

Introduction

PHILIP D. CURTIN

In the eighteenth century, an individual like Gilbert White of Selborne studied the world around him by personal observation. With today's professional specialization, he would be thought of as a dilettante, the scientific equivalent of a little old lady in running shoes. Yet many informed people, with or without professional specialization, want to understand as much as they can of the whole environment, how it came to be as it is, in the light of historical change over the long run. This book presents a broad picture of one ecosystem for the general reader, as well as for specialists who want to understand the rapidly changing body of knowledge produced in fields other than their own.

The changing relations between biological species and other aspects of the natural world within the Chesapeake watershed are the central focus of this volume. A historian, a paleobotanist, and a geologist at the Johns Hopkins University, the editors of this volume, gathered as an informal planning committee. We began by assembling representatives of as many scientific and scholarly fields as possible, asking each to explain how his or her field contributes to the whole process of understanding ecological change—and to explain these contributions to other specialists in nonspecialized language—to produce a book that speaks to the serious lay reader. Disciplines represented by the authors include climatology, anthropology and bio-anthropology, zoology and biology, archaeology, botany, geography, oceanography and marine biology. A bevy of historians also cover a variety of periods and specialties.

With the financial support of the National Science Foundation, the group held a preliminary conference to discuss draft papers. The group met again a year later to discuss revised drafts. A court reporter took notes, and the comments were then circulated for the authors' use in preparing final drafts, which form the content of this volume. If any one word describes these meetings, it is not, perhaps, collegial so much as congenial. Most academics spend vast amounts of time in meetings and writing papers, but this common enterprise was different.

One of the most intriguing aspects of our gatherings was the explication of methods. The geologist might describe the stratigraphy of hills and valleys as evidence of movement in crustal plates of the earth, the advancing and receding glaciers, or the rising and falling ocean levels that determined the formations of the present-day watershed and Bay. The climatologist, among many other things, examines the spacing of tree rings as an indicator of wet and dry periods. The paleobotanist extracts cores of sediment from the Bay or wetlands to discover pollen grains whose identities indicate a changing climate, the arrival of exotic plants, or forest fires. Marine scientists measure salinity and temperature of the waters of the Bay and study fish populations to interpret the movement and composition of the fresh and salt water that compose the Chesapeake estuary.

Examining the history of the Chesapeake ecosystem, a zoologist might use fossil evidence from middens and other deposits to establish early bird and animal populations. A bio-anthropologist would examine teeth and bones from burial sites of indigenous people to determine their diet and health, while an archaeologist might seek physical evidence about early settlement patterns and means of procuring food. An archaeologist and a botanist might excavate early gardens, looking for seeds and nuts, discovering the foods and flowers of the European settlers. Historians comb through documents, letters, and bills of lading to decipher disease patterns, uses of and attitudes toward natural resources, and changing farming practices and their impact on the land; they even analyze descriptions of witness trees, used in early surveys to identify boundaries, to deduce the distribution of species in a given time period. All these approaches (and others) were employed by the contributors to this volume.

One of the earliest and most significant lessons of this interdisciplinary effort was both the diversity of approaches and the common ground among distinct fields of study; there was also a sense of surprise at the wealth of what others knew and at how they came to know it. We learned that divergence of interpretation has traditionally existed and continues to exist among disciplines in their view of humanity's role in the natural landscape. Historians stress temporal dimensions of the landscape in human affairs, while social scientists wish to learn how particular people respond to environmental challenges in providing for their families—and how these responses influenced settlement patterns and social structures.

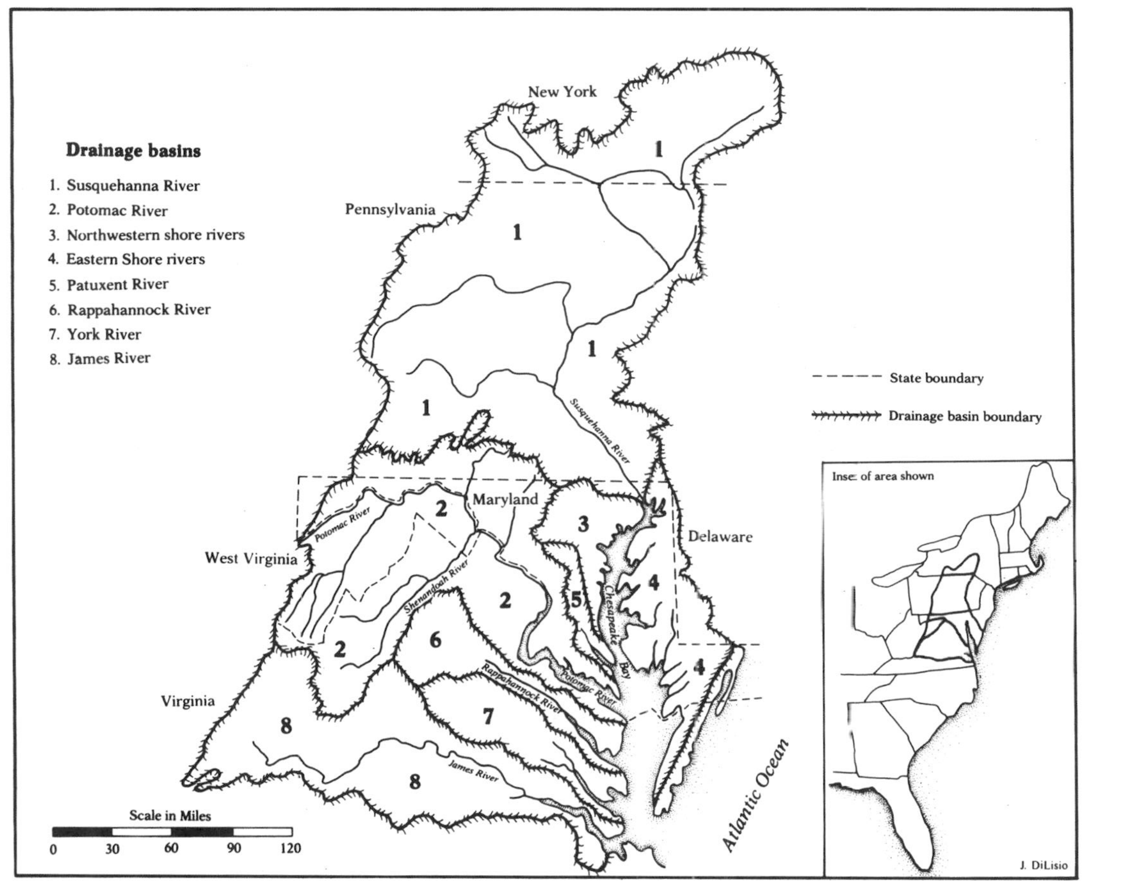

Fig. I.1. The Chesapeake Bay watershed (courtesy of James E. DiLisio).

Geographers are often concerned with spatial relationships and how society alters them over time. The scientists in our group who were concerned with the environment tended to consider processes.

This volume has been organized by both time and process. The early chapters explain natural systems of the watershed and their origins. Later chapters consider human intervention over time and in specific areas. Case studies include Indian slash-and-burn practices (for cultivation or for driving game), which increased soil erosion through deforestation; the effects of tobacco growing in what is now Tidewater Maryland; the changing farming techniques and attitudes about estate management in two generations of Maryland landowners; the introduction of pathogens, both human and plant; the role of the oyster in the aquatic system and the consequences of the oyster's depletion; the change from bottom-growing to floating vegetation in the Bay; and the response of bird and animal life to environmental factors introduced by humans.

Our knowledge goes back further for some aspects of the ecosystem than for others. The earliest phases of the geomorphology of the watershed go back to periods of glaciation, when the rivers dug a channel to the sea, whose shore was then the continental shelf. As the ice caps melted, the sea rose and invaded the channels, rising and falling over the eons of glacial and interglacial periods. Today's Chesapeake Bay is only some ten thousand years old. What a different world it was, as David Steadman comments, when the region was the home of the ground sloth, giant beaver, dire wolf, mastodon, and other megafauna. In the next few thousand years, the ice may form again and the Bay will once more be the valley of the Susquehanna, unless, of course, human-induced changes in climate create some other currently unpredictable condition.

Formed by a series of successive incursions of the sea, the Chesapeake is the largest estuary in the United States. In times of relatively high sea level, such as today, the estuary is a very large, shallow basin where fresh and salt water meet and mix. The extensive watershed of the Chesapeake includes part of six states (New York, Pennsylvania, Delaware, Maryland, Virginia, and West Virginia), which means that the Bay receives an unusual variety of upstream influence. Upon this enormous yet delicate interchange of physical materials, a vast interactive system of living creatures rests: the underwater grasses, which are food for birds and sea life; the oysters, which require a shallow, temperate environment and which filter the water as they feed; the anadromous fish, which leave the ocean each year and return to the rivers to spawn; the complex forests and marshes that surround the Bay; insect and animal life; and the humans who use the food sources and have enjoyed the clarity and productivity of the water.

The biotic network of the prehistoric watershed, other than aquatic life, consisted of the chain of plant and animal feeders—the smaller herbivores such as mice or rabbits who ingested and distributed seeds, as did the birds, which are important forest farmers, planting seeds and destroying insect pests. Deer and other ungulates browsed on forest plants and in turn fed predators, which later included the humans as well as the wolves and mountain lions.

The human population, from some ten thousand years ago until the seventeenth century, was too sparse to influence other aspects of the system in major ways, although early Indian cultures undoubtedly caused considerable local disruption of plant and animal communities. Deer populations were somewhat depleted through hunting, and forests, where Indians practiced slash-and-burn, were cleared in small, patchy areas. Over time the Indian population around the Chesapeake shifted from hunting and foraging to settlement and agriculture, which, because cultivated crops were less nutritious than the wild foods they had previously subsisted on, was supplemented with fish from the Bay. Another result of changing settlement patterns was a concentration of people, leaving the population more exposed to newly arrived European microbes, starting in the seventeenth century.

The European settlers brought new attitudes and traditions about the use of land. Their early hoe culture and their practice of abandoning fields after only a few years to move on to other ones apparently had little impact on the soil. The small European population only gradually cleared the forests, though free-ranging domesticated animals that were pastured in forests cleared patches and reduced the growth of new trees.

The settlers in one part of the watershed found the soil productive for growing tobacco, an indigenous crop the Indians had grown for their own consumption. The Europeans cultivated it far more intensively, because they had an external market in Europe. It is a highly labor intensive crop, and wealthy farmers imported African slaves as the most economical form of labor. This labor system differentiated the Tidewater population and its evolving culture from that of the watershed to the north, where small family farms rather than slave-staffed plantations predominated. Increasing deforestation and changes in plowing left the soil open to erosion. Intensive use of the soil and the cropping systems brought new pressures and pathogens to the land. As fertilizers were spread on the

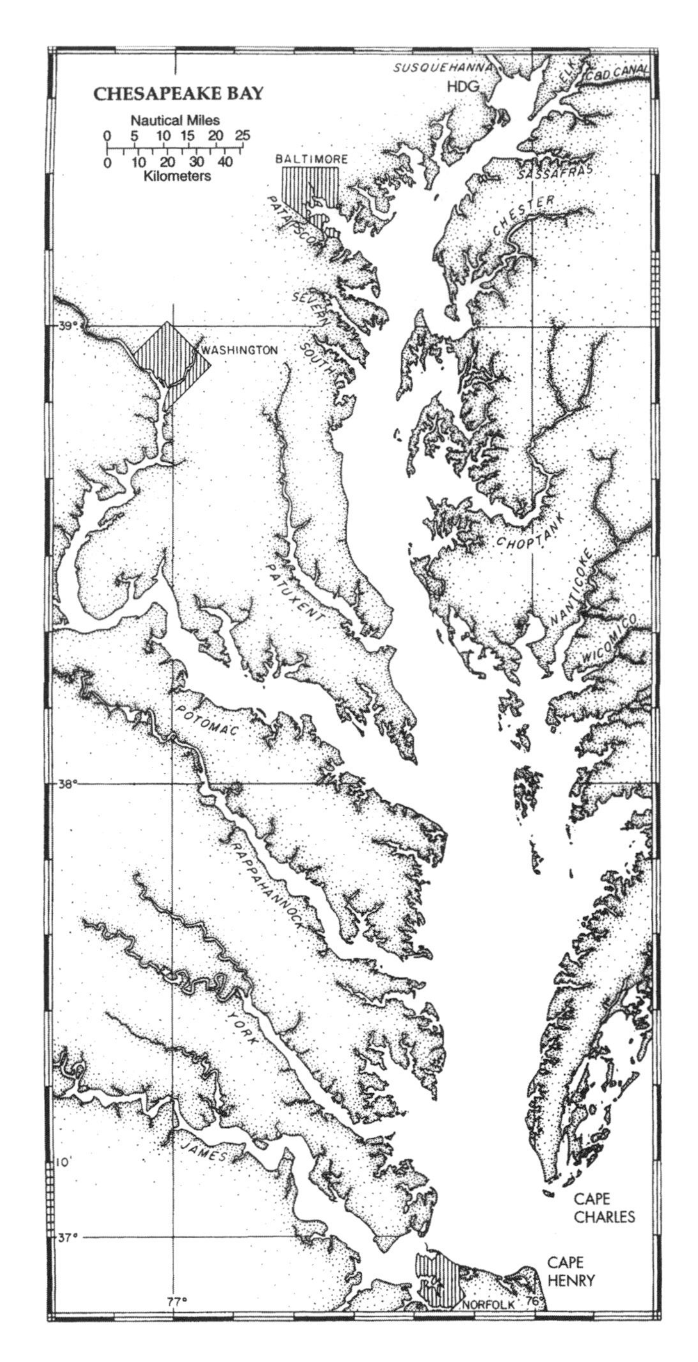

Fig. I.2. Map of the Chesapeake Bay showing the various tributary estuaries, as well as Cape Charles and Cape Henry, which mark the mouth of the Bay, and Havre de Grace (HDG), which marks the head of the Bay (Donald W. Pritchard).

land, they ended up flowing down the rivers from the upper end of the watershed and from the hills around the Bay.

Technological improvements in fishing led to the decline of fish populations in the Bay. Upstream in the watershed, dams blocked the access of anadromous fish, and the release of sawdust and chemicals into the water poisoned fish and destroyed their spawning beds. The depleted oyster population could not purify the Bay water on its own.

The Bay was once a clear-water system with plant production dominated by vast meadows of bay grasses and the microscopic plants that grew on their leaves and sediments. Driven by an increasingly "leaky" terrestrial system and nutrient loads from sewage and fertilizers, balance in the aquatic community shifted as benthic, or bottom-living, species died. In their place came the increasingly microbial food web of plant, animal, and bacterial plankton which cloud the waters today. Throughout the watershed, scores of non-native species—aquatic, terrestrial, plant, animal, and avian—were introduced, and they increased the imbalance.

Some indigenous birds and animals adapted to changing conditions, but others were locally extirpated, and a few became extinct. Some species increased through time. Deer fed in cornfields and even turned to suburban gardens. Raccoons and opossums found human refuse useful. Canada geese and other birds changed their historical range in response to the new farming practices. Beginning in the later eighteenth century, the growth of a single species—human beings—inordinately changed the conditions of life for all others. The human threat to other species in the Chesapeake ecosystem is, of course, only a local manifestation of a combination of worldwide changes in the nineteenth and twentieth centuries which marked the introduction of the industrial age.

The growing shifts in human occupations, from farming to other trades, had a telling effect on the watershed. Airborne pollutants were deposited in the Bay water, further changing its chemical composition. The human population around the Bay increased exponentially, stimulating ever-increasing rates of change. To many who had lived and worked on the estuary over a period of years, it was obvious that things were not as they had been. The water had become murky and filled with stinging jellyfish. Fish and oyster catches had plummeted, and common bird species seemed to be disappearing from people's backyards.

Several themes emerge from the chapters that follow. One is the influence of the landscape and the Bay on the people who settled here. As George Fisher and Jerry Schubel point out, it was the coast formation that brought people here in the first place; it was a good place to find a sheltered anchorage. The terrain and climate appeared to be fruitful and inviting. Beyond the coastal plain, as one moved north and west, the gravel, sand, and silt were abruptly replaced by granites and marbles, which produced a sudden "fall line," with sharp changes in elevation. People stopped and built their communities at this topographic break, as can be seen in a satellite photo or maps of population densities; Richmond, Washington, D.C., and Baltimore are all located on the fall line.

The land changed the people and, as a second theme, people changed the land. Those who came from Britain brought a land ethic that viewed the clearing of woods as necessary to "improve" the land, and the chapter on land-cover reconstruction by Robert Mitchell, Warren Hofstra, and Edward Connor deals in part with some consequences of this attitude. The settlers' concept of private property rights included the right to exploit and develop natural resources for private ends, with small regard for the wider public interest in maintaining a healthy watershed environment. Timothy Silver notes how the settlers—and the plants and animals they brought with them—changed the structure and composition of the Chesapeake's forests. The settlers created gardens, described by Anne Yentsch and James Reveal, perhaps to impose their sense of an ideal order on unruly nature, but some of the plants they brought have become today's weeds.

A third theme is the need to view the watershed from a systemwide perspective. The interdisciplinary cooperation of our approach not only is apparent in the wide range of disciplines represented but also changed the content of individual chapters as authors responded to material that had been unfamiliar to their field of specialization. The Chesapeake watershed story is a cautionary tale for those who would try to improve today's environment on the basis of a very incomplete spatial or temporal understanding of the way the ecosystem really works.

The chapters deal with many aspects of an incredibly complex natural system—many, but not all. We have consciously tried to represent parts of the ecosystem which are often underplayed in the more familiar run of environmental studies: disease, for example, and the pathogens that attack plants and animals. There are other gaps: in the disciplines, in historical periods, in geographic areas. In keeping with our watershed perspective, there is no chapter devoted to the great Susquehanna River or other major rivers of the watershed.

The book has little to say on the immense population shifts and urbanization that have occurred in the recent past. The consequences of these changes are so enormous and so generally known that we chose to neglect them in favor of less familiar aspects of change within the ecosystem and those further back in time. As early as the middle of the last century, many of the processes leading to the degradation of the Chesapeake watershed were well under way. Some chapters in this volume, when appropriate, bring that story more up-to-date.

Intense public interest in the Bay and its problems has prompted more than thirty years of environmental studies, beginning with the major Army Corps of Engineers study in 1965. A web of state and federal agencies continues to deal with watershed problems. The entire Bay is under the jurisdiction of the Environmental Protection Agency, whereas other agencies, such as the interstate commissions on the Potomac and Susquehanna basins, deal with a particular segment of the watershed. State agencies and universities have been concerned with the watershed's environmental problems, ranging from shellfish fisheries to deer hunting. Today's cooperative Chesapeake Bay Program partnership began in 1983 and includes the Environmental Protection Agency, the states of Maryland, Pennsylvania, and Virginia, and the District of Columbia, together with the Bay Commission. In addition, books written by many concerned individuals focus on the present-day Bay, its beauty, its uses, and its degradation.

The organization of environmental research and academic life is such that too few are concerned with what the diverse approach of a study such as this one can mean for broader understanding of the Chesapeake's changing ecosystem or of any other large environmental question. Amateur "naturalists" still exist, but professional "naturalists" no longer do, other than specialists in a single discipline or branch of science who try as amateurs to keep in touch with the work of specialists in other fields. As James Lynch says in his chapter on changes in bird population: "The battle to restore the Bay is nowhere near won, and we cannot be certain that all of the important problems, much less their solutions, have yet been recognized." This book is not the last word on the changing ecosystem of the watershed, but we are hopeful that it will break new ground in using an interdisciplinary approach to understand the region, not merely its present problems but historical change over a considerable interval of time.

IN ORDER TO IMPROVE READABILITY, scientific citations have been kept to a minimum. Readers seeking more information should turn to the bibliography at the end of most chapters.

Discovering the Chesapeake

CHAPTER ONE

The Chesapeake Ecosystem
Its Geologic Heritage

GEORGE W. FISHER AND JERRY R. SCHUBEL

The estuary formed, emptied, then formed again as the ocean rose, fell, then rose again with the melting and expansion of four successive ice caps. As each cap melted, rising ocean forced salt water up the valleys of the watershed to mix with the fresh water of the river, creating an immensely productive ecosystem. These processes and others operating over billions of years provided the raw materials for the farming, fishing, and manufacturing industries that shaped the culture of the Chesapeake region. The tables are now turned. Those industries and that culture now shape the ecosystem, and its life depends upon how wisely we use it.

Late on a cold December afternoon, an Eastern Shore oysterman steers his boat for home, barely a dozen bushels of oysters aboard. Discouraged, he looks over the slate gray waters of the Bay, choppy with short, steep waves, and wonders what has happened to the oysters, once so plentiful that their reefs were a hazard to navigation. In the Lancaster Valley of Pennsylvania, a young Amish farmer gazes across the fields that have supported his family for seven generations, his heart gladdened by the familiar contours of the now-barren fields, framed by a rocky ridge with bare trees stark against the winter sky. The welcome smell of wood smoke tells him that his wife has kindled a fire in the open fireplace that heats his home. A 50-year-old steelworker in the Sparrows Point plant steps into a bar a block and a half from his house, looking forward to having a beer with his friends. Some, like him, count themselves lucky to have kept their jobs at the plant, which now employs less than half as many as it did in the 1950s. Others, less lucky, hope that the contract to refit Coast Guard cutters at Bethlehem's shipyard will mean new hiring.

An elder of the Presbyterian church in a village west of Baltimore pauses to gaze at the square granite tower of her church, thinking pridefully of her great-grandfather, who, as a young stonemason fresh from Wales, had helped quarry the stone for that tower nearly 150 years ago.

Seemingly isolated events, each of these scenes is linked by a complex series of natural processes that have produced the natural resources of the Chesapeake region, processes that still shape our lives in ways of which most of us are only dimly aware. Some operate on a human time scale and their effects are easily seen. Others operate much more slowly, over the immense span of geologic time, and are much more difficult to discern.

Most chapters in this book deal with short-term processes and their links to human history. This chapter looks backward over the longer span of geologic time to unravel the geologic processes that have shaped the Chesapeake ecosystem and that still influence the system and our lives. We focus first on how the Bay and its natural resources originated, and then we discuss how we learned to use those resources and how this use has modified the ecosystem.

FROM THE PERSPECTIVE OF SPACE, we see Earth as a small blue-green planet, breathtakingly lovely but fragile and finite. Veiled by a gossamer-thin envelope of air, we see the sum total of what we have to sustain us: some good-size pools of mostly salty water, and a large ball of rock, of which we can access only the thin outer crust (a fraction of 1% of Earth's radius). In its raw form, that rocky shell is pretty useless. It is nearly half oxygen, more than 25% silicon, with smaller amounts of elements such as aluminum, iron, and magnesium and traces of others such as copper and gold. Undifferentiated, it is a typical granite. It might make an imposing tower for a church, but not much else; most of the elements on which society has come to depend are so strongly bonded to the silica in granite that the energy cost of extracting them is prohibitive. Left to itself, Earth's crust could not provide the resources needed for a steel plant, or even the soil for a farm, and Earth would be as barren as the moon.

Fortunately, though, that granite has not been left alone. A complex series of linked chemical, physical, and biological processes operating over Earth's 4.6-billion-year history has gradually refined the crust so that it is now far from uniform. In the Chesapeake region, the bedrock is mantled by a thick layer of fertile soil; minable deposits of metals like iron, copper, and chromium provide the basic raw materials for industry;

thick seams of coal provide much of the energy needed for that industry; and rain soaking into the slopes of the hinterland provides water for farms and cities and for the rich variety of brackish-water fauna in the Bay.

Some of the processes that have reworked the crust are imperceptibly slow and operate over vast reaches of geologic time. A good example is the lazy, inexorable convection of Earth's mantle, a thick layer of hot rock on which the crust floats. That convection has carried portions of the crust deep into the Earth, melted them, and subjected them to a series of complex geochemical processes that have gradually refined them into a variety of useful materials, such as oxide and sulfide ores, from which vital elements can be easily extracted.

On a much shorter time scale, streams have etched the raw crust into gently rolling hills and broad valleys, rainwater has weathered rock into fertile soil, and rhythmic variations in sea level driven by growth and shrinkage of Earth's ice caps have alternately filled and drained the Bay.

Dealing first with mantle convection and its consequences, we recognize two key kinds of convective motion. In some places, like the Red Sea, the crust is slowly being pulled apart and a nascent ocean basin is developing between two blocks of continental rock. Elsewhere, such as the western coast of South America, two slabs of crust are being rammed into one another. Where this happens, the heavier slab generally sinks into the mantle and is partially melted. Because the resulting melt is lighter than the surrounding rocks, it rises to be erupted as lava in volcanoes like those of the Andes. At the same time, compression in the overriding plate gradually thickens the crust there, producing a mountain range, again like the Andes. Over the past 600 million years, the Chesapeake region has experienced both kinds of motion, and the story of those plate motions can be read from the record preserved in the Chesapeake area today (Table 1.1).

The oldest stage that we can still discern is recorded by early Paleozoic (roughly 600-million-year-old) quartz sands and lime muds, much like those accumulating on North America's Atlantic beaches today. These sediments, now preserved as metamorphosed sandstone and marble in the horse country north of Baltimore, record deposition in a geologically quiet environment, such as that of present-day eastern North America, one well removed from tectonic plate boundaries. From the extent of these sediments and their similarity to modern Atlantic sediments, we infer that they were deposited on the margin of a 600-million-year-old ocean basin more or less on the site of the present Atlantic,

Table 1.1 Geologic history of the Chesapeake region

Years before present	Geologic period	Geologic processes	Geologic record
600 million years	Early Paleozoic	Ocean basin on site of present Atlantic gradually widens	Quartz sands and lime muds deposited
400 million years	Middle Paleozoic	Ocean basin begins to close up	Deep-sea muds, volcanic sands, and submarine slide deposits form
	Late Paleozoic	Ocean closes; North America and Africa collide	Rise of ancestral Appalachians
200 million years	Mesozoic	Breakup of North America and Africa	Fluvial sands deposited in rift valleys
	Tertiary	Atlantic Ocean basin widens	Coastal Plain sediments deposited
1 million years	Pleistocene	Chesapeake basin alternately drains and fills to form estuaries	Susquehanna valley system alternately carved and flooded
18,000 years	Recent	Present Chesapeake begins to fill	Muds and fine silt deposited in Bay

in an area undergoing crustal extension, as is the Atlantic basin today.

After roughly 200 million years of quiet sediment deposition, the character of the sediments changed dramatically. Sands rich in volcanic debris, deep-sea muds, and submarine landslide deposits, now preserved as metamorphic rocks just west of Washington, D.C., and Baltimore, show that the direction of mantle convection had reversed and the ocean basin had begun to close.

Eventually the ocean closed completely, shutting off sedimentation in the eastern Appalachians. As the ocean closed, Africa and North America collided to form a high mountain range somewhat like the modern Himalayas, where the Indian and Asian plates are colliding. As those ancestral Appalachian peaks rose, erosion cut rapidly into them, carrying coarse sand and gravel westward toward a shallow continental sea that occupied much of middle America at the time. Those sands and gravels accumulated in delta complexes at the shoreline of that sea, and extensive coastal swamps developed in backwater areas between the deltas. As the deltas shifted back and forth along the shoreline, gradually burying those swamp deposits, the weight of overlying sediment hardened the accumulated plant debris into coal—first bituminous, then anthracite. Compression continued throughout this time and eventually involved even these late sediments, wrinkling them into the spectacular fold belts that mark the western Appalachians today (see Fig. 1.1).

About 200 million years ago convection reversed once again, and the resulting extension tore North America and Africa apart, eventually forming the present Atlantic basin. The initial stages of the breakup are reflected by deposits of red continental sandstones that fill fault-bounded rift valleys dotting the Piedmont, like those at Gettysburg. Later stages are represented by both the young veneer of Coastal Plain sediments southeast of Washington and the modern Atlantic beaches. These sediments reflect a dramatic reversal of drainage patterns in the Appalachians. When the ancestral Appalachians were being eroded and contributing sediment to the continental interior, the major streams flowed west. As the present Atlantic began to open, a new set of streams draining to the east started to form. Because the crest of the range was nearer the Atlantic than the midcontinent, streams flowing east had a steeper gradient than those flowing west and thus cut more rapidly into the bedrock. As they did so, they gradually appropriated valleys cut by the earlier streams and the drainage divide shifted west, opening up the vast basins of the modern Susquehanna, Potomac, and James Rivers.

Throughout this period, erosion gradually etched the softer shales and limestones into valleys, leaving behind ridges developed on the harder rocks. Much as the weathered siding on an old barn shows the grain of the wood more clearly than a freshly sawn board, so the landscape of the Appalachians reflects many details of the underlying geologic structure (Fig. 1.1).

On the northwest, the Appalachian Plateau is characterized by dendritic drainage patterns etched into nearly flat-lying beds, with local deep gorges where major streams cut through resistant cap rocks to expose layer upon layer of sediments beneath.

Next to the southeast, the Valley and Ridge Province is developed on intensely folded rocks; consequently, the surface cuts sharply across the grain of the rock, and the resistant ridges sweep across the map in graceful arcs, which suddenly double back upon themselves as they approach crests (or troughs) of major folds. These seemingly endless ridges, so

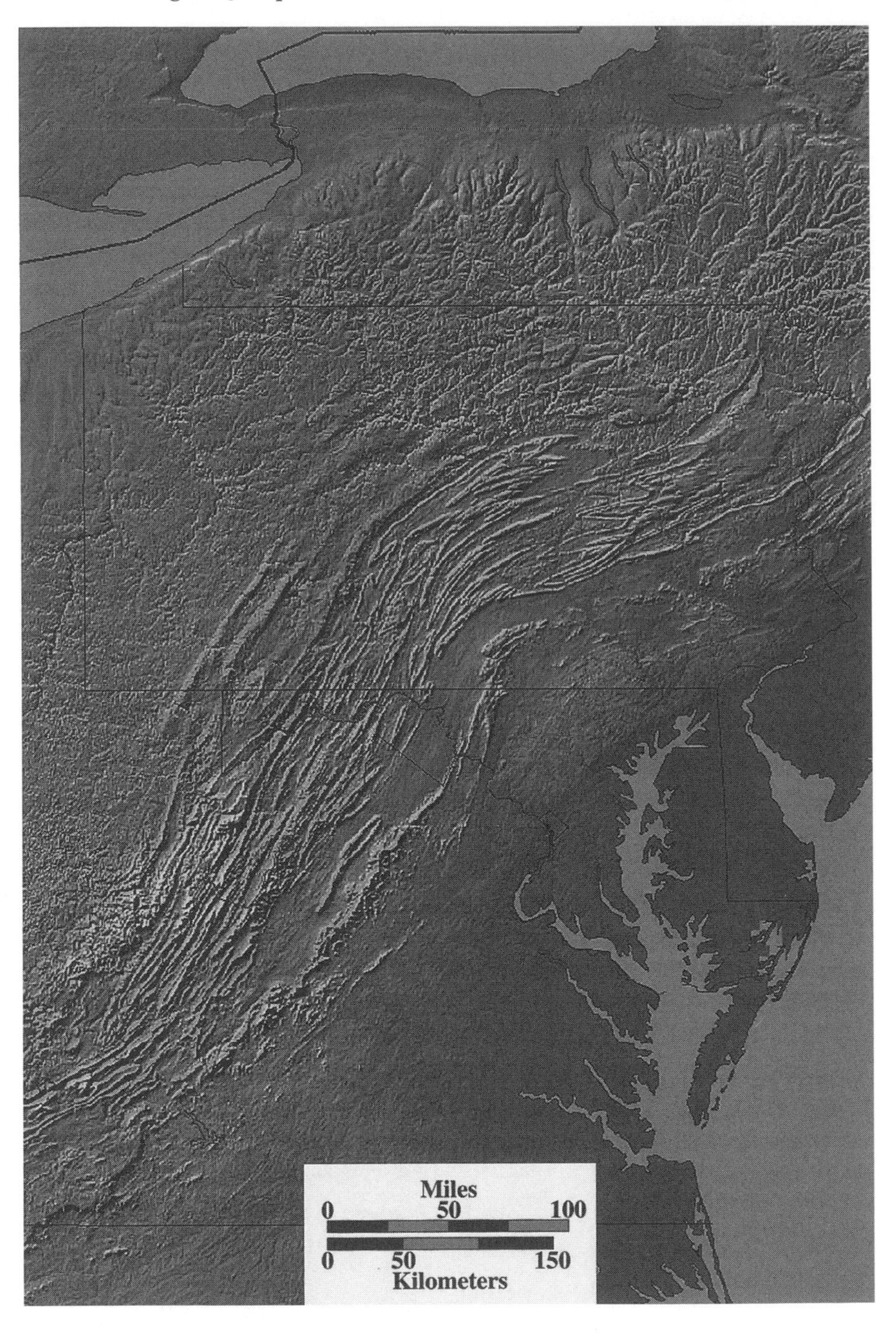
Miles
0
50
100
0
50
150
Kilometers

graceful in map view, must have been deeply frustrating to early explorers, for whom the Appalachians were a major transportation barrier. Constrained by the ridges, most streams in this province have developed a characteristic trellis-like pattern in map view. The southeasternmost part of the province is a broad valley floored by limestone formed from the lime muds accumulated on the ocean shore during the early stages of the Appalachian cycle. Variously known as the Cumberland Valley, the Shenandoah Valley, and the Lebanon Valley, it is now home to some of the richest farmland in the Chesapeake region.

Just southeast of the Valley and Ridge Province lies the Blue Ridge, an isolated region until recently, and probably the single most important barrier to westward migration during colonial days. Developed upon resistant volcanic and granitic rocks, the Blue Ridge has been a prominent feature of the landscape for the past 100 million years and has seen remarkably little denudation during that time.

The Piedmont Province, just southeast of the Blue Ridge, is different from any to the west. Developed on a variety of igneous and metamorphosed sedimentary rocks, it is a gently rolling upland, cut by streams entrenched up to 200 feet or so, in a generally dendritic pattern. Thick soils, produced by 100 million years of intense weathering, soften the contours of the land and provide a fertile base for agriculture.

Opposite: Fig. 1.1. Shaded relief map of the Chesapeake Bay drainage basin, extending from southern New York to southern Virginia. The spectacular ridges cutting diagonally across the figure are developed on rocks resistant to erosion, and they outline the major fold systems of the Appalachian Valley and Ridge and Blue Ridge Provinces. To the northwest, a dendritic network of valleys is developed upon the nearly flat-lying rocks of the Allegheny Plateau Province. Southeast of the Valley and Ridge Province is the gently rolling upland of the Piedmont Plateau, developed on metamorphic and igneous rocks in which narrow, steep-walled valleys have been cut by the major river systems. The Bay itself is cradled by soft, easily eroded sedimentary rocks of the Coastal Plain Province. The sharp transition from the broad, open river valleys of the Coastal Plain Province to the steep-walled valleys of the Piedmont defines the Fall Line, where Baltimore, Washington, and Richmond developed. (From shaded relief map of the Eastern United States compiled by Andrew D. Birrell [http://www.research.digital.com/SRC/personal/birrell/reliefMaps] from 5-degree tiles created by Ray Sterner, Johns Hopkins University Applied Physics Laboratory [http://fermi.jhuapl.edu/states]. The mosaics are © 1994, 1995, by Birrell, and the 5-degree tiles are © 1994 by Sterner.)

The sediment carried by the eastward-flowing streams that carved these geomorphic provinces is now preserved in the nearly flat-lying, barely consolidated beds that lie between the Piedmont and the Atlantic beaches and underlie the Coastal Plain Province. These soft, easily eroded rocks have produced the low-lying, virtually flat topography of the Tidewater country. The harder metamorphic and igneous rocks of the Piedmont were much less susceptible to erosion, and there the rivers are entrenched in narrow, steep-walled valleys marked by rapids and small falls. These falls mark the head of navigation for oceangoing vessels and collectively define what we have come to call the Fall Line. Though never much of a barrier to the migration of people, the Fall Line did require shifting goods from ocean vessels to land transportation. Ports naturally developed there, and the major cities of the Mid-Atlantic region—Baltimore, Washington, and Richmond—have developed along the Fall Line.

The Chesapeake Bay formed in the Coastal Plain Province during the Pleistocene period, which represents roughly the last million years of Earth's history (about one-half of 1% of the time required to open the present Atlantic basin). The Pleistocene was a time of alternating glacial periods, lasting about 100,000 years, and much briefer interglacial periods, lasting about 10,000 years.

During each glacial period, the ice caps grew and the sea level fell as much as 300 feet. At these times, the Bay was small and seaward of its present location, or even nonexistent. During the most recent glacial period, the sea level was so low that the ocean margin retreated off the edge of the continental shelf and the Bay disappeared entirely. During these glacial periods, the ancestral Susquehanna River, and its tributary streams, carved a series of broad, gentle valleys in the soft Coastal Plain sediments, each time taking a slightly different course as it sought the withdrawing sea. During the first two glacial periods, the Susquehanna cut channels through the Delmarva Peninsula which can still be recognized in seismic sections through the Bay sediments. During the last glacial period, the shelf was exposed long enough to develop boreal forests and tundra, and it was home to both large mammals, including mammoths, and the Paleo-Indian hunter-gatherers who subsisted on those mammals.

During the interglacial periods, the ice caps melted, the meltwater returned to the sea, and the sea level rose to invade the Susquehanna valley and its tributaries. The sea first moved up the deeper portions of the valley cut during the previous low-stand of sea level. Then, as the sea continued to rise, it spilled out of the deeper portions of the valley system onto the gently sloping valley sides, which became the sandy floor of the Bay. Each rise in sea level transformed the Bay environment from a fresh-water river system into an estuarine system. Each transformation began at the Bay mouth and gradually worked its way north as the heavier sea water pushed its way under the lighter river water, forcing the boundary with fresh water farther and farther up the growing Bay. As the Bay filled with water, it served as a trap to catch sediment carried by the tributary river systems, and the sediment gradually filled the river channel cut during the previous low-stand of sea level.

The lowest part of each sedimentary sequence is sand, deposited in the channel system while sea level was low. The sands were covered first by fine estuarine and Bay bottom sediments, and then, at the southern end of the Bay, by near-shore marine deposits. These marine deposits represent sands deposited in barrier islands and spits, like those along the Atlantic coast today. Long-shore currents drove these barrier-spit complexes relentlessly southward, shifting the mouth of the Bay and the position of the active river channel progressively farther south. The sediments filling the older channel systems record the history of the Bay during the Pleistocene. The sediments filling the current channel provide a detailed record of the evolution of the ecosystem during the last few hundred years.

The history of the Chesapeake Bay basin, then, has been one of draining and filling, first with water and later with sediment. The history of the Chesapeake Bay estuary has been one of comings and goings, of beginnings and endings. The present Bay has already lasted far longer than its ancestors. Its story began about 18,000 years ago, with the latest rise in sea level, but the sea has been high enough to recapture the Bay (and most other coastal embayments throughout the world) only for the last 10,000 years or so.

One oddity deserves mention. The major rivers that carved the valley system now occupied by the Bay converge near the Bay's mouth, directly above a meteorite crater 50 miles across. The crater formed 35 million years ago, long before the Bay itself, when a large meteorite struck the Atlantic and buried itself in the thickening wedge of Coastal Plain sediments. The impact thoroughly disrupted the sediment pile and even shattered some of the metamorphic rocks beneath. Although the crater is much older than the Bay, compaction of the disrupted sediments in the crater could nonetheless have formed a long-lasting depression, which in turn localized the confluence of the region's river systems and helped to control the position of the Bay's mouth.

So the landscape of the Chesapeake region is the product of processes that have worked their way with Earth's crust over the last 100 million years or so. The Tidewater landscape—broad rivers navigable by ocean-going vessels, flanked by nearly flat, fertile land ideal for cultivation—reflects the shape of the valley system that the glacial Susquehanna cut in the soft sediments of the Coastal Plain. That landscape proved ideal for the plantation economy that still gives the Eastern Shore its special character. The terrain that developed on the harder rocks of the Piedmont above the Fall Line—gently rolling hills separated by deeply incised streams and rapids that prevent navigation—required a totally different economic and political style, based on smaller farms, milling of grain, marketing of lumber, and the manufacturing of goods using local resources.

But the influence of our geologic past goes far beyond the landscape and the soil that provides our agricultural base. Virtually every natural resource that the early settlers learned to use reflects some part of the complex process of geochemical refining that affected the rocks of the Chesapeake region. Use of those resources set in motion economic forces that still exert a profound influence on our lives. Prime examples are chromium, iron, and coal.

The chromium story is particularly illustrative. Chromium comes from the mineral chromite, found in the rock serpentine. Serpentine has a pretty checkered history. It was formed from a dense igneous rock that lay just beneath the floor of the pre-Atlantic Ocean, then thrust up onto the continent during the Appalachian compressive cycle, and finally buried and metamorphosed. Serpentine contains little or no aluminum, so it makes a thin, poor-quality soil. Isaac Tyson, a farmer in Baltimore County in the early 1800s, had a good deal of serpentine on his farm and probably tried to figure out what do with that poor soil. Somehow he discovered chromite in the serpentine, recognized that it was an ore of chrome, and determined that other chrome deposits could be found by looking for that same kind of worthless soil. He did so, developed mines to extract the chromite he found, and established the Baltimore Chrome works to refine the chromite into chromium. That plant became the major chromium producer in the world, and Maryland was the world leader in chromite mining until the discovery of rich deposits in Turkey in the 1850s. Chrome continued to be refined in Baltimore long after the Maryland mines were abandoned, and today we still pay the price for the rather crude way in which that process was carried out. The cleanup of the refinery site, until recently owned by Allied Signal, is a major environmental problem, one the company and the city have worked to solve for 35 years, because the site sits squarely atop one of the Inner Harbor's prime development sites.

The example of iron again illustrates how the original natural resources of a region influence the economy and the region long after the initial exploitation of those resources has died out. Iron was, of course, a mainstay of the colonial economy—needed for cannons and muskets to fend off the British, for shoes for horses to plow the fields, for nails to build the boats to carry grain to market, and for a host of other uses. The iron industry began around 1720 as a cottage industry, using iron carbonate ores mostly derived from the Arundel Formation—one of those coastal plain sediments deposited during the current cycle of crustal extension centered on the Atlantic. These carbonate nodules were also used as whetstones, and they account for the name Whetstone Point, the original site of Baltimore. By the middle 1700s those ores were superseded by limonite ores formed where the Piedmont marbles and quartz-rich metamorphic rocks, above and below, made contact. Many local iron furnaces developed at this stage: Catoctin Furnace in western Maryland, Principio Furnace near Baltimore, and numerous smaller ones. The last continued in use until about 1900.

As a result of all this activity, Baltimore developed a substantial business exporting iron and iron products. When other ores eventually became available at lower cost (first in Cuba, then in northern Minnesota), those ores were shipped to Baltimore for refining into steel. Baltimore was favored in steel making because of its proximity to the two other essential ingredients: coal, found in the upper Susquehanna basin of Pennsylvania, and marble, found in the Piedmont. By the time of World War II, these advantages and the presence of a deep-water port had led Bethlehem Steel to build its huge plant at Sparrows Point and to branch out into ship building, two industries which were until recently the mainstays of Baltimore's economy.

As industry grew in the Mid-Atlantic region and the population began to concentrate in cities, the need for energy to run steel mills, iron foundries, and factories and for fuel to heat homes far outstripped the waterpower provided by the region's streams and the wood provided by its forests. People turned eagerly to coal to solve both problems. The northern part of the Susquehanna basin includes the world's largest reserves of anthracite, formed from the plant material which accumulated in the Late Paleozoic swamps along the western fringe of the Appalachians 300 million years ago. Appalachian coal was discovered as early as 1698, and the first mine was opened by Colonel Jacob Weiss following discovery of anthracite at Mauch Chunk (now Jim Thorpe, Pa.) in 1781. Production was inhibited by inadequate transportation until the Schuylkill Canal opened in 1825. After that, coal production increased steadily to 99.6 million tons in 1917, before declining to a low of 2.9 million tons in 1983.

During the early years, anthracite was extracted by deep-mining techniques. When thinking of the coal mining industry, most people still visualize deep mines. Now, however, roughly 65% of U.S. coal production is from open-pit surface mines, with only 17% from deep mines. Deep mining has declined because it is expensive and dangerous. For example, a mine belonging to the Pennsylvania Coal Company was being extended beneath the bed of the Susquehanna River in 1959 when a cave-in flooded first that mine and eventually nearly all the mines in the Northern Anthracite Field, eliminating most mining there.

The coal mined in the upper Susquehanna basin and elsewhere in the central Appalachians provided a strong economic stimulus to business in the Chesapeake region. The need to transport the coal to industrial centers led to the development of the region's railroads, and in a strange turn of events the railroads themselves became the largest consumers of coal. After the Civil War, most of the mines were owned or controlled by the railroads, and the partnership between the railroads and the mining interests became one of the strongest political forces in Pennsylvania, regularly winning substantial concessions from the state. This partnership even managed to obtain exemptions from the state's clean-stream laws until the mid-1960s, by which time many streams in the Susquehanna basin were seriously contaminated.

One of the most serious consequences of coal mining for the environment is referred to as acid mine drainage. In the natural landscape, the bedrock is mantled by a thick layer of soil, which isolates most of the surface water from raw, unweathered bedrock. As the soil mantle thickens, the bedrock is weathered by water percolating down through the soil cover. This process is extremely slow, so the products of the weathering are added to the groundwater and the streams very slowly. Natural weathering rates are commonly on the order of a few feet of rock per million years. Mining changes the situation dramatically, however. In a few decades a vast amount of unweathered rock is suddenly exposed to groundwater, both in the mine itself and in piles of waste material at the head of the mine. In most coal mines, that unweathered rock contains a substantial amount of the mineral pyrite (FeS_2), and reaction between pyrite and oxidizing water increases the acidity of the water. Groundwater reacts rapidly with this fresh rock, substantially increasing the water's acidity and that of the local stream system. The effect of this acid mine drainage on ecosystems of the upper Potomac and the upper Susquehanna has been nearly catastrophic. Once again, a natural resource formed by geologic processes operating hundreds of millions of years ago has been used by industry in a way that has had drastic effects on the ecosystems of the region today.

Like mining, nearly every human activity has imposed costs on the local ecosystem. Lumbering on the slopes of the Appalachian Mountains has increased the rate of runoff, reducing percolation of rain into the groundwater system and increasing the rate of erosion and sediment deposition in the Bay. Clearing of land for farming and construction has had a similar effect. Use of too much fertilizer in farming and landscaping has increased the plant growth in the Bay and its tributaries. As those plants die and decay, they sharply reduce the amount of dissolved oxygen and threaten the health of aquatic ecosystems. Runoff from factory farms has exacerbated this problem and perhaps led to fishkills and outbreaks of pfisteria. Overproduction of oysters has reduced the harvest to a mere fraction of what it was in the 1920s.

THE GEOLOGIC PROCESSES THAT SHAPE our planet have played a fundamental role in defining many aspects of the Chesapeake ecosystem and our interaction with it. They refined Earth's crust over the past 600 million years to give us the raw materials which have largely determined the direction of economic life in the Bay region. But careless use of those resources has left us with a legacy of pollution and ecological damage with which we must cope.

The waxing and waning of Pleistocene ice caps defined the form and size of the Bay itself and gave shape to the seafood industries that depend upon the Bay's immense biological productivity. Ultimately those same processes will lead to the end of the Bay. If one were to use the events of the Pleistocene to forecast the remaining geologic and oceanographic life span of the Bay—the time until the next draining—one might conclude that it has already outlived its ancestors and that the end must be imminent. But there are no signs of an early death by draining. Sea level is rising more rapidly than the Bay is filling with sediment, and its domain is growing as it spills out of its traditional basin and claims new land. If the greenhouse effect and global change produce an increase in Earth's average temperature—as many scientists believe is inevitable

—the rate of sea level rise will increase, and the geologic life of the Chesapeake Bay estuary will be extended even further.

So the question that we face is not how long the Bay will live but how well. The answer is in our hands.

CHAPTER TWO

Climate and Climate History in the Chesapeake Bay Region

JOHN E. KUTZBACH AND THOMPSON WEBB III

The climate of the eastern seaboard of North America is influenced by global-scale controls, such as the greenhouse effect of atmospheric carbon dioxide, and local controls, such as the prevailing winds and the proximity to the ocean. Analyses of written records (e.g., weather diaries of former presidents) and geologic records (e.g., pollen in lake sediments, fossil air in polar ice cores, and plant and animal fossils in rock strata) tell us that the climate has changed. As continental glaciers waxed and waned, the Atlantic Ocean alternately fell and rose, thereby dramatically altering the size, shape, and hydrology of the Chesapeake Bay. At the time of the late Paleozoic supercontinent Pangea, there was no Atlantic and the regional and global climate controls were drastically altered. Simulations of the climate produced with the help of computationally intensive supercomputers provide glimpses of past climates and predictions of future climates and of the future level of the sea.

The mild, relatively equable climates of the Mid-Atlantic region have long attracted humans. The region has an environment rich in vegetation, terrestrial wildlife, and marine life. The current climatic conditions, however, have existed for only a short time. The climate has varied considerably over past centuries and drastically over geologic time. These changes may be traced to external events, such as variations in the sun's intensity, or to variations in Earth's orbit about the sun, or to complex interactions among the atmosphere, the ocean, the biosphere, and the land surface, and even human activities. The changes in climate have affected the vegetation, and these vegetation changes have in turn affected the climate. Hundreds of millions of years ago, under climatic conditions vastly different from today, vegetation grew in the area of the Appalachians from which coal was later produced. Today this coal is being burned, and one by-product of this combustion is contributing to the recent increase in atmospheric concentrations of carbon dioxide which may lead to changes in regional climates worldwide. This example illustrates how the climate and vegetation of the past, in concert with recent human activities, may affect future climates.

The Climate Now

The Chesapeake Bay area, located along the eastern seaboard of North America, is generally described as having a temperate and humid climate. Several factors combine to produce these conditions. One important controlling factor is the warmth of the western North Atlantic Ocean with its strong offshore current, the Gulf Stream, conveying warm tropical waters northward and thereby keeping coastal water temperatures between about 15°C (60°F) (winter) and 25°C (75°F) (summer). Another controlling factor is the predominant air circulation associated with the subtropical high-pressure system centered over the North Atlantic (Fig. 2.1). In the summer half year, April through September, the clockwise circulation of air around this center produces, to the west of the center, winds from a southerly direction that carry moist air masses from the Atlantic or from the Gulf of Mexico to the Bay area. These airstreams bring generally warm and humid conditions and frequent showers and thundershowers. In the winter half year, October through March, the subtropical high-pressure system is weaker and centered farther south and the controlling factor for the Bay area climate becomes the prevailing westerly winds of middle latitudes. Westerly airstreams cross the continent bringing cooler and slightly drier conditions to the Bay area. This season is marked by the passage of cyclonic storms, either traveling along the coast from southwest to northeast or moving east from the continental interior. Weather associated with these storms includes rain or drizzle and, more rarely, snow or freezing rain, followed by brief periods of freezing temperatures after the passage of cold fronts. On average, however, the major weather boundary or frontal zone marking the southern limit of polar and Arctic air masses lies along the U.S.-Canadian border, 500 km or more to the north of Chesapeake Bay. This geographic separation of the Bay area from polar and Arctic air masses accounts for the region's relatively mild winters.

Climatic statistics for Washington, D.C. (Table 2.1) provide an example of the regional conditions. The mean temperature is about 26°C

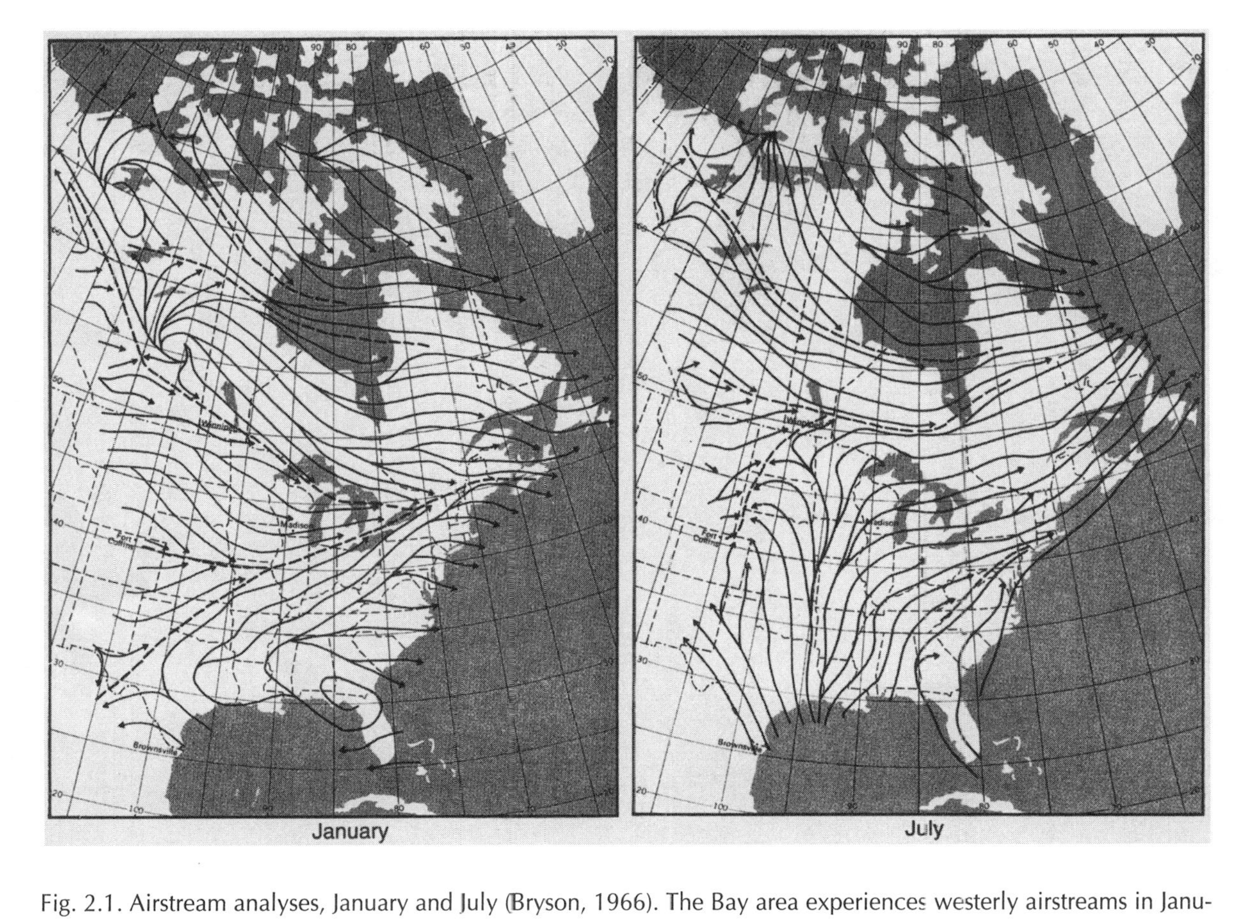

Fig. 2.1. Airstream analyses, January and July (Bryson, 1966). The Bay area experiences westerly airstreams in January and more southerly airstreams in July.

Table 2.1 Washington, D.C., climate statistics
(latitude 38°51′N, longitude 77°03′W, elevation 4m [34 m])

Month	Temperature (°C)				Precipitation (mm)		Mean snowfall (cm)
	Daily mean	Mean daily range	Extremes: Maximum	Extremes: Minimum	Mean	Maximum in 24 hrs	
January	2.7	8.2	26	−15	77	44	11
February	3.2	9.3	28	−15	63	39	10
March	7.1	10.0	32	−12	82	87	7
April	13.2	11.2	35	−4	80	45	trace
May	18.8	10.8	34	1	105	110	0
June	23.4	10.3	38	8	82	93	0
July	25.7	9.8	39	13	105	75	0
August	24.7	9.5	38	12	124	162	0
September	20.9	10.0	38	4	97	92	0
October	15.0	10.4	34	−1	78	126	trace
November	8.7	9.8	29	−9	72	64	2
December	3.4	8.4	24	−17	71	47	8
Annual	13.9	9.8	39	−17	1,036	162	38

Month	Precipitation (days with >0.25 mm)	Thunderstorms (days)	Mean cloudiness (tenths)	Mean sunshine (hrs)	Wind: Most frequent direction	Wind: Mean speed (m/sec)
January	11	0	6.8	45	NW	5
February	8	0	6.4	52	NW	5
March	12	1	6.1	55	NW	5
April	10	3	6.5	55	S	5
May	12	6	6.4	56	S	4
June	9	5	5.6	65	S	4
July	10	7	5.9	64	SSW	4
August	10	5	5.6	61	S	4
September	8	3	5.4	62	S	4
October	8	1	5.3	58	SSW	4
November	8	1	5.9	55	SSW	4
December	9	0	6.2	49	NW	4
Annual	115	32	6.0	56	S	4

Source: Shortened and adapted from Court (1974).

(79°F) in July and about 3°C (37°F) in January, and the annual average is about 14°C (57°F). Daily extremes range from a high of 39°C (102°F) in July to a low of −17°C (1°F) in December. There is no distinct dry season. Rainfall is about 75 mm (3 inches)/month in winter and about 100 mm (4 inches)/month in summer; the annual total is about 1,000 mm (40 inches). Extremes of rainfall are sometimes associated with hurricanes that track north over the Bay area in their decaying phase as tropical or extratropical storms. These storms, usually occurring August through September, have lost their truly destructive winds in their journey north over land but can still drop 6 inches or more of rain in one day. The passage of Tropical Storm Agnes over the Bay area in 1972 caused severe flooding, sediment transport, and significant changes in the Bay's salinity. On average about one of these decaying hurricanes passes over or near the Bay area each year. Measurable snowfalls occur November through March. Although the long-term average snowfall is only about 10 cm (4 inches) per month during this season, memorable individual storms, such as in the winter of 1922, have dumped 24 inches or more of snow in the area.

The Bay itself modifies the local climate (Fig. 2.2). The annual average temperature along the Bay may be 1–2°C (2–4°F) higher than the temperature a few tens of miles inland. The average number of days with temperatures below freezing varies from about 100–110 in Washington, D.C., and in the region immediately to the north of the Bay to about 70–80 along the Bay itself; this sharp gradient occurs over distances of only 20–30 miles. The unique microclimate of the Bay area may help explain regional historical patterns of diseases such as malaria (see Chap. 7) and regional concentrations of plant and animal pathogens (see Chap. 14).

The Climate of Recent Centuries

The eastern seaboard is fortunate in having the longest-running weather records and climate reports in North America, some dating from the early 1600s (Barron, 1992; Bradley and Jones, 1992). In 1612 Captain John Smith wrote in *A Map of Virginia*: "The summer is as hot as in Spain; the winter cold as in France or England. The heat of summer is in June, July, and August, but commonly the cool breezes assuage the vehemence of the heat. The chief of the winter is half December, January, February, and half March. The cold is extremely sharp, but here the proverb is true, that no extreme long continues" (quoted in Ludlam, 1966: 32).

The list of early weather observers in the Bay area includes George Washington, James Madison, and Thomas Jefferson (Ludlam, 1966). Jefferson (1825) published a summary of his meteorological journals at Monticello for 1810–16 based upon 3,905 observations, normally taken twice per day: before sunrise and between 3 and 4 P.M. July was the

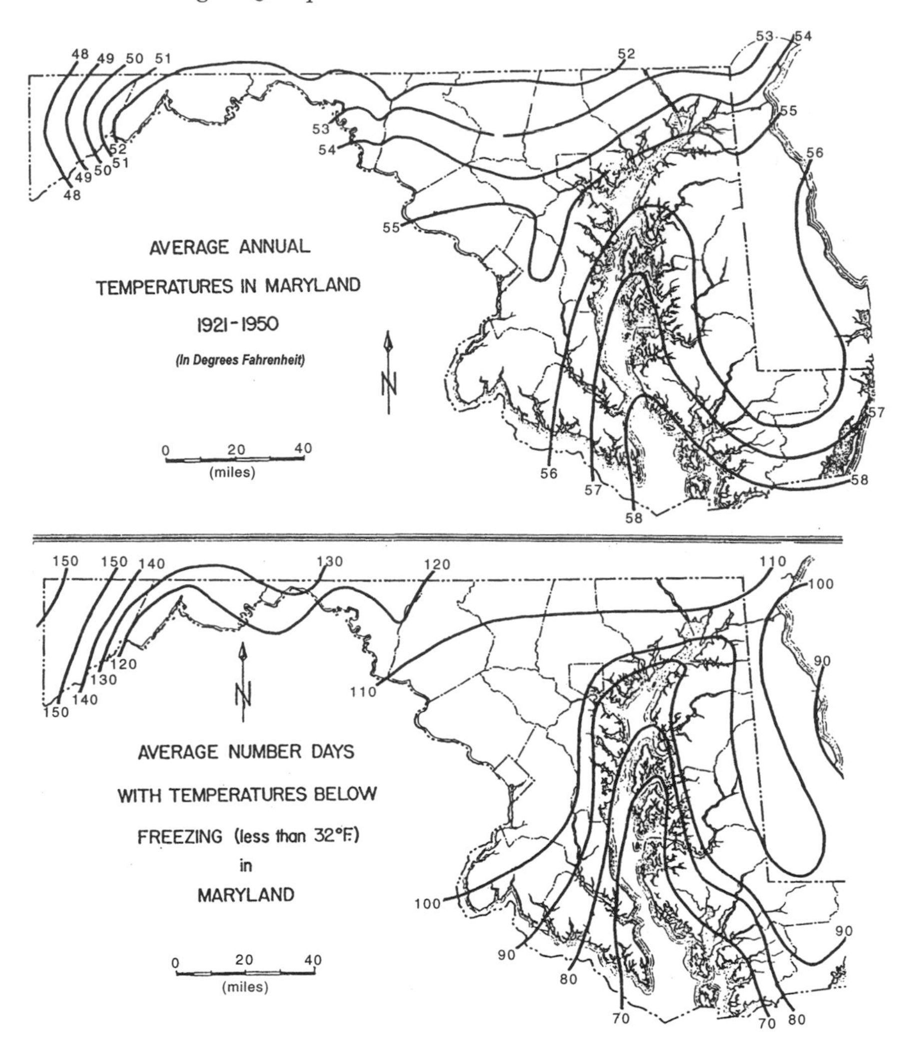

Fig. 2.2. Regional climates of Maryland, 1921–1950 (Vokes, 1957).

warmest month (75°F), January the coldest (36°F); and the annual average temperature was 55.5°F. Jefferson's records describe a thermal climate rather similar to that of today (see Table 2.1). In addition to detailed tables of temperature, precipitation, and wind, Jefferson noted the following example of a practical application of his statistics: "It is generally observed that when the thermometer is below 55°[F], we have need of fire in our apartments to be comfortable." Over the seven years, there were freezing temperatures an average of 190 mornings per year and 120 afternoons per year. "Whence we conclude that we need constant fires four months in the year, and in the mornings and evenings a little more than a month preceding and following that time." Jefferson kept phenological records of the first dates of cherry, lilac, and dogwood blossom, and of the arrivals of martins, robins, and fireflies. It would be interesting to know if these "natural" calendar events have shifted. Jefferson also published a summary of weather observations from Williamsburg, Virginia, 1772–77.

How representative were these early weather records and how variable was the climate from decade to decade and century to century? H. E. Landsberg assembled weather records from several cities, including Boston-Cambridge, Mass., New Haven and Albany, N.Y., Morristown, Pa., Baltimore and Woodstock, Md., Williamsburg and Charlottesville-Monticello, Va., and Charleston, S.C. From these diverse and sometimes incomplete records, Landsberg produced an average or composite climatic chronology for the eastern seaboard of the United States, centered in Philadelphia. This chronology, starting in 1730 and extending to 1967, is the best long climatic record that is relatively close to the Bay area. We have added to Landsberg's long composite records the temperature and precipitation records for Philadelphia for 1948–93 (Fig. 2.3). For the period of overlap (1948–67) the results are similar, but we must caution that the two sections of the record are not strictly comparable.

The Philadelphia chronology illustrates the sometimes chaotic nature of weather and climate. The annual average temperature changes considerably from year to year, with typical variability being on the order of 2°F and with a range of about 5°F. Monthly and seasonal variability is larger. Decades and centuries also differ. Conditions were relatively warm in the late eighteenth century and relatively cold in the nineteenth century. There was a warming trend in the late twentieth century, but there were also cold winters, such as 1976–77, when parts of the Bay froze. The precipitation record displays interannual and interdecadal variability of about ±5–10% compared to the long period average, with some of the wettest conditions occurring during the cold periods of the nineteenth century. The climate of the 1990s included several very wet summers. The modern Philadelphia precipitation records indicate an average value of 41 inches/year, which is very close to the average of 42 inches/year for Landsberg's long composite record.

Some corroboration for the cool, wet conditions of the mid-nineteenth century as depicted in this local record is available from continent-

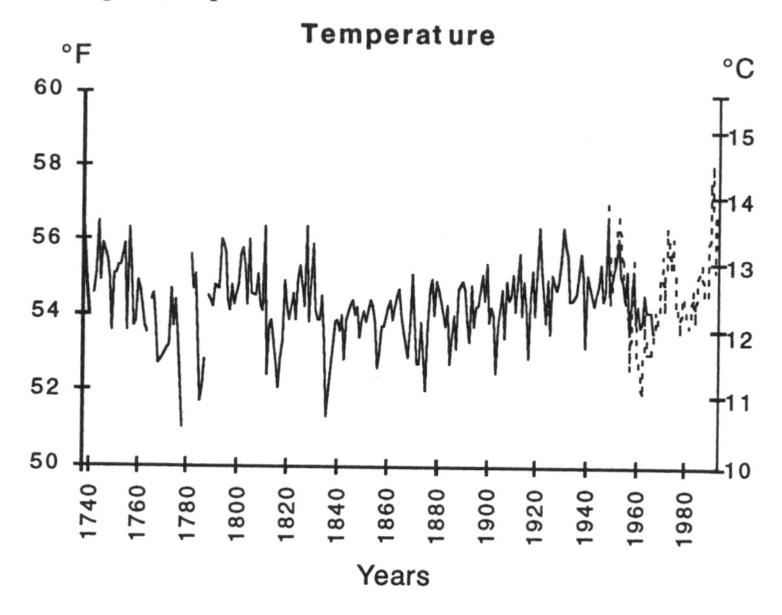

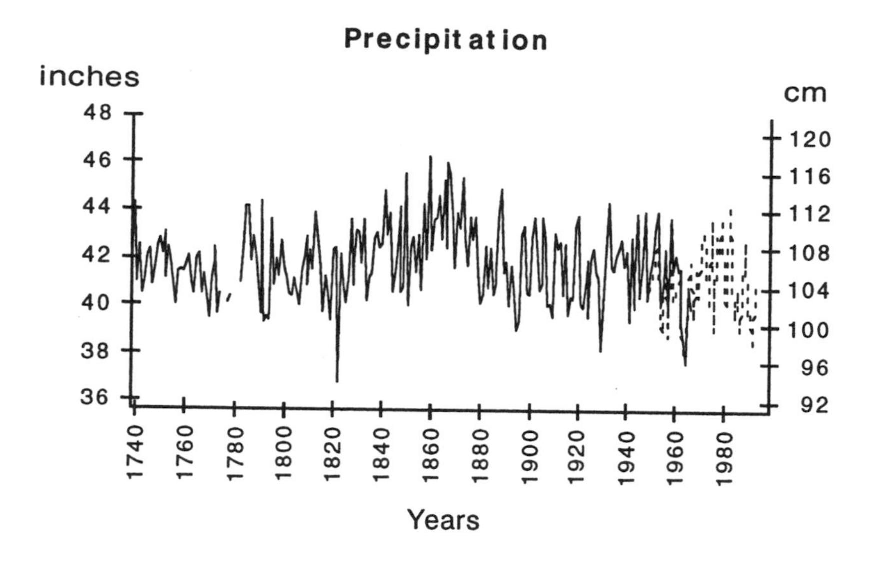

Fig. 2.3. Climate time series for the eastern seaboard, centered on Philadelphia (adapted from Landsberg, 1970; © American Association for the Advancement of Science). The solid curves are from Landsberg for 1738–1967. The dashed curves are from climate records for Philadelphia, 1948–93. The amplitude of the single-station Philadelphia record (1948–93) has been reduced by a factor of four

wide maps of the climate for the 1850s and 1860s, assembled from the reports of many stations (Fig. 2.4). These maps show that in the middle of the nineteenth century, the temperatures on the eastern seaboard were about 1°C (2°F) lower and precipitation was about 5% higher than at present. A southward shift of the westerlies and the prevailing storm track in late fall and winter appears to account for the wetter and cooler conditions during the 1850s and 1860s. Changes in circulation and storm tracks are no doubt associated with recent climate changes as well (Hayden, 1981). For example, systematic records of storm frequency, begun in 1885, show a trend toward increased numbers of East Coast cyclones in recent decades.

Records of tree growth provide another source of year-to-year climatic information about regional environmental variability (Fig. 2.5). Tree-ring-width chronologies from the Potomac River basin of northern Virginia have been analyzed for 1730 to about 1980. The tree growth records were calibrated against streamflow in the Potomac in recent decades and then this calibration was used to infer river flow since 1730. These tree growth records (and inferred river flow records), like the instrumental records, show a high degree of interannual and interdecadal variability. However, the estimated streamflow does not agree in detail with the Landsberg-derived precipitation record, perhaps because of possible sampling biases in both records or because streamflow is related to a combination of factors, including not only annual rainfall but also its seasonal distribution, as well as the seasonal cycle of evaporation. Changes in land use could also influence streamflow.

The Climate of the Past Millennium

The late nineteenth century marked the end of several centuries of generally cold conditions in northern mid-latitudes, a time known for large extensions of mountain glaciers in Europe and western North America (Grove, 1988). It is to the growth of these mountain glaciers that the period owes its somewhat exaggerated name of "The Little Ice Age." Although the temporal limits of the Little Ice Age are not well delin-

to make it more comparable with the amplitude of the multistation record (1758–1967). The average values are as follows: temperature (1738–1967), 54.6°F; temperature (1948–93), 55.9°F; precipitation (1738–1967), 42 inches, precipitation (1948–93), 40.9 inches.

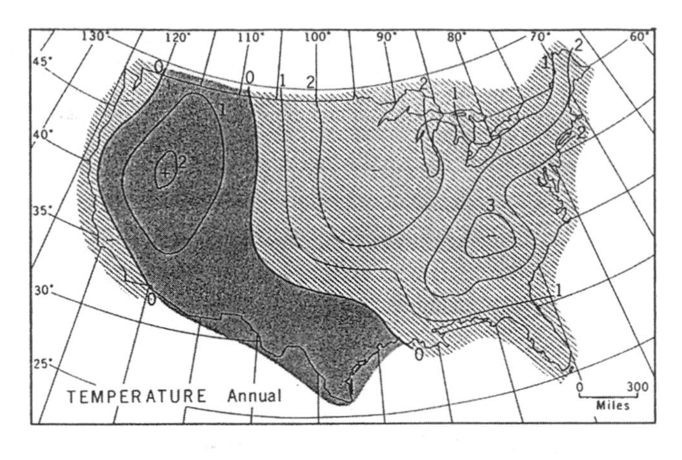
130°
120°
110°
100°
90°
80°
70°
60°
45°
40°
35°
30°
25°
TEMPERATURE Annual
0 300
Miles

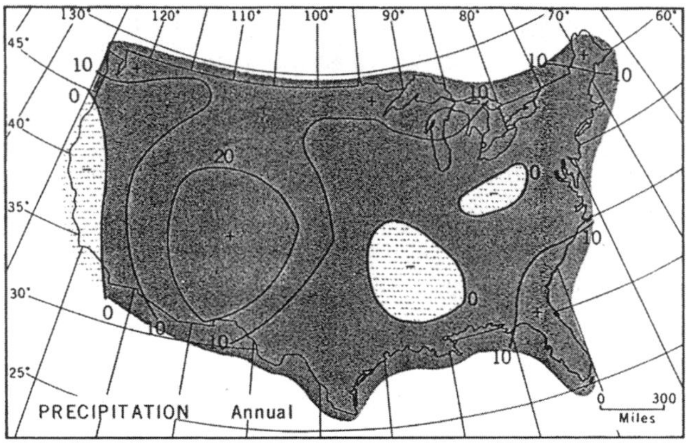
130°
120°
110°
100°
90°
80°
70°
60°
45°
40°
35°
30°
25°
PRECIPITATION Annual
0 300
Miles

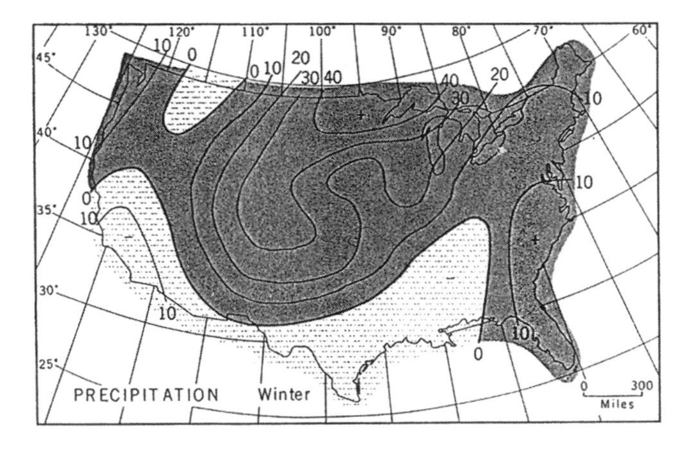
130°
120°
110°
100°
90°
80°
70°
60°
45°
40°
35°
30°
25°
PRECIPITATION Winter
0 300
Miles

eated, it is often described as commencing around 1300 A.D. and persisting until about 1880. Thus only the last 150 years of the Little Ice Age are documented with instrumental records. The Little Ice Age was preceded by a warmer period centered around 1000 and known in Europe as the Medieval Warm Period. Historical and archaeological records indicate that this period was marked by the colonization of Greenland and possibly parts of the east coast of North America by Vikings (Bryson and Murray, 1977).

The Bay area lacks thousand-year historical records, but aspects of its climatic history can be inferred from environmental records. For example, pollen, charcoal, and trace metals from a sediment core in the Nanticoke River indicate a relatively dry climate around 700 to 900 years ago; that is, during the Medieval Warm Period (see Chap. 3). The end of this dry period, around 1300, presumably marks the onset of the moister and cooler conditions of the Little Ice Age. Far north of the Bay area, pollen analyses from sediments in Conroy Lake in Maine show a marked increase in spruce pollen around 1300, and climatic estimates based upon these (and other) pollen indicators suggest that summer temperature in Maine was perhaps 1–2°C (2–4°F) lower and annual precipitation 5% higher in the Little Ice Age than it was in 1000 or the present (Gajewski, 1988). Because the shifts in climatic pattern associated with the Little Ice Age are of large scale (see Fig. 2.4) and probably reflect large changes in airstreams, the Bay area may have experienced a shift toward cool and wet conditions of similar magnitude. There were certainly examples of individual cold winters. Much of the Chesapeake Bay was frozen in the winters of 1641–42, 1645–46, 1779–80, and 1783–84 (Ludlam, 1966: 33, 115). The weather diaries of both George Washington and Thomas Jefferson record the snowstorm of 26–29 January 1772, which left three feet of snow at Mt. Vernon and near Monticello (1966: 144–45).

The causes of climate variability on the time scale of years to centuries are poorly understood. At least four possible mechanisms have been suggested (Hansen et al., 1984; Rind and Overpeck, 1993): (1) small variations in solar radiation, (2) episodes of increased or decreased volcanic

---

Opposite: Fig. 2.4. Climate of the 1850s and 1860s (Wahl and Lawson, 1970): (a) temperature deviations (°F) of the data in the 1850s and 1860s from climatic averages 1931–60, annual average; (b) precipitation departures of the data in the 1850s and 1860s from climatic averages 1931–60 (as % of 1931–60 average amount), annual averages; (c) same as (b), winter (January through March).

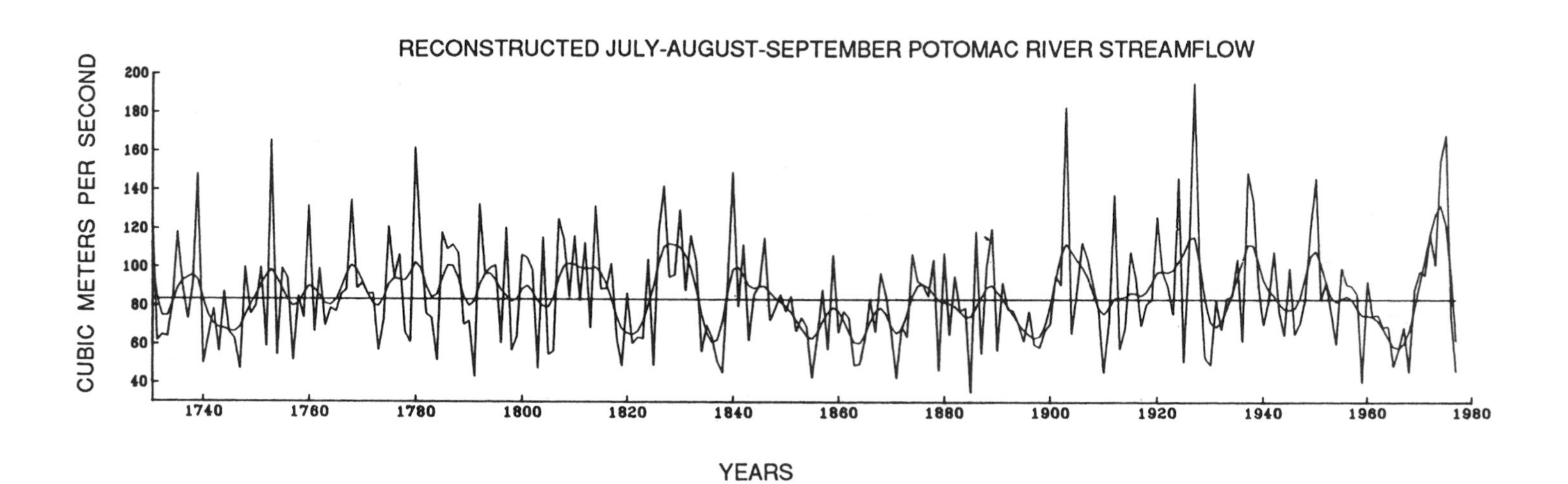

Fig. 2.5. Tree-ring chronology, calibrated in terms of Potomac River streamflow, July–September (Cook and Jacoby, 1983). The smooth estimates emphasize periods greater than eight years. The long-term median discharge is indicated by the straight line.

activity, (3) natural internal variability of the atmosphere and ocean and specifically of the circulation in the Atlantic Ocean and in the tropical Pacific Ocean, and (4) human-caused changes in vegetation and, more recently, changes in the carbon dioxide concentration of the atmosphere.

The output of the sun has varied by about 0.1% through the most recent 11-year sunspot cycle, as measured by Earth-orbiting satellites, and century-scale variations as large as 0.4–0.5% have been suggested. A decrease in solar radiation of 0.5% might be expected to lower the global average temperature about 2°F.

Large volcanic eruptions such as Tambora (1815) and Krakatoa (1883) in Indonesia and Pinatubo (1991) in the Philippines are capable of cooling the climate by as much as 1–2°F globally for several years (Hansen et al., 1992). The primary mechanism for this cooling is the eruptive injection of sulphate aerosols into the stratosphere. These aerosols reflect sunlight and thus cool the planet for several years, the typical lifetime of the stratospheric aerosol from an individual eruption. For example, 1816 was a very cold year in the Philadelphia record (see Fig. 2.3), especially the summer, and this cold event was no doubt related to the Tambora eruption. In contrast, other dips in the temperature record are not obviously related to volcanic eruptions. It is plausible that decades or centuries of slightly higher or lower volcanic activity could bias the climate toward cooler or warmer conditions.

Natural oscillations of the atmosphere and ocean are also a powerful source of internally generated variability within the climate system. The relatively slow circulation and the long travel time of internal ocean waves can produce climate variability on time scales ranging from years to centuries, and the El Niño / La Niña oscillations are a well-known example of this type of variation. These oceanic changes in the tropical Pacific influence the weather and climate along the eastern seaboard of North America through wave-like perturbations of the westerly jet stream and storm tracks.

The possible climatic effects of mid-latitude deforestation, occurring earlier in Europe than in North America, are generally thought to be small but have never been thoroughly investigated. Deforestation changes the local temperature, the surface brightness (albedo) and roughness, the recycling of water vapor by means of transpiration, and the amount and timing of runoff of surface water following storms.

None of these sources of climate variations—whether solar, volcanic, oceanic, or human-induced—are known with sufficient accuracy that we can establish the cause or causes of all the variability that we observe in our climate. The variability is real, however, and natural systems of the Bay area have responded, and will continue to respond, to these changes.

The Climate since the Last Glacial Maximum and Earlier Glacial-Interglacial Cycles

Paradoxically, as we look into the more distant past where climatic changes were much larger than in the recent past, we are more certain of the causes. A huge dome of ice reaching more than 2 km in elevation covered the northeastern United States and much of eastern Canada 21,000 years ago and was associated with concomitant changes in vegetation and sea level along its margin (Fig. 2.6). This glaciation was the most recent of perhaps ten major glacial advances in the Northern Hemisphere in the past million years. Although not covered by glacial ice, the Chesapeake Bay area had a climate drastically different from today. The main ice front lay across Pennsylvania and New Jersey, about 100 km north of the northernmost point of the Bay, and occupied the northernmost sector of the Chesapeake watershed. The buildup of ice on the continents lowered global sea level by about 100 m (more than 300 feet), so that present-day Chesapeake Bay was then the valley of the Susquehanna River; the river cut its channel during this and previous glacials and emptied into the Atlantic far to the east along the continental shelf (see Chaps. 1 and 3). As testimony to the harsh glacial environment, pollen records from Maryland and Virginia indicate landscapes of tundra and spruce (see Chap. 3).

These large differences between the glacial and modern vegetation provide a useful source of data for estimating past temperature and moisture conditions along the eastern seaboard at glacial maximum (Webb et al., 1993). The estimates (Fig. 2.7) suggest much lower temperatures than at present and precipitation 40% less than present; that is, a much colder and drier climate. For the Bay area, these pollen-based estimates imply summer temperatures of 10–15°C (50–60°F) (rather than 75–80°F), winter temperatures of 15°F (rather than 35–40°F), and annual rainfall closer to 25 inches than to the present-day 40 inches. These harsh climatic conditions are fairly typical of conditions in central Canada today near the tundra-forest border (although winter temperatures are colder than 15°F in Canada now). The climatic inferences from pollen indicate conditions similar to those simulated by climate models (COHMAP, 1988; Kutzbach and Webb, 1991). Climate models show that a glacial-age atmospheric jet stream, much stronger than today's, flowed from west to east along the southern margin of the ice sheet. This jet stream was present

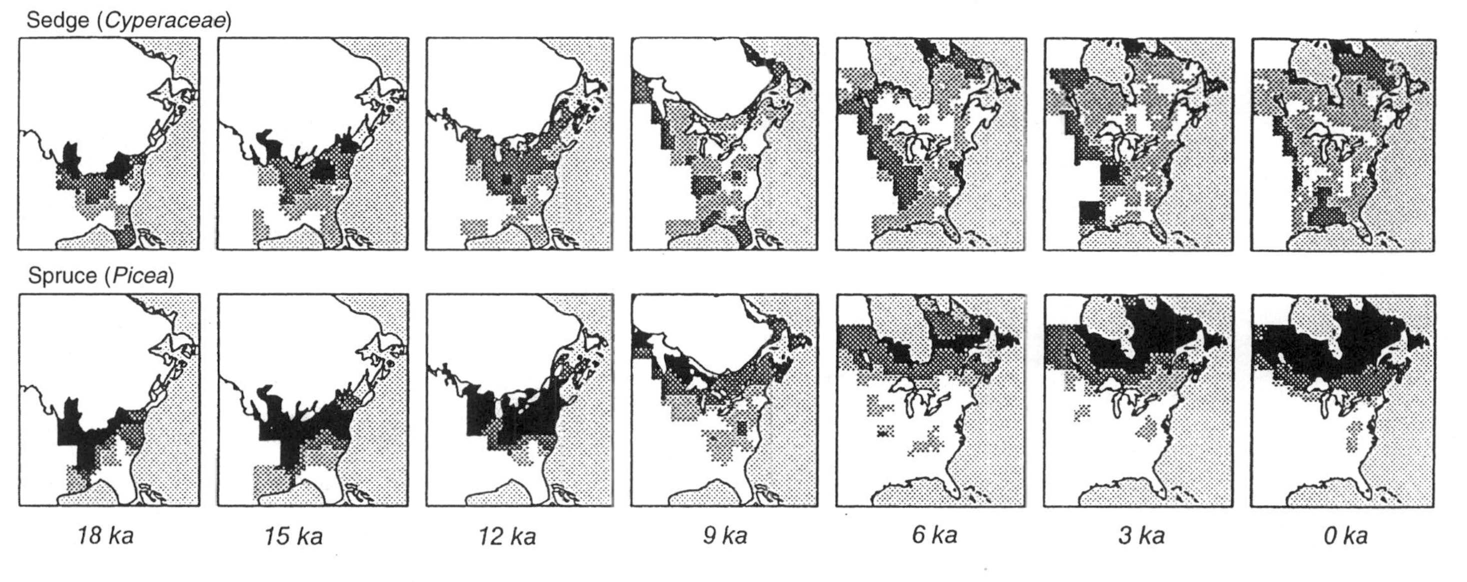

Fig. 2.6. Maps of ice sheet extent, shown in white, and of pollen abundance (modified from T. Webb et al., 1993). The ice boundary retreat occurs from 18,000 radiocarbon years ago (18 ka) to 9,000 radiocarbon years ago (9 ka). The pollen abundance (as % of a sum of 24 major taxa) is shown for sedge and spruce from 18 ka to the present (0 ka). For sedge and spruce pollen, white (south of the ice sheet) is <1%; light gray, 1–5%; dark gray, 5–20%; black, >20%. The changes in the coastline and the Great Lakes reflect changes in water levels from 18 ka to the present.

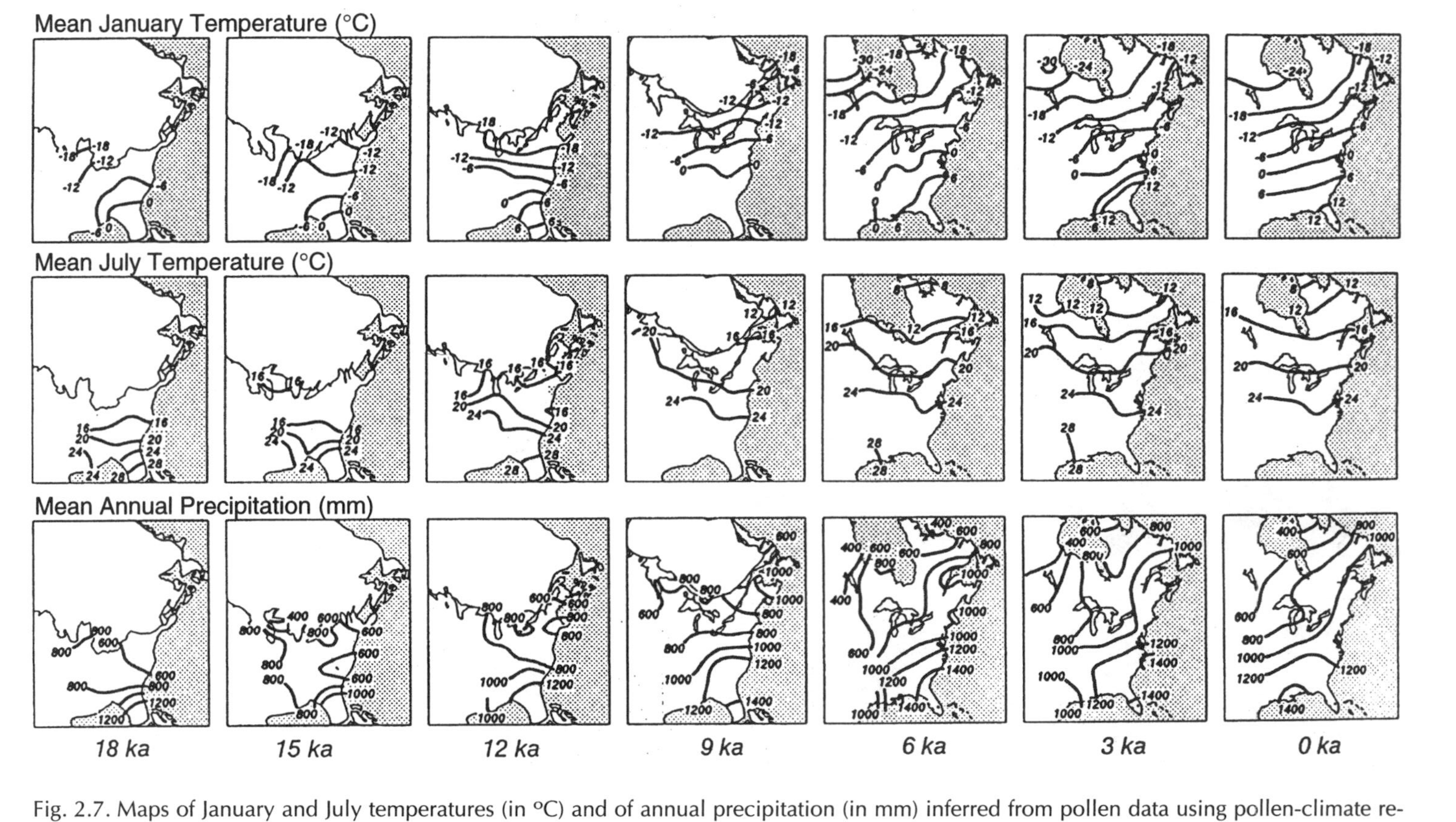

Fig. 2.7. Maps of January and July temperatures (in °C) and of annual precipitation (in mm) inferred from pollen data using pollen-climate response surfaces (T. Webb et al., 1993). The map for today (0 ka) is the modern observed climate values. The ice sheet, shown in white in the north, is present from 18,000 radiocarbon years ago (18 ka) to 9,000 radiocarbon years ago (9 ka). The changes in the coastline and the Great Lakes reflect changes in water levels from 18 ka to present.

in both summer and winter and was associated with strong westerly winds at the surface. In summer, these strong westerlies contrasted sharply with the modern-day warm, moist, southerly airstream. In winter, strong northwesterly winds would have blown off from the ice sheet, along the coast, and out over the Atlantic (Wells, 1983). Wind-formed coastal dunes and basins that date from this period exist in the Bay area and as far south as the Carolinas. These frigid conditions must have lasted for 10,000 or more years, from 25,000 to 15,000 years ago. Then, first slowly and later with great speed, the ice melted and vegetation changed (Gaudreau and Webb, 1985).

The postglacial trend toward warmer conditions was interrupted at least once by an abrupt cooling around 10,500 years ago that persisted for several centuries. This brief cold period was first identified in Europe and called the Younger Dryas. It may have been caused by large changes in the circulation of the North Atlantic Ocean associated with the flooding of the surface waters by low-density meltwater from the melting ice sheets. This cold event was marked in the Bay area by a brief return to higher concentrations of spruce and birch pollen as recorded in sediments (see Chap. 3) and is evident in coastal pollen records in southern New England and New Jersey (Peteet et al., 1993).

Short-term climate oscillations like the Younger Dryas interval are classified as suborbital variations, because the major advances and retreats of glacial ice are now widely recognized as having been initiated by subtle changes in Earth's orbital parameters. These include the tilt of the axis of rotation, which varies from about 22° to 25° with a period of 41,000 years and directly affects the angle of the sun with respect to the horizon; the eccentricity of the Earth's orbit around the sun, which varies from near circular to about 5% eccentricity with a period of about 100,000 years; and the precession of the Earth's rotational axis, which changes the season of perihelion from summer to winter and back again with a period of about 23,000 years (Imbrie and Imbrie, 1979). Ice accumulates and advances when orbital conditions cause cool summers, thereby favoring the persistence of snow cover; ice melts and retreats when orbital conditions cause summers to warm. However, the details of these processes remain poorly understood.

The most recent great melting commenced about 15,000 years ago. By 12,000 years ago, the ice border had retreated to the St. Lawrence River (see Fig. 2.6), and by 9,000 years ago, when the ice border was well north in Canada, sea level was rising rapidly (Fig. 2.8), reflooding the continental shelf and beginning to flood the deeper parts (>20 m) of coastal

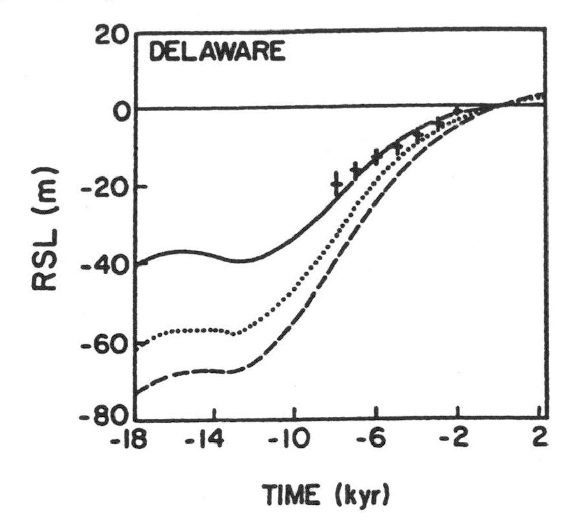

Fig. 2.8. Observed (+'s) and simulated sea-level curves for Delaware; the three simulations were based upon a range of model parameters (modified and adapted from Peltier, 1987). Time is in thousands of years (kyr), with negative values before present. RSL is relative sea level in meters, with negative values indicating that the water was below current sea level.

river valleys, such as those in the Chesapeake Bay area. Summer conditions may have warmed to about 70°F, but both summers and winters remained colder than at present (see Fig. 2.7). By 6,000 years ago summertime solar radiation was about 6% greater than at present, most of the North American ice sheet had melted, and summer temperatures in eastern North America, as estimated from climate model simulations, were perhaps 2–4°F higher than at present (Wright et al., 1993; Webb, 1998). Increased amounts of oak pollen and charcoal (indicating fires) and lower lake levels indicate that 9,000 to 6,000 years ago it was warmer and drier than at present in the northeast (see Chap. 3). Bay basins in Delaware were dry. Climate models also simulate this tendency for drier conditions in the northern mid-latitude continents owing to increased evaporation. Since 6,000 years ago the orbital conditions moved slowly toward modern values and summers became slightly cooler.

One further aspect of glacial-interglacial climate change is that glacial periods have lower levels of atmospheric carbon dioxide than interglacial periods. This difference in CO_2 levels, detected through analysis of fossil air trapped in Greenland and Antarctic ice, is thought to be a response to changes in Earth's orbit. The change in greenhouse gas concentration *amplifies* the orbital effects in that it cools the cold periods (and warms the warm periods) more than they would be cooled (warmed) by orbital changes alone. Thus, discussions of greenhouse warming have echoes in the climatic records of the past.

Ancient Climates and the Long Cooling Trend

According to the rules of celestial mechanics, Earth's orbital parameters have always been changing, influenced by the gravitational tugs of the sun, moon, and larger planets. One might therefore infer that glacial-interglacial cycles extend back continuously and indefinitely into the past. But this is not the case. Sediments recovered from the central North Atlantic Ocean, thousands of kilometers northeast of the Bay area, have helped pinpoint the onset of glacial cycles in the Northern Hemisphere 2.4 million years ago. Before this time, the sediments displayed a rather uniform accumulation of carbonate derived from the shells of planktonic forams; after this time, the carbonates are mixed with ice-rafted debris brought south on icebergs that calved from high-latitude glaciers. The characteristic periods of ice-rafting cycles match the orbital periods of about 23,000, 41,000, and 100,000 years.

Where do we look for the cause of the Northern Hemisphere glacial cycles that began about 2.4 million years ago? The answers to this question are still shrouded in mystery but certainly involve plate tectonics, that is, continental drift and mountain uplift.

We will start with Pangea, the supercontinent, which existed from around 320 million years ago to about 180 million years ago. Around 280 million years ago Africa was colliding with North America and ancient Appalachian mountains were forming and rising along this collision boundary in an almost west-east, near-equatorial mountain chain (see Chap. 1). The "Chesapeake Bay area" was on the equator near these mountains. Monsoon winds switched from north to south and back again across this region in response to the seasonal thermal extremes in Laurasia (the northern continent) and Gondwana (the southern continent) (Kutzbach and Ziegler, 1993). The near equatorial highlands were a focal point for heavy precipitation and probably help to explain the formation of lowland swamps and (later) coals (Otto-Bliesner, 1993).

About 180 million years ago, as North America and Africa moved north, they began to separate. This separation marked the birth of the Atlantic Ocean (see Chap. 1). The rapid seafloor spreading and the development of the Atlantic basin lasted almost 100 million years and corresponded to a very warm period in geologic history. One possible explanation for this warmth is that a massive increase in atmospheric carbon dioxide was associated with the volcanism that accompanied the seafloor spreading and the distancing of Chesapeake Bay from Gibraltar (van Andel, 1985).

Several factors have been involved in the transition from this warm epoch, the Mesozoic, to the cool epoch, the Cenozoic, in which we now live. Over the past 100 million years, the spreading of the seafloor has slowed, CO_2 levels have dropped, and consequently the climate has cooled. In addition, northern continents have continued to move slowly poleward, lowering temperature further. The breaking apart of North America, Greenland, and Europe made three relatively small land masses from one large continent. These smaller land masses experienced decreased midcontinent summer temperature extremes owing to the increased penetration of maritime air masses into their interiors (North, Mengel, and Short, 1983). With lower summer temperatures, the possibility increased that snow and ice would persist through the summer months.

Finally, during the past 10 million years, a new phase of mountain building commenced in the Himalayas and Tibet, and to a lesser extent in western North America. This recent uplift of mountains and plateaus may have further cooled the climate in two ways. First, the high mountains and plateaus provide a focus for upslope precipitation on their windward sides, and increased precipitation could increase mechanical and chemical weathering rates; in turn, the weathering of silicate rocks is associated with the net removal of atmospheric carbon dioxide and lowering of temperature (Raymo, Ruddiman, and Froelich, 1988). Second, the emergence of high mountains and plateaus causes shifts in air currents, bringing polar air masses farther south in some regions and tropical air masses farther north in others (Kutzbach, Prell, and Ruddiman, 1993). For example, experiments with climate models show that uplift of mountains and plateaus in western North America causes cold air masses to push farther south in eastern North America (Ruddiman and Kutzbach, 1989). Other factors might be involved, such as the closing of the isthmus of Panama several million years ago, which led to changes in the strength of the Gulf Stream. Perhaps all of these factors contributed to the development of ice sheets in the Northern Hemisphere around 2.4 million years ago and the onset of orbitally forced glacial-interglacial cycles of increased magnitude.

The Future Climate of the Chesapeake Bay Area

What will the future bring? History indicates that climate is restless and always changing on time scales of years to millions of years. These natural processes, active in the past, will likely continue. And, for the first time, humans may bring a new factor to the sometimes cyclic, sometimes chaotic, climate equation. We are adding huge amounts of carbon dioxide to the atmosphere, largely from burning of fossil fuels but supplemented by deforestation. As stated by Roger Revelle, "human beings are now carrying out a large scale geophysical experiment of a kind that could not have happened in the past nor be reproduced in the future" (Revelle and Suess, 1957). The increase began slowly in the eighteenth and nineteenth centuries and has accelerated in the twentieth century. The present-day concentration could double by the middle of the next century (IPCC, 1996). This rising CO_2 concentration is increasing the atmospheric absorption of infrared radiation upwelling from the Earth's surface and should, if acting by itself, produce a warmer climate. Many climatic records, including those in eastern North America, show a slight warming trend in the twentieth century (see Fig. 2.3). However, it is not yet possible to conclude whether this trend is an indication of greenhouse warming or just part of a natural cycle related to variations such as the yet unexplained Little Ice Age, though evidence of greenhouse warming continues to mount (IPCC, 1996).

The predictions of changes in regional climates because of greenhouse warming are not yet considered to be reliable estimates of future conditions. One example of this relative unreliability is that different models simulate somewhat different climatic futures. Several models, however, indicate warming of 2–3°C (4–6°F) in the middle latitudes of North America. Various attempts have been made to estimate the potential effects of this magnitude and rate of warming on the vegetation of eastern North America (Davis and Zabinski, 1992). Modern-day pollen-climate relationships indicate that under warmer conditions, forests of southern pines would expand northward, replacing deciduous forests near the Chesapeake Bay. The future rate of change of vegetation might equal or exceed the maximum rates of vegetation change observed in the transition from glacial to interglacial climates, resulting in a severe disequilibrium between vegetation and climate (Overpeck, Bartlein, and Webb, 1991; Webb, 1992).

If global warming occurs, it may also be accompanied by a continuing rise in sea level by perhaps 25–50 cm (10–20 inches) over the next 100 years (sea level has risen 10 cm [4 inches] in the past century). These small but significant rises of sea level are predicted as a consequence of thermal expansion of sea water and the melting of some mountain glaciers. Chesapeake Bay will enlarge slightly.

Whatever the effects may be, human-induced climate change owing to combustion of fossil fuels will likely run its course over the next several centuries and millennia. By that time, natural forces, in particular orbital changes, can be expected to once again pace the long-term climatic trends. According to the orbital theory of climate change, which has been so successful in correctly marking the timing of past glacials, our world will move slowly toward a cold interval culminating in about 25,000 years and then, after a small recovery, a major glacial period in about 55,000 years. By then sea level will be lower and the Chesapeake Bay will have drained once again and been replaced by the valley of the Susquehanna River. And the river will once again extend itself out onto the exposed continental shelf before emptying into the Atlantic.

BIBLIOGRAPHY

Barron, W. R. 1992. Historical climatic records from the northeastern United States, 1640 to 1900. In *Climate since A.D. 1500*, ed. R. S. Bradley and P. D. Jones (London: Routledge).

Bradley, R. S., and P. D. Jones, eds. 1992. *Climate since A.D. 1500* (London: Routledge).

Bryson, R. A. 1966. Air masses, streamlines, and the Boreal Forest. *Geograph. Bull. (Can.)* 8: 228–69.

Bryson, R. A., and F. K. Hare. 1974. The climates of North America. In *World Survey of Climatology*, vol. 11, *Climates of North America*, ed. H. E. Landsberg (Amsterdam: Elsevier), 1–47.

Bryson, R. A., and T. J. Murray. 1977. *Climates of Hunger* (Madison: University of Wisconsin Press).

COHMAP. 1988. Climatic changes of the last 18,000 years: Observations and model simulations. *Science* 241: 1043–52.

Cook, E. R., and G. C. Jacoby. 1983. Potomac river streamflow since 1730 as reconstructed from tree rings. *Journal of Climate and Applied Meteorology* 22: 1659–72.

Court, A. 1974. The climates of the conterminous United States. In *World Survey of Climatology*, vol. 11, *Climates of North America*, ed. H. E. Landsberg (Amsterdam: Elsevier), 193–343.

Davis, M. B., and C. Zabinski. 1992. Changes in geographical range resulting from greenhouse warming: Effects on biodiversity in forests. In *Global Warming and Biological Diversity,* ed. R. L. Peters and T. E. Lovejoy (New Haven: Yale University Press), 297–308.

Gajewski, K. 1988. Late Holocene climate changes in eastern North America estimated from pollen data. *Quaternary Research* 29: 255–62.

Gaudreau, D. C., and T. Webb III. 1985. Late Quaternary pollen stratigraphy and isochrone maps for the northeastern United States. In *Pollen Records of Late-Quaternary North American Sediments,* ed. V. Bryant and R. G. Holloway (Dallas: American Association of Stratigraphic Palynologists Foundation), 247–80.

Grove, J. M. 1988. *The Little Ice Age* (New York: Methuen).

Hansen, J., A. Lacis, D. Rind, G. Russell, P. Stone, I. Fung, R. Ruedy, and J. Lerner. 1984. Climate sensitivity: Analysis of feedback and mechanisms. In *Climate Processes and Climate Sensitivity,* ed. J. E. Hansen and T. Takahashi. American Geophysical Union Geophysical Monogram Series 29: 130–63.

Hansen, J., A. Lacis, R. Ruedy, and M. Sato. 1992. Potential climate impact of Mount Pinatubo eruption. *Geophysical Research Letters* 19: 215–18.

Hayden, B. P. 1981. Secular variation in Atlantic coast extratropical cyclones. *Monthly Weather Review* 109: 159–66.

Imbrie, J., and K. P. Imbrie. 1979. *Ice Ages: Solving the Mystery* (Hillside, N.J.: Enslow Publishers).

IPCC [Intergovernmental Panel on Climate Change]. 1996. *Climate Change 1995: The Science of Climate Change. The Second Assessment Report of the IPCC,* ed. J. T. Houghton and others (Cambridge: Cambridge University Press).

Jefferson, T. 1825. *Notes on the State of Virginia* (Philadelphia: H. C. Carey and I. Lea).

Kutzbach, J. E., W. L. Prell, and W. F. Ruddiman. 1993. Sensitivity of Eurasian climate to surface uplift of the Tibetan Plateau. *Journal of Geology* 101: 177–90.

Kutzbach, J. E., and T. Webb III. 1991. Late Quaternary climatic and vegetational change in eastern North America: Concepts, models, and data. In *Quaternary Landscapes,* ed. E. J. Cushing and L.C.K. Shane (Minneapolis: University of Minnesota Press), 175–218.

Kutzbach, J. E., and A. M. Ziegler. 1993. Simulation of late Permian climate and biomes with an atmosphere/ocean model: Comparisons with observations. *Philosophical Transactions of the Royal Society of London. Series B* 341: 327–40.

Landsberg, H. E. 1970. Man-made climatic changes. *Science* 170: 1265–74.

Ludlam, D. M. 1966. *Early American Winters, 1604–1820* (Boston: American Meteorological Society).

North, G. R., J. G. Mengel, and D. A. Short. 1983. Simple energy balance model resolving the seasons and continents: Application to the astronomical theory of the ice ages. *Journal of Geophysical Research* 88:6576–86.

Otto-Bliesner, B. L. 1993. Tropical mountains and coal formation: A climate model study of the Westphalian (306 MA). *Geophysical Research Letters* 20: 1947–50.

Overpeck, J. T., P. J. Bartlein, and T. Webb III. 1991. Potential magnitude of future vegetation change in eastern North America: Comparisons with the past. *Science* 254: 692–95.

Peltier, W. R. 1987. Glacial isostasy, mantle viscosity, and Pleistocene climatic change. In *North America and Adjacent Oceans during the Last Deglaciation*, ed. W. F. Ruddiman and H. E. Wright Jr. Decade of North American Geology (Boulder: Geological Society of America), 155–82.

Peteet, D. M., R. A. Daniels, L. E. Heusser, J. S. Vogel, J. R. Southon, and D. E. Nelson. 1993. Late-glacial pollen, macrofossils and fish remains in northeastern USA: The Younger Dryas oscillations. *Quaternary Science Reviews* 12: 597–612.

Raymo, M. E., W. F. Ruddiman, and P. N. Froelich. 1988. Influence of late Cenozoic mountain building on ocean geochemical cycles. *Geology* 16: 649–53.

Revelle, R., and H. E. Suess. 1957. Carbon dioxide exchange between atmosphere and ocean and the question of an increase in atmospheric CO_2 during the past decade. *Tellus* 9: 18–27.

Rind, D., and J. Overpeck. 1993. Hypothesized causes of decade-to-century-scale climate variability: Climate model results. *Quaternary Science Reviews* 12: 357–74.

Ruddiman, W. F., and J. E. Kutzbach. 1989. Forcing of late Cenozoic northern hemisphere climate by plateau uplift in Southern Asia and the American West. *Journal of Geophysical Research* 94(D15): 18409–27.

van Andel, T. H. 1985. *New Views of an Old Planet* (Cambridge: Cambridge University Press).

Vokes, H. E. 1957. *Geography and Geology of Maryland.* Bulletin 19 (Baltimore: State of Maryland Board of Natural Resources, Department of Geology, Mines and Water Resources, J. T. Singewald Jr., Dir.).

Wahl, E., and T. L. Lawson. 1970. The climate of the mid nineteenth century United States compared to the current normals. *Monthly Weather Review* 98: 259–64.

Webb, R. S., K. H. Anderson, and T. Webb III. 1993. Pollen response-surface estimates of Late-Quaternary changes in the moisture balance of the northeastern. *Quaternary Research* 40: 213–27.

Webb, T., III. 1992. Past changes in vegetation and climate: Lessons for the future. In *Global Warming and Biological Diversity,* ed. R. L. Peters and T. E. Lovejoy (New Haven: Yale University Press), 59–75.

———, ed. 1998. Late Quaternary climates: Data syntheses and model experiments. *Quaternary Science Reviews* 17: 465–688.

Webb, T., III, P. J. Bartlein, S. P. Harrison, and K. A. Anderson. 1993. Vegetation, lake-levels, and climate in eastern United States. In *Global Climates since the Last Glacial Maximum,* ed. H. E. Wright Jr., J. E. Kutzbach, T. Webb III, W. F. Ruddiman, F. A. Street-Perrott, and P. J. Bartlein (Minneapolis: University of Minnesota Press), 415–67.

Wells, G. L. 1983. Late-glacial circulation over central North America revealed by aeolian features. In *Variations in the Global Water Budget,* ed. A. Street-Perrott, M. Beran, and R. Ratcliffe (Dordrecht, Netherlands: D. Reidel), 317–30.

Wright, H. E., Jr., J. E. Kutzbach, T. Webb III, W. F. Ruddiman, F. A. Street-Perrott, and P. J. Bartlein, eds. 1993. *Global Climates since the Last Glacial Maximum* (Minneapolis: University of Minnesota Press).

CHAPTER THREE

Forests before and after the Colonial Encounter

GRACE S. BRUSH

Sediment cores from the Chesapeake estuary, tributaries, and marshes contain pollen grains, seeds, charcoal, and other materials that are used as evidence for changes in forest composition and the environment. Changes in climate are reflected by increased charcoal in sediments which indicate warm, dry conditions. Widespread disturbance of the land, such as deforestation and cultivation for agriculture, gives rise to increased ragweed pollen and sediment accumulation from soil erosion.

The present landscape surrounding the Chesapeake Bay is a patchy mosaic of forests, cultivated and abandoned agricultural fields, wetlands, residential lawns, gardens, urban asphalt, and other kinds of land cover. The entire drainage area, except for tidal wetlands, serpentine barrens, and scattered Native American dwellings, was forested prior to European settlement. Precolonial forests were all removed at one time or another within the last 200 to 300 years for lumbering, mining, agriculture, and road and railroad building. At present, the region surrounding the Chesapeake is about 40% forested. Most existing forests represent new growth on abandoned fields or lumbered areas; few are older than a century and most range in age from 30 to 70 years.

The history of the forests in the Chesapeake drainage area includes response to climate change over several millennia before European settlement as well as sudden and massive deforestation after colonization. Both sets of changes had profound effects on the landscape. Because land and water are intimately linked, land use also transformed parts of the estuary from a system of mostly bottom dwellers to one dominated by floating and swimming organisms. The effect of climate change is well documented on the land, but less so in the estuary.

The history of land use and deforestation has resulted in ecologically different landscapes through time. For example, early tobacco farming, in which small fields were planted for a few years and then abandoned to become fallow fields, resulted in fragmentation into a fine mosaic of forest patches interspersed with patches of young trees, herbs, and shrubs. In contrast, large-scale agriculture, particularly as it became more highly mechanized, stripped extensive areas of the landscape of all vegetation, leaving only a few trees in locations unsuitable for cultivation and in hedgerows used to separate property and as windbreaks. Deep plow fur rows altered soil structure and likely destroyed many native perennial species that reproduce by root and rhizome growth, such as woodland orchids and violets. This in turn would have affected the microorganism populations in the soil, many of which grow symbiotically on the roots of perennials and make nitrogen available for plant consumption. Intensive fertilization of agricultural fields also changed soil nutrients and microstructure. The most extensive land clearance in the region of the Chesapeake occurred in the late 1800s and early 1900s, when 80% of the land was cleared of forests. Later, wetlands were drained for arable land. The net result of these alterations to the land was a more homogeneous landscape. There has been some afforestation since the 1930s, as increased numbers of abandoned farms were vegetated primarily by seedlings of natural plants. Over time these abandoned fields have become forest stands.

Species composition has changed since European colonization as a result of disease. The chestnut blight, introduced in the late 1920s, brought about the demise of the American chestnut, which in many areas in Maryland constituted about 30% of forests. More recently, American elm, common throughout the region, has been greatly reduced by the Dutch elm disease, and Eastern hemlock by the woolly adelgid.

Distributions of existing forests in Maryland were compiled into a map based on field data collected from 1973 to 1976 (Brush, Lenk, and Smith, 1980). The different species associations are related to geologic substrate and soils (Fig. 3.1). The region surrounding the Chesapeake includes parts of the Coastal Plain and Piedmont Provinces. Extensive floodplains and numerous tidal marshes characterize the topographically flat Coastal Plain, where crystalline metamorphic and sedimentary bedrock is overlain by partially unconsolidated sedimentary deposits. Soils

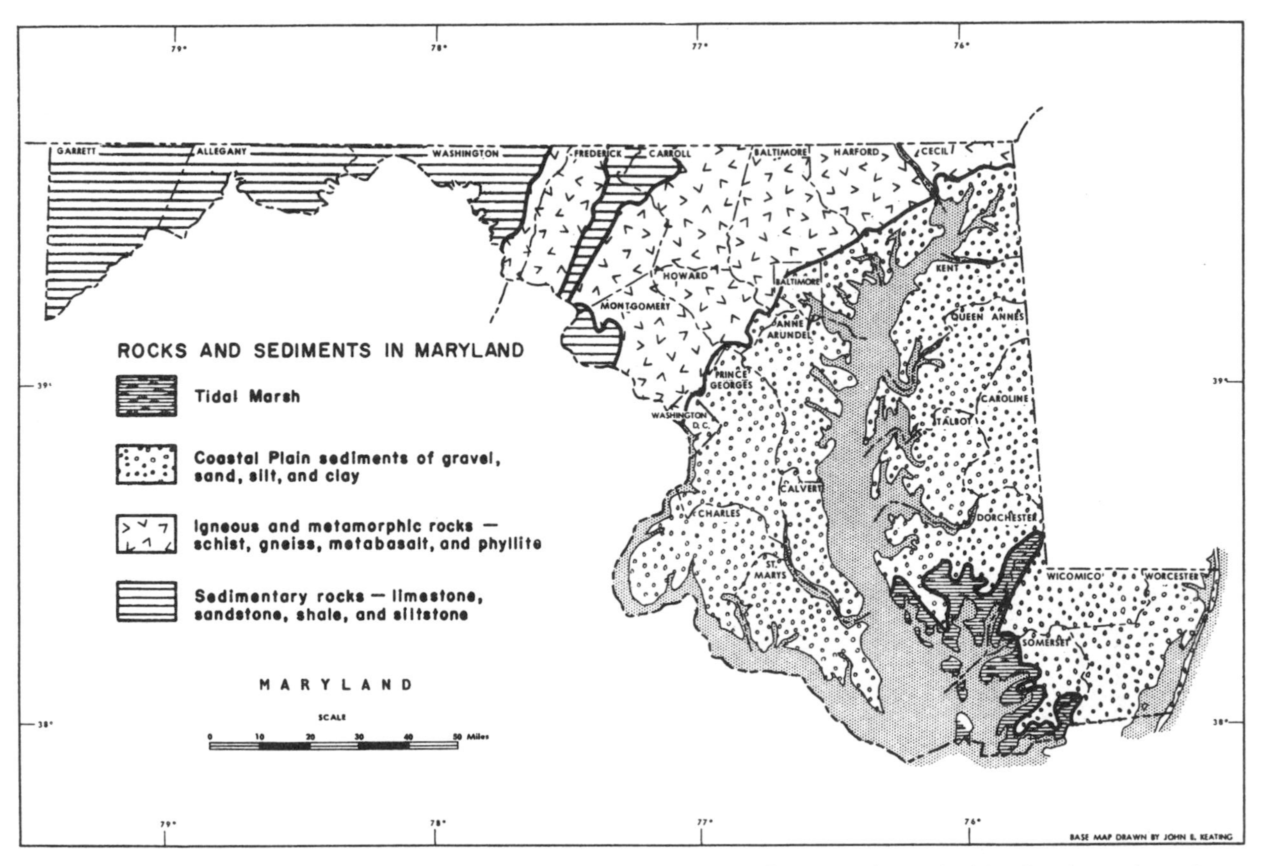

Fig. 3.1. A map of the major rock and soil types in the Coastal Plain, Piedmont, and Appalachian Provinces throughout Maryland (from Pomerening, 1967).

include fine silt and clay, with some gravel adjacent to the streams, where flooding is frequent; mesic silt-loam, with a high water-holding capacity and sufficient drainage to avoid waterlogging; well-drained sand and gravel, with a low water-holding capacity; and soils, called fragipan, that are functionally shallow because of an impermeable layer close to the surface.

The hillier Piedmont Province to the west is separated from the Coastal Plain by the Fall Zone, the headwater region of tidal streams. Piedmont soils consist of thick silt-clay soils called saprolites, weathered from underlying igneous and metamorphic bedrock and often reaching depths of 12 m. Intermingled with the thick saprolites are areas of serpentine and quartzite rock generally lacking any soil. Serpentine occurs sporadically throughout the area. Soldier's Delight, in the Reisterstown area, northwest of Baltimore City, was mined for chromium in the late nineteenth century. Quartzite rocks in general occur on the tops of ridges and contain boulders and broken-up rock, with very little silt and clay. Floodplains bordering streams are periodically flooded and underlain by silt and clay transported there by the streams when they flood the adjacent land. This water-transported soil is called alluvium.

Areas underlain by serpentine, quartzite, gravel, and fragipan represent the driest substrates in the drainage area, because there is very little if any soil to hold water. In the case of gravel, the substrate is so porous that water is rapidly drained into the groundwater, which is sometimes too deep to be accessible to plants. Consequently, the only water available for plants is that derived directly from precipitation. Various species of oak, including blackjack, post, and chestnut oak, characterize these regions. In the most extreme of these environments, such as the serpentine, trees exhibit the stunted growth and thick leaves characteristic of arid areas. Floodplain alluvium contains the wettest and often most nutrient-rich substrates and supports trees such as sycamore, silver maple, box elder, and green ash in the Piedmont Province and also river birch in the Coastal Plain Province. The more mesic areas, underlain by thick saprolite in the Piedmont and clay-loam soils in the Coastal Plain, support forests characterized by about 20 species of oak, black gum and sweet gum, species of hickory, beech, and a number of other species. Altogether there are about 150 woody species in the Maryland portion of the Chesapeake drainage area. The close relationship between existing remnant forests and the substrate's water-holding capacity indicates the overriding influence of geology and soils on species distributions.

Are these fragmented forests—the offspring of a long history of forest dissection and land use—anything like the forests the colonists encountered when they first landed here? Some historical documents contain descriptions of the landscape, including the forests, when the settlers arrived (Semmes, 1929; Stetson, 1956), but none describe environmental conditions prior to European colonization. The post-European history itself is fragmentary at best in documentation of environmental change. The dearth of historical records has led some to the conclusion that change results only from human activity; the corollary is that in the absence of human activity there was a never-changing forest primeval. Thus, humans alone are held responsible for changes to the landscape. It is certainly true that humans have changed the landscape. But did the precolonial forests change independent of human intervention, and how would one find this out? Those two questions form the body of this chapter.

History Contained in Sediments (Stratigraphy)

How does the environmental historian acquire data in the absence of written documents? Estuaries, like lakes and other open bodies of water, are depositional features that eventually become filled with sediment. The sediment comes from the land, and the rate at which it is deposited in the basin—be it in an estuary, lake, or pond—depends on whether the land that drains into the basin is covered by vegetation or cleared for lumbering, mining, or agriculture. The degree to which the basin becomes filled with sediment depends also on the size of the basin, and in the case of estuaries also on the rate of rise of sea level and land subsidence.

The present Chesapeake Bay is the most recent estuary to have formed along the Coastal Plain of Maryland and Virginia during the Pleistocene, which includes the last several million years. For most of the Pleistocene, glaciers covered much of the Earth, for as long as 100,000 years at a time. Sea level was much lower than today because water was locked in ice, and much of the continental shelf was terrestrial. In the intervening warm periods, each of which lasted about 10,000 years, river valleys extending from the shelf inland were submerged by rising sea level as the glaciers melted and receded northward. Estuaries formed as seawater flowing landward met river water flowing to the ocean. Some 10,000 to 15,000 years ago, the head of the Chesapeake Bay was where the mouth is today. As the ice melted and sea level continued to rise, the Chesapeake Bay reached its present configuration.

Much of the sediment deposited in the estuary is silt and clay, to which is attached nutrients, trace metals, and other substances derived either from weathering rock or decaying vegetation. Also attached to the silt and clay particles are nutrients from fertilizers and chemicals either applied to the landscape or discharged directly into the estuary. In addition to sediment carried by streams from the land, pollen grains, seeds, and other parts of plants, such as leaf litter, are transported to and deposited in the estuary. How far the material will be transported into the estuary depends on the size and weight of the exported particles and the depth and turbulence of the receiving water. Materials originating in the atmosphere are also deposited in the estuary. These include windblown pollen grains of many tree species, as well as chemicals such as nitrogen, lead, and sulfuric acid, many of which are discharged from industrial sources. Along with these external materials, organisms that live in the estuary itself—as evidenced by the outer coverings (silica frustules) of diatoms, spores of dinoflagellates, parts of insects, invertebrates, and fish scales—settle to the bottom and are buried by the incoming sediment. Not everything deposited in the estuary is preserved. Only those organisms or parts of organisms resistant to physical or biological decay, and more or less inert chemicals are preserved in the accumulating sediment.

Organisms, chemicals, and other identifiable materials preserved in the sediment represent the particular conditions of climate and land use at the time of deposition. Thus, when a plastic or metal tube is inserted into the sediment, the core of sediment extracted from the depositional basin contains a history of environmental conditions over the period of time that the sediment was deposited. The history derived from the sediment is referred to as stratigraphy. Just as the value of a manuscript is based on a correct and complete sequence of pages, so the stratigraphic record is useful as a historical document only if deposition has been continuous and the sediments once deposited have not been reworked either by physical scouring or biological activity.

To guarantee stratigraphic integrity and to read the historical record contained therein, the sediment core is analyzed first to identify any features that signify disturbance, such as evidence of scouring and burrows left by worms or other organisms that could turn the sediment over or mix it up after it has been deposited. X-rays are used to identify features not visible to the naked eye. If the core shows any evidence of disturbance, it is not used for historical reconstruction.

Dating of Sediments and Development of Chronologies

In order to convert core depth to time, sediment cores are dated. All cores must be individually dated, no matter how close the collection sites, because sediment deposition is highly variable in the estuary. For example, 1 m of sediment might be deposited over 100 years at one site, and at another location a few meters distant, a meter of sediment may span 1,000 years. This variation in sedimentation rates (the time required for 1 cm of sediment to be deposited) is a result of the hydrodynamic complexity of the shallow estuary, where wind and storms greatly affect fluid motion and the transport of materials.

Several methods are used to date sediments, including measuring decay rates of radioactive isotopes and changes in pollen that reflect historically dated changes in vegetation (that is, changes already established and dated by other means). Carbon-14, with a half-life of 5,730 years, is used for dating sediments deposited between 500 and 50,000 years ago. Lead-210, with a half-life of 27 years, is used for sediments less than 100 years old. Finding a high amount of cesium-137 in sediments generally dates the sample to 1963 or 1964, when the testing of atomic weapons, begun in 1955, was most intense. Historic changes in vegetation are often reflected in changes in the pollen record. In the Chesapeake Bay region, historically dated agricultural activities are recognized in the sediment by an increase in ragweed pollen and a change in the ratio of oak pollen to ragweed pollen. The demise of the American chestnut owing to disease in the 1920s and 1930s can be seen in some Chesapeake sediment cores by the absence of chestnut pollen above a certain depth, thereby providing a datable horizon. If it is known when exotic plants were introduced in a particular area, the presence of the plant's pollen in the sediment can be used as a time horizon. Historically dated exotics have provided a number of dated horizons for sediment cores collected in estuaries and marshes in California, for example. Historical horizons used for dating Chesapeake Bay cores are listed in Table 3.1.

Chronologies are constructed for each sediment core by adjusting the average sedimentation rates between dated horizons according to the pollen concentration in each subsample. For example, the average sedimentation rate between the 1780 agricultural horizon and the horizon where chestnut pollen is absent (1930) might be 1 cm yr^{-1}, but as mentioned earlier, sedimentation is highly variable in the estuary, and the sedimentation rates will not be similar throughout the period from 1780 to 1930. Therefore, in order to compile an accurate chronology, it is necessary to determine sedimentation rates for all subsamples of a core. This is done by assuming that for short periods of time the total pollen rain is more or less uniform, even though there are yearly seasonal variations, and that most of the pollen is wind transported and enters the estuary

Table 3.1 Pollen horizons used for dating sediments

Time horizon (range)	Change in land use/vegetation	Indicator in sediment core
1930 (1923–27)	decline of American chestnut	decrease in chestnut pollen to <1%
1840 (1820–60) 1780 (1760–1800)	40–50% of land cleared	% ragweed pollen >10; ratio of oak to ragweed pollen <5 (generally <1)
1730 (1720–40) 1650 (1640–60)	<20% of land cleared (generally <5% of land cleared for initial and tobacco farming)	% ragweed pollen >1 to <10; ratio of oak to ragweed pollen >5 (generally >10)
Pre-European	Indian agriculture	% ragweed pollen <1 or absent

from the atmosphere. This assumption is valid because many trees and shrubs in this region are wind pollinated and they produce by far the largest amount of pollen.

Pollen grains are small particles that have transport properties similar to fine silt and clay. Thus, if most pollen introduced into the estuary is windblown from terrestrial vegetation and independent of the sediment introduced with river runoff, and if its influx is more or less uniform (e.g., there is no significant change in vegetation cover and species composition), then the pollen concentration in any level of the sediment core will reflect the rate of accumulation of the other particles that make up the sediment. If the rate of sediment accumulation increases, the concentration of pollen will be correspondingly less, and if sediment accumulation decreases, pollen concentrations will be correspondingly greater.

Using these assumptions, sedimentation rates for each sample of a core can be calculated. Once a sedimentation rate is derived for each sampled level of the core, the number of years required for the deposition of that layer can be calculated by dividing the sedimentation rate for the particular sample by the depth of the sample. For example, if the depth of the sample is 1 cm and the sedimentation rate is 0.1 cm per year, the number of years required for that cm of sediment to be deposited is 1/0.1 = 10 years. Working from the top of the core down, a chronology can then be constructed which assigns specific years to each layer of sediment. For example, Table 3.2 shows the reconstructed chronology of part of a sediment core taken in 1980 (the sedimentation rate is expressed as *R*). Sedimentation rates determine the resolution of the chronology.

Table 3.2 Chronology of a portion of a sediment core collected in 1980

Depth (cm)	R (cm yr^{-1})	Number of years	Chronology (A.D.)
0–2	0.96	2.1	1980–1978
2–4	1.20	1.7	1978–1976
4–6	0.98	2.0	1976–1974
6–8	0.79	2.5	1974–1972
8–10	1.33	1.5	1972–1970
10–12	0.85	2.3	1970–1968
12–14	1.80	1.1	1968–1967
14–16	1.02	2.0	1967–1965
16–18	0.72	2.8	1965–1962
etc.			

Stratigraphic Indicators

Input rates, called influxes, of organisms and chemicals, which are the number of organisms or amount of chemicals deposited per square cm per year, can be calculated for any component in any interval, or sample, of the core by dividing the concentration of the component (the number of organisms in or the weight of 1 cubic cm or one ml) by the sedimentation rate (cm/year) for the interval.

Like many other ancient records written in archaic language, stratigraphy must be deciphered and translated into history. Remember, the stratigraphic record contains a history of only those components that are preservable. Of those that are preserved, not all are identifiable. And of those that can be identified, many are organisms or chemicals that are found under various environmental conditions. Organisms preserved in the sediment for which ecological requirements and tolerances are known can be used as indicators of environmental conditions. Distributions of chemicals in the sediment can also be used to indicate environmental conditions if the processes that govern their introduction and distribution in the estuary are known. Examples of some indicators used to reconstruct history from stratigraphy are shown in Table 3.3.

Differential pollen and seed production by different species necessitates calibrating pollen and seeds to the abundance or vegetation cover

Table 3.3 Some paleoecological indicators of environmental conditions

Environmental conditions	Fossil indicator
Regional climate	
Dry	
Carya (hickory), *Pinus* (pine), *Quercus* (oak), *Castanea* (chestnut), *Ilex* (holly)	pollen
Wet	
Liquidambar (sweet gum), *Nyssa* (black gum)	pollen
Local hydrologic conditions	
Standing water, continual submergence	
Zannichellia palustris (horned pondweed), *Najas* spp. (naiads), *Potamogeton* spp. (pondweeds), *Vallisneria americana* (wild celery), *Elodea canadensis* (waterweed)	seeds
Intertidal, flooding twice daily, 9–12 hrs/tidal cycle	
Zizania aquatica (wild rice)	seeds
Pontederia cordata (pickerelweed), *Peltandra virginica* (arrow arum), *Nuphar advena* (spatterdock)	pollen
Channel bank, flooding twice daily, 3–5 hrs/tidal cycle	
Acnida cannabina (water hemp), *Polygonum punctatum* (dotted smartweed)	seeds
High marsh, flooding may occur 0–4 hrs/tidal cycle	
Typha spp. (cattails)	seeds, pollen, stems
Bidens laevis (smooth bur-marigold), *Leersia oryzoides* (rice cutgrass), *Cyperaceae* spp. (sedges)	seeds
Swamp, flooding 100% of the growing season	
Salix nigra (black willow)	seeds, leaves
Riparian floodplain forest, flooding ~25% of the growing season	
Acer negundo (box elder)	seeds, pollen
Fraxinus pennsylvanica (green ash), *Betula nigra* (river birch)	pollen
European agriculture	
Increase in *Ambrosia* spp. (ragweed)	pollen
Erosion from deforestation and agriculture	
Increased sedimentation rates	
Nutrient loading	
Total organic carbon, nitrogen, phosphorus, and biogenic silica	
Oligotrophic (low nutrient) conditions: 4 to 6 species of submerged macrophytes	seeds
Eutrophic (high nutrient) conditions or high turbidity: *Zannichellia palustris* (horned pondweed)	seeds
Anoxia	
Degree of pyritization of iron	ratio of pyritic iron to the sum of acid-soluble iron and pyritic iron
Fire	
charcoal	
Pteridium (bracken fern)	spores

of the source plants. Pollen and seeds in surface sediments are compared with the number and size of plants measured in the same area. A comparison of pollen percentages in surface sediments of the Potomac River with the percentage basal area of trees along a broad band adjacent to the river shows differences in pollen representation of forests in this area. For example, birch and elm are closely represented by pollen, whereas tulip poplar, sweet gum, red maple, and beech are underrepresented, and sycamore, walnut, and hemlock are overrepresented.

History of Forests in the Chesapeake Region

The history of forests in areas drained by the Chesapeake is reconstructed from pollen and seed stratigraphy of sediments deposited in the Chesapeake Bay and tributaries since the last glaciation. Cores from different locations span different periods of time, depending on the depositional pattern. Some records extend back several thousand years to the time when glaciers retreated northward, sea level rose because of melting ice, and the Chesapeake Bay was formed as the sea invaded the land. Others include shorter periods of time, when climate varied around changes in precipitation rather than temperature. The record contained in the sediments allows for comparisons of the effect of climate on forest composition, as well as the effects of human activity.

The last glaciation, some 15,000 years ago, extended just north of Maryland into Pennsylvania, reaching the vicinity of Scranton in the east. During full glacial time, sea level was on the order of 100 m lower than it is today and the estuary that is now the Chesapeake Bay was the valley of the Susquehanna River. Much of the presently submerged continental shelf was terrestrial during that time. Sediment cores collected at the mouth of the Chesapeake Bay show that the region surrounding the Bay (which was then the Susquehanna River valley) supported an assemblage of plants including the northern taxa fir and spruce. Pine, birch, and alder were also present (Harrison et al., 1965).

A 20,000-year sequence of pollen grains contained in Buckles Bog in the unglaciated part of the Appalachian Plateau in western Maryland shows relatively rapid changes in species composition after the full glacial period (Maxwell and Davis, 1972). During full glacial time, before the glaciers began to melt and recede northward, tundra grew at high elevations. About 13,000 years ago tundra was replaced by open spruce forests consisting of a mixture of spruce and jack or red pine. The open spruce forests lasted about 2,500 years until a mixed coniferous-deciduous forest of white pine, hemlock, birch, fir, ash, hornbeam, and oak replaced the boreal forest 10,500 years ago. Oak-chestnut forests did not become prominent until 5,000 years ago. Beech and chestnut increased quickly and were followed by large increases in hickory. The top 3 cm of the sediment, which was deposited during the last few centuries, contains up to 17% ragweed, reflecting the post-European deforestation of the Appalachian Plateau.

The longest continuous history of postglacial vegetation in the Mid-Atlantic region is contained in sediments deposited in the floodplain of Indian Creek, a tributary of the Anacostia River close to Washington, D.C. (Fig. 3.2). The sequence begins 12,000 years ago, with a coniferous forest of fir, spruce, and pine present for about 1,000 years. Alder, ash, birch, hornbeam, and hazelnut were also present. There was some oak but no hickory, and sweet gum and black gum were absent. Among the non-tree taxa, grasses and sedges were dominant. During the next 500 years, fir and spruce declined and pine and ash increased, indicating a possible warming trend. This was followed by a reversal, with increased amounts of spruce and birch and a large decline in ash, suggesting a relatively sudden switch to cooler conditions.

The reversal to cooler conditions following initial glacial retreat has been observed in many parts of the world; it is known as the Younger Dryas period. It is attributed to increased runoff of very cold water from the melting glaciers into warmer water in unglaciated regions. After 10,000 years, spruce and fir diminished and never returned to the area. Hemlock dominated the landscape for 5,000 years. Black gum appeared and gradually increased. The dominance of pine and hemlock, with some deciduous taxa (such as black gum, alder, and birch) but with few herbaceous plants indicates a closed canopy, mixed coniferous-deciduous forest. Large amounts of charcoal, accompanied by oak and hickory, turn up in the sediment after 5,000 years, indicating a warmer and drier climate characterized by frequent fires. The role of fire in providing open space for herbs and shrubs in the forest is indicated by an increase in non-tree pollen, particularly blueberry and arrowwood, coincident with the increase in pollen of oak and hickory. A similar understory was observed in sediments from Buckles Bog in the Appalachian Province. The oak-hickory forest spanned 3,500 years prior to European settlement. During this interval, while oak and hickory were increasing, pine decreased and sweet gum appeared on the landscape. Herbaceous plants, like goldenrod, increased, and members of the blueberry family emerged to form a major component of the vegetation.

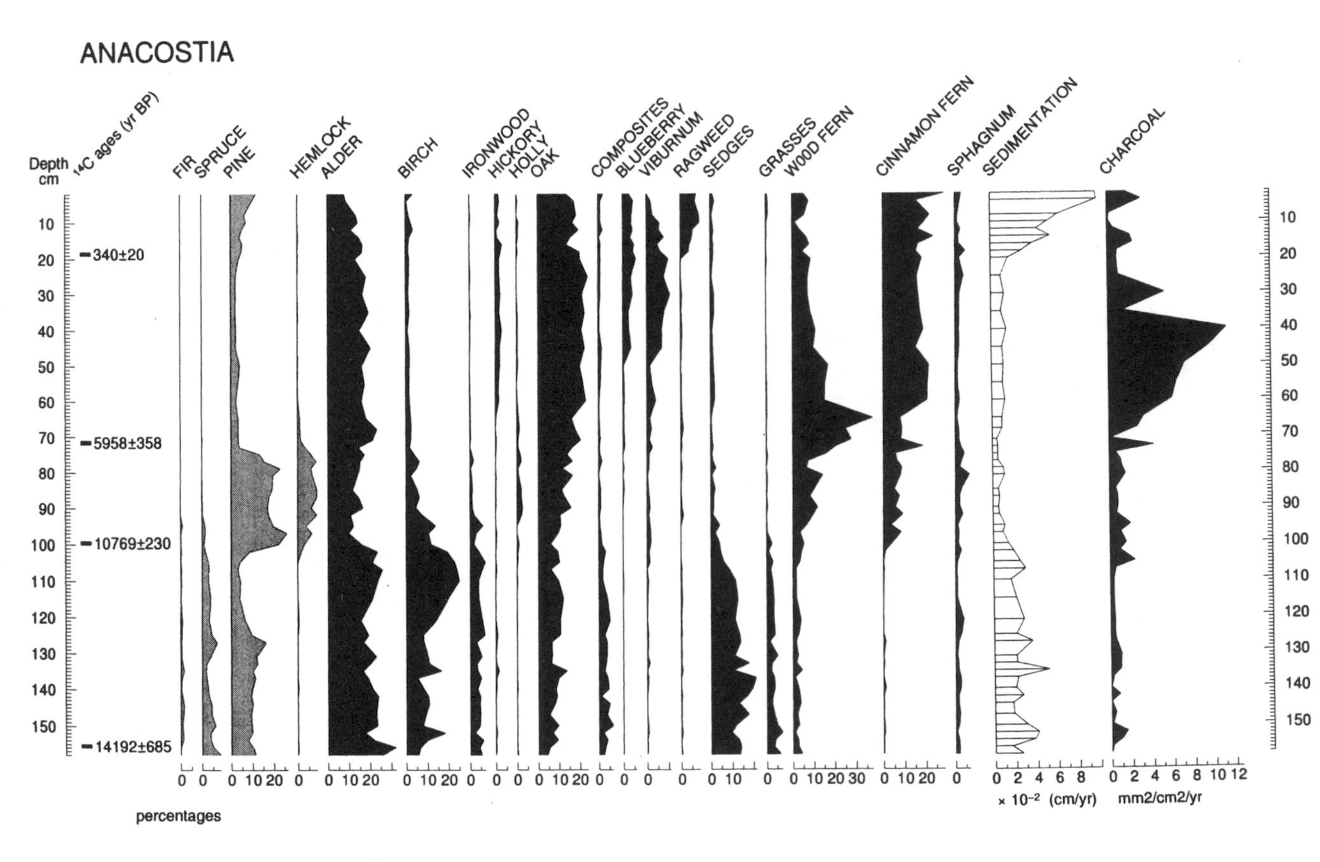

Fig. 3.2. A 12,000-year pollen profile from a sediment core collected in the floodplain of the Anacostia River.

Ragweed is a major portion of the pollen flora in sediment deposited in the last 350 years at the Anacostia site. The pollen records a largely non-arboreal vegetation and reflects the change in the landscape from forested to herbaceous vegetation after European settlement.

A similar pattern of growth is seen north of Indian Creek in the Magothy River. Pollen extracted from sediment cores collected in the headwaters of the Magothy River provides a continuous record of the forests adjacent to the river for the last 4,000 years (Fig. 3.3). Major shifts in plants occupying wet and dry habitats are shown to have occurred over that time period. From 4,000 to 2,500 years ago, black gum and sweet gum were the dominant trees and were accompanied by river birch and cinnamon fern. These taxa indicate wetter conditions. After 2,500 years, sweet gum and black gum were replaced by chestnut, American holly, and blueberry, which thrive in drier conditions and which remained dominant until European settlement. Within this dry period were two intervals of exceptionally dry conditions, about 1,500 years ago and later about 800 to 900 years ago.

Pollen, charcoal, and metals extracted from a sediment core collected in the Nanticoke River with a 1,500-year history show a pronounced dry period from 700 to 900 years ago, synchronous with the dry episode in the Magothy River (Fig. 3.4). This warm, dry horizon is also present in cores collected from marshes in the same region. The period is distinguished by a high ratio of plants that occupy dry habitats (such as oak, hickory, and pine) to those that occupy wet habitats (such as river birch, sweet gum, and black gum). The 700- to 900-year-old period is contemporaneous with the historically documented Medieval Warm Period. Along with large amounts of charcoal, the interval is characterized by high sedimentation rates. It is interpreted as representing a warm, dry period where frequent fires, particularly on Coastal Plain sandy soils, resulted in a burned landscape with episodically high rates of erosion and sediment runoff.

A sediment core from Furnace Bay in the upper Chesapeake Bay—which includes a 1,500-year history prior to European settlement, as well as the post-European interval—provides a comparison of the pollen representation of forests prior to colonization, at the time of the colonial encounter, and during the different phases of land use in postcolonial time (Fig. 3.5). During the 1,500 years prior to colonization, pine, hemlock, oak, and hickory remained relatively uniform, but there were fluctuations in plants that occupy wet habitats, such as ash, walnut, alder, and birch. Tree pollen in sediments deposited since European settlement follows

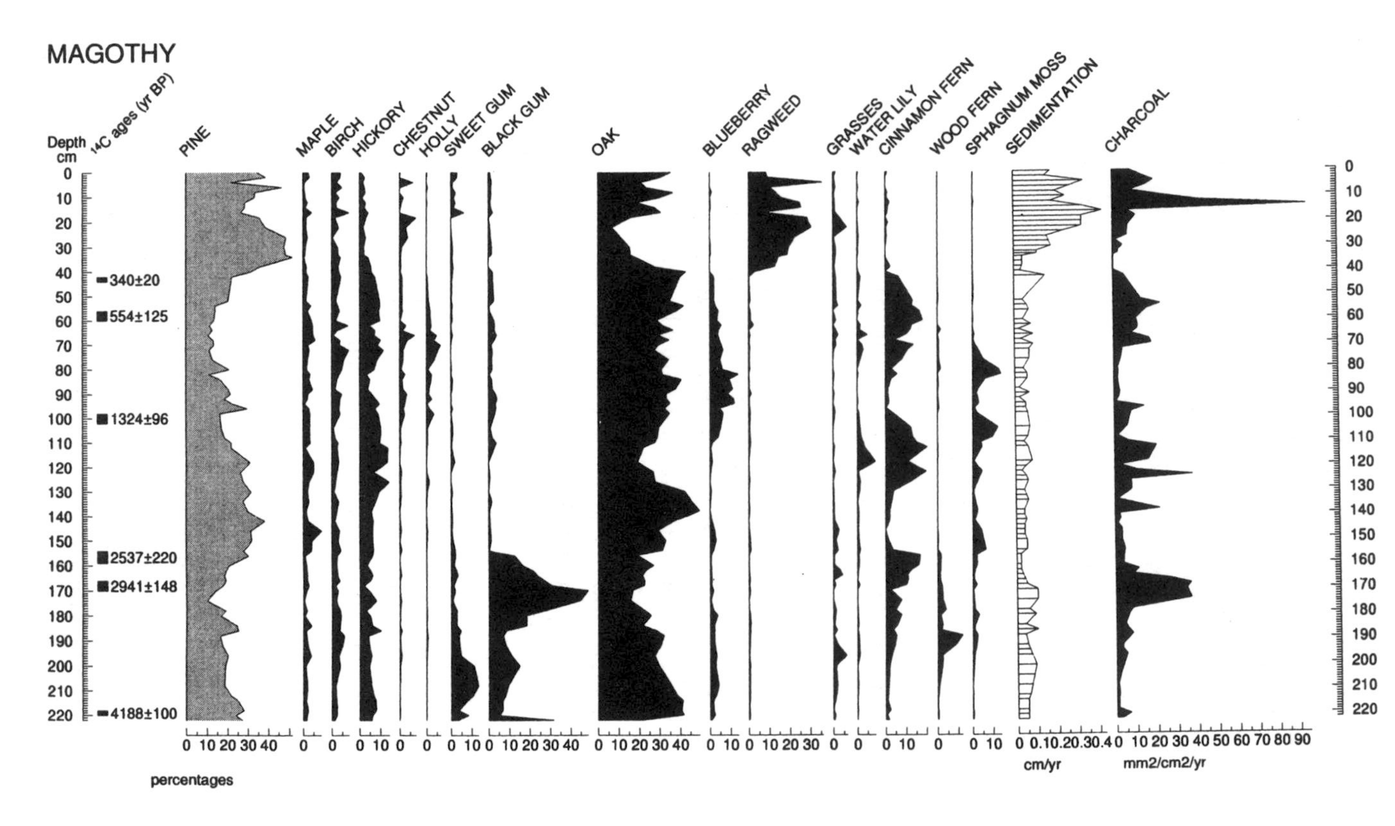

Fig. 3.3. A 4,000-year pollen profile of the major forest taxa in the Magothy River.

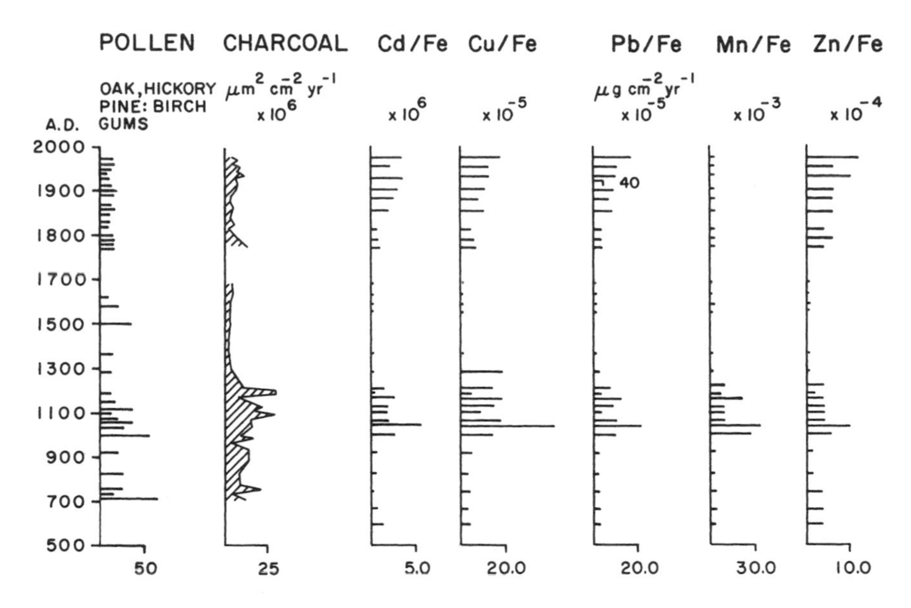

Fig. 3.4. A 1,200-year profile of pollen, charcoal, and metals in the Nanticoke River (from Brush, 1986).

fairly closely the pattern of deforestation. The demise of the American chestnut through disease in the 1920s is represented in the pollen profile by the absence of chestnut pollen. Post-1960 chestnut represents the Oriental chestnut, which was planted to replace the American chestnut in the 1940s. Flowering probably did not begin until 15 or more years after planting. Basswood (not shown in Fig. 3.5) was present prior to colonization, but it does not appear in the pollen record in postcolonial time.

Summary and Discussion

Pollen and seed stratigraphy in sediments deposited in the Chesapeake Bay system indicates that forest composition shifted dramatically over the past several millennia because of changes in climate. Major shifts occurred even within the last several centuries. All of these changes took place prior to European settlement and are synchronous with climate change documented from stratigraphy and history elsewhere on Earth. Pollen profiles indicate that the patterns of forest types at the time

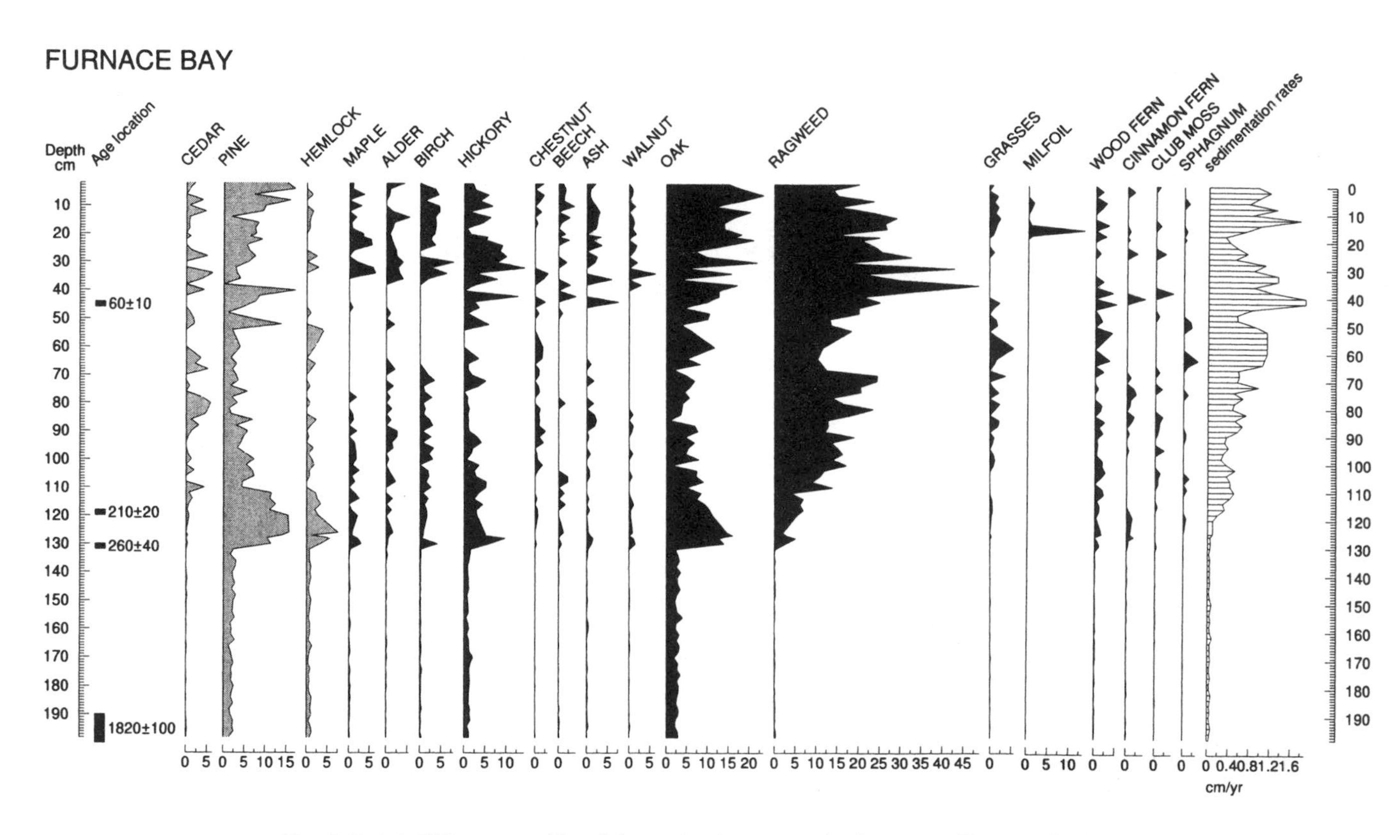

Fig. 3.5. A 1,700-year profile of the major forest taxa in the upper Chesapeake Bay.

of colonization were similar to what they would be today if the entire landscape were still forested, as it was when the colonists arrived. There have been a few changes in species distributions. Two species, American chestnut and American elm, have been reduced practically to extinction by disease. Virginia pine appears to be replacing red cedar in serpentine areas, for reasons unknown but possibly related to post-European fire control. The real legacy of European colonization has been the transformation of a highly variable, almost completely forested landscape into a herbaceous-dominated system, interspersed with different-sized fragments of secondary forests in different stages of succession. The result has been a vast reduction of forest habitat for wildlife. It is difficult to predict what would happen to today's largely deforested landscape if it were subjected to the kind of climate changes seen in the past.

The land and water are inextricably connected. The transformation of the landscape from forests to mostly agricultural and urban land has been accompanied by a change in the Chesapeake biota from bottom-dwelling organisms in precolonial time to a dominance of floating and swimming organisms today. Thus, the effects of European land use are most dramatically realized in the loss of much of the bottom-dwelling estuarine resource from decreased water clarity, increased nutrients, and decreased oxygen directly related to land use. This fundamental change is illustrated in stratigraphy by changes in sedimentation rates, chemistry, submerged aquatic vegetation, and diatoms in sediment deposited over the past few centuries.

Sedimentation rates doubled to quadrupled in the shallow waters of the Bay and in the upper tributaries, more or less following the pattern of land use, with more sediment being deposited during periods of most extensive deforestation. Estuarine waters became muddier, with increased sedimentation; the resulting turbidity limited the amount of light reaching the bottom of the water column. Thus bottom-dwelling, or benthic, plants, including algae and submerged grasses, were stressed. The sediment coming into the estuary carried nutrients from fertilizers, providing additional food for aquatic plants (mainly algae), particularly those living in the upper parts of the water column where light limitation was not a factor. As the waters became more nutrient-rich, or eutrophic, these organisms, many of which are diatoms, proliferated and reduced still further the amount of light reaching bottom waters. Eventually, many of the benthic populations disappeared. This was most dramatically illustrated by widespread extinctions of submerged grasses throughout much of the Chesapeake shallow waters in the early 1970s, and it is recorded in seed stratigraphy. Diatom stratigraphy also shows a change from larger benthic species to very small free-floating (planktonic) forms.

High sedimentation rates accompanied by the preservation of large amounts of carbon, nitrogen, sulfur, and iron in sediments deposited after European settlement indicates that low-oxygen conditions may have become more persistent as water nutrient levels increased, thus further stressing the benthic populations. In estuaries, lighter fresh water flows over the incoming heavier sea water, creating a stratification that is particularly strong in spring when the amount of fresh water discharged into the estuary is highest. It is also strong during years of high spring precipitation. As a result there is very little mixing of the top and bottom waters, and the bottom, more saline, water can become depleted of oxygen if the demand exceeds the available oxygen. This can happen when, along with being consumed by bottom-dwelling organisms, oxygen in the bottom water is depleted as large numbers of algal cells produced by excess nutrients in the top waters die and settle to the bottom, using oxygen as they decay, particularly during spring and summer when stratification is greatest.

History, whether derived from written records or stratigraphy, is essential in understanding how complex ecosystems function. The close connection between deforestation and the decline of the estuarine resource is clearly illustrated in the stratigraphic record. Stratigraphy has also shown that this system is vastly more dynamic and complicated than previously imagined. Climate change over the last few millennia has produced very different assortments of forest species than were present at the time of colonization. New research on estuarine stratigraphy will explore the nature of the estuary during those periods of different forest composition.

BIBLIOGRAPHY

Boynton, W. R. 1997. Estuarine ecosystem issues on the Chesapeake Bay. In *Ecosystem Function and Human Activities: Reconciling Economics and Ecology*, ed. R. D. Simpson and N. L. Christensen Jr. (Holladay, Utah: Chapman), 71–94.

Brush, G. S. 1982. An environmental analysis of forest patterns. *American Scientist* 70: 18–25.

———. 1986. Geology and paleoecology of Chesapeake estuaries. *Journal of the Washington Academy of Sciences* 76(3): 146–60.

Brush, G. S., C. Lenk, and J. Smith. 1980. The natural forests of Maryland: An explanation of the vegetation map of Maryland (with 1:250,000 map). *Ecological Monographs* 50: 77–92.

Byrne, R., J. Michaelsen, and A. Soutar. 1977. Fossil charcoals as a measure of wildlife frequency in southern California. In *Proceedings of the Symposium on the Environmental Consequences of Fire and Fuel Management in Mediterranean Ecosystems,* ed. H. A. Mooney and C. E. Conrad. General Technical Report WO-3 (Washington, D.C.: USDA Forest Service), 361–67.

Cooper, S. R., and G. S. Brush. 1991. Long-term history of Chesapeake Bay anoxia. *Science* 254: 992–96.

Foresman, T. W., S. T. A. Pickett, and W. C. Zipperer. 1997. Methods for spatial and temporal land cover assessment for urban ecosystems and application in the greater Baltimore-Chesapeake region. *Urban Ecosystems* 1(4): 201–16.

Harrison, W. R., J. Malloy, G. A. Rusnack, and J. Terasmae. 1965. Possible late Pleistocene uplift at the Chesapeake Bay entrance. *Journal of Geology* 73: 201–29.

Maxwell, J. A., and M. B. Davis. 1972. Pollen evidence of Pleistocene and Holocene vegetation on the Allegheny Plateau, Maryland. *Quaternary Research* 2: 506–30.

Pomerening, J. A. (revised by F. P. Miller). 1967. *General soil map of Maryland.* University of Maryland Extension Bulletin 2121 (College Park: Maryland Agricultural Experiment Station and Soil Conservation Service).

Raiswell, R., F. Buckley, R. A. Berner, and T. F. Anderson. 1988. Degree of pyritization of iron as a paleoenvironmental indicator of bottom-water oxygenation. *Journal of Sedimentary Petrology* 58: 812–19.

Semmes, R. 1929. Aboriginal Maryland, 1608–1689, 2: The Western Shore, Md. *Historical Magazine* 24: 195–209.

Shreve, F., M. A. Chrysler, F. H. Blodgett, and F. W. Besley. 1910. *The Plant Life of Maryland.* Maryland Weather Service v. 3 (Baltimore: Johns Hopkins Press).

Stetson, C. W. 1956. *Washington and His Neighbors.* Richmond, Va.: Garrett and Massie.

Thornton, Peter. 1991. A paleoecological study of salt marsh sediments from the NOAA Estuarine Research Reserve site at Monie Bay, Maryland: The application of principal components analysis to fossil seed data. M.A. thesis, Johns Hopkins University.

Yuan, S., and G. S. Brush. 1992. Postglacial vegetation history of the Coastal Plain of Maryland. *Bulletin of Ecological Society of America* 73(2): 396.

CHAPTER FOUR

Human Influences on the Physical Characteristics of the Chesapeake Bay

DONALD W. PRITCHARD AND JERRY R. SCHUBEL

The dynamic balance of tides and river flow controls the motion and mixing of the Bay's water and the responses of the Bay to human influences. Understanding the physical characteristics of the Bay helps us distinguish the effects humans have on currents and mixing from those resulting from natural processes. Currents and mixing control patterns of siltation, vertical stratification, and low-oxygen zones. These patterns, in turn, affect the Bay's living organisms.

Chesapeake Bay and its network of tributaries provided early explorers and colonists with protected waters, bountiful seafood, and passageways into the interior. These same tributaries would later be used to convey the colonists' products back downstream for shipment to the mother country. They also would carry the unintended by-products of human activities back downstream, causing deterioration of the natural environment and its living resources. Human influences on the Bay and its tributaries increased as the population grew and with greater agricultural activity, as well as with the later industrialization.

Throughout human history, the effects of human activities have been concentrated in the upper reaches of the Bay proper and in the upper reaches of its tributaries, with the specific range controlled by the physics of the system—the mixing and motion of the waters. Although human influences on the physics of the system have been modest and localized, the physics has been a controlling force in determining the location and extent of human influences on the Chesapeake Bay estuarine system.

The physics of the Bay—the motion and mixing of its waters—are controlled by the interplay of winds, tides, and river flow with the geometry—the size and shape—of the basin that holds the Chesapeake Bay estuarine system. Humans have had no measurable influence on the winds and the tides and only modest influence on the river flow. Although humans have altered the geometry through dredging and filling, the alterations and the effects of these alterations on the physics have been localized and modest. Perhaps the most significant change has been the creation of the Chesapeake and Delaware Canal. The modest changes in the physics of the Chesapeake Bay by humans set it apart from many other major estuaries.

However, one human activity, the overharvesting of oysters, and particularly the dredging of the upthrusting reefs of dead oyster shells that had accumulated over thousands of years, could have had an effect on the duration and intensity of hypoxia (low levels of dissolved oxygen) and anoxia (complete lack of dissolved oxygen) in the deeper waters of the Bay proper and the lower reaches of the major tributaries.

The Geometry of the Chesapeake Bay

The Chesapeake Bay, the largest estuary in the United States and one of the largest estuaries in the world, stretches for about 320 km from its seaward end at the Virginia capes, Cape Charles and Cape Henry, to the mouth of the Susquehanna River at Havre de Grace, Maryland. The Chesapeake Bay estuarine system is made up of the Bay proper and its tributary estuaries. Some of the essential statistics are summarized in Table 4.1.

The modern Chesapeake Bay estuary fills an ancestral dendritic river valley system that was carved into the soft coastal plain sediments during the last low-stand of sea level. As recently as 18,000 years ago, sea level was more than 100 m lower than today and the Susquehanna flowed down the main axis of the Bay, running more than 175 km across the exposed continental shelf before it finally reached the sea. It was during this period that most of the characteristics of the Bay basin were formed. By about 18,000 years ago the climate had begun to warm and the glaciers began to melt and retreat. The glaciers' meltwater ran into the ocean, causing it to rise. By about 10,000 years ago the sea had risen high enough to poke its nose into the Bay basin and begin converting the river valley system into an estuarine system by drowning it. Like all other estuaries of the world, the Chesapeake Bay estuarine system is less than 10,000 years old. It differs from most in the size and complexity of its basin.

Table 4.1 Dimensions of the Chesapeake Bay estuarine system

Waterway	Length (km)	Average width (km)	Maximum width (km)	Tidal mean depth (m)	Tidal mean area (km^2)	Tidal mean volume (km^3)
Bay proper[1]	320	20.0	45.0	8.28	6.5×10^3	53.8
Estuarine system[2]	NA	NA	45.0	6.46	11.8×10^3	76.3

NA = not applicable.

[1]The Chesapeake Bay proper, excluding major tributaries.

[2]The Chesapeake Bay estuarine system, which includes the Bay proper and all tributaries.

Discharge of Fresh Water to the Estuary

Although the Bay has more than 50 tributary rivers, only 3—the Susquehanna, Potomac, and James—account for more than 80% of the total freshwater input, with the Susquehanna accounting for nearly half of the total (49%). The Eastern Shore river with the largest discharge, the Choptank, discharges only 1.2% of the total freshwater input. All Eastern Shore rivers combined contribute less than 4% of the total. The contribution of each river corresponds closely to the proportion of the total drainage basin it represents (Table 4.2).

In the average year, the total amount of fresh water discharged into the Bay (71 km^3) is roughly equivalent to the tidal mean volume of the Chesapeake Bay estuarine system (76 km^3)—the combined volume of the Bay proper and its tributary estuaries. In years of low flow the total discharge of fresh water may fall to about half of the tidal mean volume of the Chesapeake Bay estuarine system (e.g., in 1965 it fell to about 52%), while in periods of high flow the total freshwater discharge may exceed the volume of the entire estuarine system by more than 50%. For example, in 1972, the year of Tropical Storm Agnes, the Susquehanna's annual discharge of 61 km^3 exceeded the volume of the main stem of the Bay by some 7 km^3. In contrast, in 1965 the river's annual discharge of 19 km^3 accounted for only 35% of the volume of the Bay proper.

Only the Susquehanna discharges directly into the main body of the Bay; all other rivers discharge into their own estuaries well upstream from their junctures with the Bay. This pattern reflects the drowning of the ancestral river valley system. When the Bay basin was carved during the earlier low-stand of sea level, the marginal rivers were tributary to the Susquehanna, which flowed down the length of the Bay and out across the continental shelf.

Table 4.2 Drainage basin areas and freshwater discharges for the Chesapeake Bay and tributary estuaries

	Drainage basin area		Freshwater discharge		
Waterway	km²	% of total	m³/sec	km³/yr	% of total
Chesapeake Bay system	165,200	100.0	2,238	70.6	100.0
Susquehanna River[1]	71,300	43.2	1,103	34.8	49.3
Potomac River[2]	36,100	21.9	448	14.1	20.0
James River[2]	26,330	15.9	313	9.9	14.0
Rappahannock River[2]	6,762	4.1	70	2.2	3.1
York River[2]	6,764	4.1	68	2.1	3.0
Patuxent River[2]	2,288	1.4	27	0.9	1.2
Choptank River[2]	1,788	1.1	27	0.9	1.2
Nanticoke River[2]	1,770	1.1	26	0.8	1.2
All other Eastern Shore rivers[3]	6,930	4.2	81	2.6	3.6
All other Western Shore rivers[4]	5,168	3.1	75	2.4	3.4

[1]Drainage area and freshwater input above mouth at Havre de Grace.

[2]Drainage area and freshwater input above confluence of the river plus subestuary with the main stem of the Chesapeake Bay.

[3]Includes all named tributary rivers and named and unnamed small streams that enter the Bay along its Eastern Shore, excluding the Choptank and the Nanticoke.

[4]Includes all named tributary rivers and named and unnamed small streams that enter the Bay along its Western Shore, excluding the six Western Shore tributary rivers listed by name in the table.

The flows of the Bay's major rivers are typical of mid-latitude rivers: high discharge in spring, produced by snow melt and spring rains; low flows in late summer and early fall, followed by moderate flows throughout the remainder of the year. This pattern is revealed clearly by averaging the flows, by month, across the more than 65 years of record (1929–93) for the Susquehanna (Fig. 4.1). The range in average flow from month to month is large, with the maximum average monthly flow exceeding the minimum by about 13.5 times.

Although the general annual pattern is repeated from year to year—with the maximum discharge occurring during spring (March through May) and the minimum discharge occurring in autumn (August through October)—there is no such thing as an average year for the Chesapeake

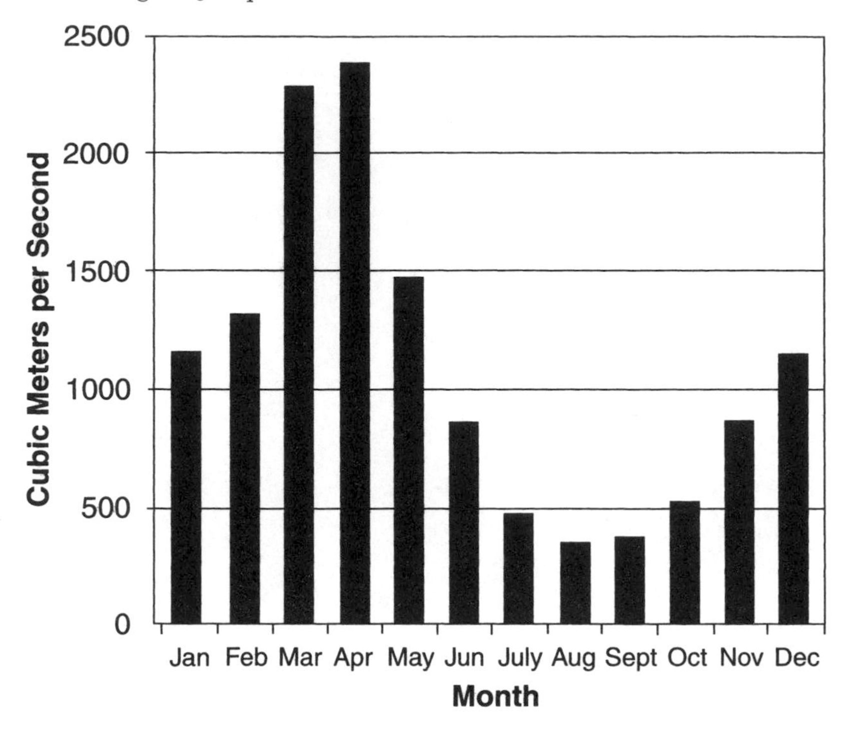

Fig. 4.1. The ensemble-averaged monthly flow of the Susquehanna River at Havre de Grace, for the 65-years from 1929 to 1993.

Bay. There are large year-to-year variations, and occasionally a tropical storm or hurricane comes through that completely changes the typical annual pattern of river flow. For example, the maximum monthly discharge of fresh water to the Bay for 1975 occurred in September, a month that usually contributes the second-lowest level of freshwater discharge.

The large variations in the total river flow into the Chesapeake Bay are important in determining the rate of flushing of the estuary, the degree of vertical stratification (mixing) of its waters, and the amount of nutrients (and other materials) discharged into the Bay. These three factors—the rate of flushing, degree of vertical stratification, and the amount of nutrients—are important in determining the intensity and extent of hypoxic and anoxic conditions in the deeper sections of the middle reaches of the Bay proper and the lower reaches of the major tributaries. The first two factors relate to the physics; the third is an input into the Bay. It is the interaction of the physics and the input of nutrients—their amount, forms, and timing—that control the Bay's response.

The timing of the variations in river flow is also important. A large discharge in the fall does not have the same effects as a large discharge in the spring or early summer. The magnitude and timing of the variations in flow are important to the ecological health of the system. Animals have been dealing with this variability since the Bay was formed and have adapted to it. Indeed, the variations are important in maintaining the balance. The variations in flow of the Susquehanna also are important in controlling the physics that drives the exchange of water between the Bay and the estuaries tributary to the upper Bay, such as the Gunpowder and Bush.

Humans have had only a modest impact on the flows of the rivers that enter the Bay. Prior to European settlement, the watershed of the Chesapeake Bay was almost completely forested. In precolonial days, the peak spring discharges resulting from snow melt and spring rains would have been somewhat dampened and delayed and spread out over longer time intervals compared to the present. Runoff from cultivated lands, and especially from roads and urban and suburban areas, occurs over shorter time intervals with little delay and less infiltration into the soil. Consequently, stratification of the Bay waters may have been somewhat less severe in precolonial days than at present but the differences were probably modest.

The reservoirs along the lower Susquehanna have an even smaller impact on river input. Since these reservoirs have little storage capacity, they have little influence on river flow. It is only during the summer periods of very low flow that discharge may stop entirely on weekends, when the hydroelectric plants shut down, but the effects of these actions on the Bay are small.

Although humans have had little impact on the input of fresh water to the Bay, they have had a large impact on the input of nutrients and other materials transported by rivers flowing into the Bay, as will be discussed later.

The Scales of Motion and Mixing in the Bay

The Tides

The most obvious motions of the Bay's waters are the oscillatory tidal currents which have average speeds of 1.25 knots to 2.00 knots. In general, the average and maximum current speeds are lower in the lower reaches of the Bay, increase in the central reaches of the Bay, and drop off again near the head of the Bay. Within each of these zones, current speeds are higher where cross sections are smaller and lower where cross sections are larger. Average tidal current speeds increase by about 30–40% on spring tides and decrease by about the same amount on neap tides.

Approximately every 12 hours a new tidal wave enters the Bay at its mouth and progresses up the Bay. Since the travel time to reach the head of the Bay at Havre de Grace is approximately 14 lunar hours, a crest does not quite reach the head of the Bay before the following crest enters its mouth. In this respect, the Bay is unusual in that it is able to contain a complete semidiurnal tidal wave at all times. The Chesapeake Bay is indeed the mother of all estuaries.

The oscillatory motions of the tides are accompanied by a rise and fall of the water level in the Bay. The tide range decreases from about 0.9 m at the Bay's mouth to a minimum of about 0.3 m at Annapolis, and then rises again to 0.7 m at the head of the Bay. The maximum range in the entire system is 1.2 m and occurs at Walkerton, Virginia, on the Mattaponi River.

Motions and mixing of the Bay's waters occur on scales that range from random molecular motions to turbulent scales occurring over distances from centimeters to meters vertically and to 100 meters horizontally. The time scales of motion and mixing range from less than a second to several hundred seconds. These turbulent diffusion processes have a major role in controlling the dispersion of suspended and dissolved substances, both those naturally occurring (such as plankton, including the eggs and larvae of shellfish and finfish) and anthropogenic pollutants. It is the tidal currents that provide the energy to drive these turbulent motions that lead to dispersion.

If there were no rivers flowing into the Bay and no winds, the tidal currents would flow in one direction for about 6 hours, then reverse and flow in the other direction for about 6 hours. It would thus seem that there would be no net motion. However, there are complex interactions between the tidal currents and the tidal rise and fall of the water surface, and between the local tidal currents and the local topography, that would lead to net transport even in the absence of rivers or wind.

The first of these complex, nonlinear interactions results from the fact that the tide advances up the Bay as a progressive wave. Maximum flood current speeds occur near the time of high water, when the area of the local cross section of the Bay is greater than average, and maximum ebb current speeds occur near the time of low water, when the area of the local cross section of the Bay is smaller than average. Because of this asymmetry of flow, there is a transport of mass up the estuary, which is called the Stokes transport. This transport causes a buildup of the tidal mean water surface relative to distance up the Bay. This slope of the tidal mean water surface drives a tidal mean residual flow down the estuary, such that over the length of the Bay the up-estuary–directed Stokes transport is balanced out by the down-estuary, slope-driven residual flow. However, over local reaches of the Bay having lengths extending over several tidal excursions, such a balance does not occur, resulting in a longitudinal dispersion mechanism that exceeds that which would otherwise be expected from the simple oscillatory tidal current motions.

The second nonlinear interaction—the interaction of the oscillatory tidal currents and the local topography—results in the development of tidal mean residual eddy circulations, particularly in the vicinity of headlands. These eddies can cause both longitudinal and lateral transports of dissolved and suspended materials which significantly exceed those resulting from the simple oscillatory tidal current motions.

The Residual, Tidally Averaged Flows

Of greater importance than these complex, nonlinear tidal interactions are the motions resulting from the inflow of fresh water from the tributary rivers and streams, those driven by winds that blow over the surface of the Bay, and those driven by winds that blow over the surface of the adjacent continental-shelf waters. The interactions of these processes with the tides produce residual motions—motions leftover after averaging out the oscillatory tidal motions—that give the Bay its distinctive characteristics and make the Bay an estuary. Let us consider the river input first.

Because fresh water is being delivered by the rivers all the time and because the Bay is neither getting fuller nor emptier on average over periods of from 5 to 20 tidal cycles long, there must be an average net flow directed toward the ocean equal to the long-term average of the volume rate of inflow of fresh water to the Bay. Anything less and the Bay would be getting fuller; anything more, and it would be getting emptier. Since the fresh water is lighter than the salt water that enters from the ocean, it tends to move along the top of the heavier sea water. The vigorous sloshing back and forth produced by the tides mixes the fresh water and the salt water, producing gradients in density both longitudinally, along

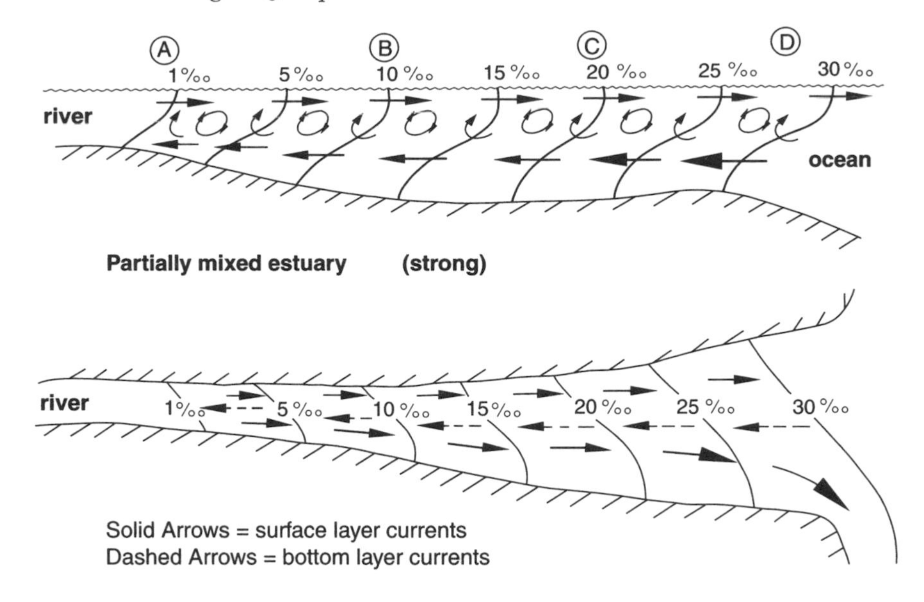

Fig. 4.2. Schematic depiction of the distribution of salinity and of the long-term (greater than 10 tidal cycle–averaged) residual circulation in a partially mixed estuary, characteristic of the northern half of the Chesapeake Bay under nearly all conditions for all seasons, and of the lower Chesapeake Bay and of the major tributaries under most conditions during late winter, spring, and summer. The upper diagram is a longitudinal vertical section taken along the thalweg of such an estuary; the lower diagram is a plan view.

the axis of the estuary, and vertically, throughout the depth of the water column.

These density gradients drive a tidally averaged residual circulation in which the lower salinity, lower density water of the upper layers moves seaward and the higher salinity, higher density deeper layers move up the estuary. Vertical mixing driven by the oscillatory tidal currents maintains the longitudinal salinity gradient, and hence density gradient, along the length of the estuary. This circulation pattern, which has been given the name "estuarine circulation pattern," is shown schematically in the upper diagram of Fig. 4.2.

Here is what happens: If there were no tide, a heavy torpid layer of salty water would lie in the bottom of the Bay and over it a lighter layer of essentially fresh water would run to the sea. Because water has some internal friction, that is, some viscosity (though not much), some sea salt would be mixed into the upper layer but in such small amounts that the upper layer would remain essentially fresh most of the way to the Virginia capes. But the tide *is* there. The great mass of water in the Bay sloshes back and forth at speeds that reach 2 knots. The flows are turbulent and tear great globs from both layers. From the upper layer, globs are carried down into the lower layer, and globs from the lower layer are transferred to the upper. There they mix so the "fresh" upper layer becomes increasingly salty toward the mouth of the Bay, while the "salty" lower layer becomes increasingly fresh toward the head, yet at any given place in the Bay the lower layer is always saltier and denser than the upper layer. The difference in salinity between the two layers is nearly the same over much of the length of the Bay.

The sea salt in the upper layer is discharged, mixed with the fresh river water, back into the sea. If this attrition of the lower layer were to go on without compensation, the entire Bay would soon be "fresh" from top to bottom. The sea would have been eliminated. But this has not happened. Although the total inventory of salt in the system varies seasonally and somewhat from year to year, its average salinity has remained essentially unchanged for thousands of years. There must be a supply of salt, an inexhaustible supply; and there is: the sea. Compared to the Bay, the sea is infinite; its volume is 20 million times greater. A slow, persistent current moves silently up the Bay in the lower layer to resupply the salt that has been flushed out to sea in the upper layer.

The rivers must discharge their water to the ocean. On average, the rivers entering the Bay must get rid of some 2,000 m^3 every second. However, when we measure the discharge to the ocean in the upper layer through the mouth of the Bay, we find the discharge is not 2,000 m^3/sec but rather nearly 10 times that amount, or 20,000 m^3/sec. This excess water is water transported from the lower layer to the upper layer by a steady vertical flow that occurs along the entire length of the estuary. This vertical motion, called the "entrainment velocity," transports water and salt from the lower layer into the upper layer. Random, turbulent motions also occur. On average, they do not produce a net transport of water, but they do lead to a net transport of salt from the higher salinity lower layer to the fresher upper layer.

The amount of water in the Bay has changed very little from week to week, from year to year, or even from century to century over at least the last few thousand years, so it is clear that the net difference in flows of the two layers through the mouth of the Bay must equal the total freshwater input to the Bay; no more, no less. If the flow outward in the upper layer is 10 times the total river flow, then the flow inward in the lower layer must equal 9 times the total river flow. The net difference must equal the total freshwater inflow.

Given that the total freshwater input to the Bay varies seasonally and from year to year, the total net discharge through the mouth of the Bay must also change, but not by a great deal. The ratio of the discharge of the upper layer to the freshwater input also varies seasonally and from year to year but roughly inversely to the variation in freshwater input. As a result, the volume rate of outflow of the upper layer does not vary greatly over the year.

Because of the effects of the Earth's rotation, the seaward flow of the near-surface layers is stronger, and extends deeper, on the right-hand side of the estuary (looking seaward) than along the left, while the up-estuary–directed flow of the deeper layers is stronger, and extends to shallower depths, on the left-hand side of the estuary than on the right. Thus, in the Chesapeake Bay, the residual seaward-directed flow of the upper layers is stronger and extends to greater depths along the western shore, whereas the up-estuary–directed flow in the lower layers is stronger and extends to shallower depths along the eastern shore. The division between the upper seaward-flowing layer and the up-estuary–flowing lower layer is thus not level. Rather, it slopes upward to the east. As a result of these lateral variations in the flow pattern, salinities are lower along the western shore of the Bay than along the eastern shore. The lower diagram in Fig. 4.2 schematically demonstrates these lateral and longitudinal features of the estuarine circulation pattern.

Motions Driven by the Wind

Winds blowing over the surface of the Bay and its major tributaries (near-field wind forcing) and winds blowing over the nearby open ocean (far-field wind forcing) set up complex differences in the elevation of the water surface, and hence complex patterns in the speed and direction of surface currents. These wind-driven motions usually last longer than those caused by tides. The complex responses of the Bay and its tributaries to wind forcing have been adequately studied only in the past 30 years.

Winds blowing up the axis of the Bay weaken both the seaward-directed, density-driven flow of the near-surface layers and the up-estuary–directed, density-driven flow in the deeper layers. If the wind is strong and persistent enough, the classical estuarine circulation pattern can be reversed, with the near-surface layers flowing up the estuary and the deeper layers flowing seaward. Winds blowing down the axis of the Bay strengthen the classical circulation pattern, with both the seaward-directed flow in the near-surface layers and the up-estuary flow in the deeper layers becoming stronger.

Because of the effects of the Earth's rotation, the net transport of wind-driven flows of the waters of the continental shelf are directed to the right of the wind direction. Therefore, winds blowing over the waters of the continental shelf adjacent to the mouth of the Bay and directed generally parallel to the coastline from the northeast toward the southwest cause a rise in tidal mean sea level at the mouth of the Bay. This rise in sea level advances up the Bay as a long wave, and consequently the tidal mean sea level at the mouths of the tributary estuaries to the Bay also rise. This set-up of water results in a flow of water into the Bay through its mouth and a flow into each of the tributaries. As a result, the tidal mean volume of water in the Bay is increased over the duration of such a wind system, which may last for several days. When the offshore wind subsides or changes direction, the excess volume of water accumulated in the Bay during the wind event must flow out through the mouth of the Bay. Such outflows have been observed to last for as long as three full days, during which no flood tidal flow through the mouth of the Bay occurred.

The combination of near-field and far-field wind-driven motions with the density-driven motions results in very complex time-varying circulation patterns in the major tributaries to the Bay. For example, records taken over a one-year period in the mid-reaches of the Potomac River estuary, when averaged over 10 days or more, show the classical estuarine flow pattern of seaward flow in the upper layers and up-estuary flow in the deeper layers. But when these records were averaged over successive 25-hour (two-tidal cycle) periods, the classic two-layered estuarine circulation pattern occurred only 40–50% of the time, with an average duration of about 2.5 days. A reverse, two-layered pattern, with up-estuary flow in the surface layers and down-estuary flow in the lower layers, occurs about 20% of the time, with an average duration of one to two days.

Other circulation patterns that were observed a statistically significant number of times in yearlong records are three-layered flow patterns, with surface- and bottom-layer flows directed up the estuary and a mid-depth-layer flow directed down the estuary; reverse three-layered patterns, with surface- and bottom-layer flows directed seaward and a mid-depth-layer flow directed up the estuary; flow directed out of the estuary at all depths (discharge); and flow directed up-estuary at all depths (storage). Similar variations in the residual (tidal averaged) flow patterns occur in the main stem of the Bay and in other major tributary estuaries, but when averaged over the long term, the classic two-layered estuarine flow pattern always emerges.

It is unlikely that human activities have influenced the wind-forced motions in the Chesapeake Bay estuarine system. Unless global warming increases the frequency and intensity of storms, it is also unlikely that humans will have any effect in the future on the meteorologically induced motions of the Bay.

In summary, we have the oscillatory tidal flow as the dominant circulation pattern in the Bay. Part of the potential energy of this highly organized, periodic flow is dissipated through the kinetic energy of the highly disorganized, aperiodic, random turbulent motions that mix the fresh water and sea water. The resulting density field drives a second highly organized flow pattern—the estuarine circulation. It is this slower, more subtle circulation—moving water persistently seaward in the upper layer and landward in the lower layer at speeds of only about one-fifth those of the stronger tidal currents—that is in many respects the more important circulation pattern. The play of local winds on the Bay and of far-field winds on shelf waters can add to, subtract from, or completely reverse these classic estuarine circulation patterns, but their effects are short-lived.

Human Influences on the Topography of the Bay

In 1608 Captain John Smith prepared the first chart that outlined the entire Bay proper and all of its major tributaries from the mouth of the Bay at the Virginia capes to the head of tide. Although the chart did not have any bathymetric (depth) information, it was a remarkably accurate characterization of the major features of the Bay's shoreline. Bathymetric information began to be added to charts in the mid to late 1600s, and by the early to mid 1700s navigation charts began to resemble modern charts.

A comparison of early navigation charts indicates that the main body of the Bay and its major tributaries changed little in depth over the first 125 years or so of European colonization. With the intensification of agriculture beginning in the last quarter of the eighteenth century—more land being deforested and put into cultivation, deeper plowing, and the abandonment of exhausted fields—soil erosion in the watershed soared. Between 1760 and 1860 the influx of sediments to the Bay from its rivers was at least four times that of precolonial levels and for part of this period may have been as much as eight to ten times as high. All the rivers except the Susquehanna discharged this sediment into the upper reaches of their estuaries—far upstream from their junctures with the Bay. Most of it was then trapped by the circulation patterns of the tributary estuaries. Note that the trapping of sediment in the upper reaches of the Bay and of each of its tributary estuaries is a characteristic not only of the Chesapeake Bay but of most estuaries because of their circulation patterns.

The most dramatic shoaling occurred in the upper reaches of some of the smaller tributaries to the western shore of the upper Chesapeake Bay, the Gunpowder and the Bush, and in the 50-km stretch of the main body of the upper Bay from the mouth of the Susquehanna River down to about Love Point. Joppa Town, at the mouth of the Little Gunpowder River, is the classic example of a former port left stranded. Joppa Town was the county seat of Baltimore County and Maryland's most prosperous seaport in the early 1700s. By 1750, shoaling seriously interfered with the operation of the port and its importance waned. In 1763 the county seat was moved to Baltimore. Today the only reminder that Joppa Town was once a seaport are the exposed upper sections of stone mooring posts that once held the hawsers of seagoing vessels and are now nearly hidden in the underbrush more than 3 km from navigable waters. Upper Marlboro on the upper Patuxent and Port Tobacco on the upper Potomac experienced the same fate.

From about 1760 to 1860 the Susquehanna Flats, the bulbous-shaped delta extending from the mouth of the Susquehanna River at Havre de Grace for 8 km downstream, shoaled by 2 to 3 m (6 to 10 ft) to an average depth of just over 1 m. Farther seaward, the segment from Sandy Point on Spesutie Island to the latitude of Swan Point (just above the Bay Bridge at Annapolis) shoaled over this same 100 years by about 2 m. The present average depth of this section of the Bay is 4 m. Depths increase markedly seaward from this segment. The next 40-km stretch of the Bay has an average depth of just under 10 m.

Throughout the period of record there has been a steady erosion of the shoreline, particularly along the eastern shore of the Bay, and of the various islands found along the east side of the Bay. Erosion has reduced the area of many of these islands to small fractions of what their sizes were just one hundred years ago. About two-thirds of the sediment eroded from the shoreline and from the islands is sand and gravel, which is deposited as a narrow shelf near the shoreline and around the islands.

There has been some reforestation of abandoned farmlands over the last 100 years, and the several dams built on the lower reach of the Susquehanna River some 65 years ago have intercepted a significant part of the sediment influx during years of low to average river discharge. There has also been a rise in relative sea level over this same time period, such that at the present and for several prior decades even the upper reaches of the Bay and of the major tributaries have not shoaled significantly.

In summary, the size and shape of the Chesapeake Bay basin has changed little since European settlement and most of the changes have happened around the margins, particularly where rivers enter the system. Humans cut channels to accommodate larger ships, but the channels are limited in number and extent, affect only a very small percentage of the area of the Bay bottom, and have had little effect on the physics of the Bay. There may be one exception.

Although there has been little change in the gross shape of the shoreline and the distribution of depths in the Chesapeake Bay and its tributary estuaries, there have been significant changes in the detailed topography of the shallow shelves extending from the shorelines of the Bay and its tributaries. This includes the shelves off the shorelines of many islands along the eastern side of the middle reaches of the Bay and in tributary embayments such as Tangier Sound and Pocomoke Sound. During the 4,000 years or so before the settlement of the Bay region by Europeans, these shelf areas, with depths of less than about 6 m and exposed to overlying waters having a salt concentration greater than five parts per thousand, were ideal for the growth of the American oyster. Under conditions of very limited harvest pressure from American Indians, successive layers of oyster shell built up into upthrusting reefs of various shapes, often extending up to Mean Low Water.

Reports by early writers, both colonists and travelers passing through the Chesapeake Bay region during colonial times, attest to the abundance of oysters. These accounts frequently refer to the navigation dangers posed by extensive oyster reefs. Except for the local depletion of nearshore beds by settlers for their own use, there was no significant exploitation of this accumulated resource until the early 1800s. At that time, demand from the more populous New England states, where overfishing had seriously depleted local oyster beds, led to the start of a steady increase in fishing effort in the Chesapeake Bay and its tributaries.

Data on oyster harvests were not reported until 1839, when some 700,000 bushels were landed in Maryland. The exploitation of Chesapeake Bay oyster beds then increased rapidly, such that the annual harvest in the Bay as a whole was estimated to be some 17 million bushels in 1875, with a peak production of perhaps 20 million bushels in 1885. It is likely that between 1800 and 1920 some billion bushels of oysters were harvested from the Bay and its tributaries. Only in the last 50 years, after the harvest was much reduced, have there been programs in both Maryland and Virginia to return any portion of the shells removed. Such a return of shells is necessary to provide a clean substrate for the settling of oyster spat.

In fact, over most of this period heavy use was made of the oyster shells after removal of the meat. Crushed shells were used as construction material for roads and as grit for chickens and were burned to produce lime for fertilizer. Significant portions of the town of Crisfield, on Maryland's Eastern Shore, and parts of the City of Baltimore are built on oyster shells. These uses of shell led to the dredging of the dead remnants of the once productive upthrusting oyster reefs. One comparison of soundings taken on oyster beds in the James River in 1847 with those taken recently show that the mean depths over these beds increased in that period by about 1.5 m. Probably most of that increase in depth occurred prior to 1900.

Summer Hypoxic Events and the Physical Hydrography of the Bay

The development of hypoxia and anoxia in waters deeper than about 7 m in the central reaches of the Chesapeake Bay has been considered to reflect the ecological condition of the Bay. Typically, low-oxygen conditions start in late spring and extend through summer. A study in which the oxidation states of sulfur and iron in pollen-dated sedimentary layers were determined showed that anoxic conditions, and probably strongly hypoxic conditions, did not occur in the Bay prior to about 1760, a period sometimes called the agriculture horizon. An important management question is the degree to which increases in the human inputs of nutrients and other substances have increased the frequency, duration, extent, and intensity of hypoxic and anoxic waters in the Bay.

Unfortunately, prior to 1949 there had been only occasional observations of dissolved-oxygen distributions in the deeper waters of the mid-Bay. It was not until 1949, when the Chesapeake Bay Institute (CBI) initiated a continuing series of surveys of temperature, salinity, and dissolved-oxygen measurements throughout the Bay and its tributary estuaries, that there were sufficient data to determine the spatial extent and intensity of midsummer hypoxia. CBI's data collection program continued for more than 30 years. Other scientific organizations on the Bay have augmented and extended the program.

An analysis of these data reveals that small volumes of hypoxic waters appear in the deep trough in the middle reaches of the Bay in mid-May; the volume of hypoxic (and anoxic) waters increases rapidly after about the middle of June, reaching a maximum in late July or early August; the hypoxic volume then decreases to small values by late September; and complete reaeration usually occurs by mid-October. Over most of June through September, hypoxic conditions are found in the deep trough of the Bay from just north of the Bay Bridge (39°10′ N) southward to the latitude of the mouth of the Rappahannock River (37°40′ N). All of the studies of hypoxia in the Bay have shown large year-to-year variations in the maximum volume of hypoxic water. Although there has not been complete agreement among all investigators as to the time trends revealed by the data set, the evidence indicates that there is a strong direct relationship between the volume of freshwater discharge from the Susquehanna during the spring freshet and the volume of hypoxic and anoxic waters in the following midsummer.

For 1949–85, there is a strong correlation between the volume of fresh water entering the upper Bay during the spring months of April and May and the volume of hypoxic water in the subsequent midsummer months of July and August. About 80% of the variance in the hypoxic volume over this period can be explained by the variance in the spring freshet flow. In 1958 the freshwater inflow to the upper Bay during the spring freshet months of April and May totaled 17.13 km^3, the highest flow in these months for 1949–85. The subsequent midsummer hypoxic volume of 8.6 km^3 was also the largest observed during the same interval. Compare these values to those for 1965, which had a spring freshet of 6.52 km^3 (38% of the 1958 value) and a midsummer hypoxic volume of just 1 km^3 (12% of the 1958 value). Because of the large year-to-year variation in the spring freshwater inflow, there is a concurrent large variation in the midsummer hypoxic volume that masks any long-term trend that may exist.

All investigators who have studied hypoxia in the Chesapeake Bay agree that there is a strong correlation between the degree of stratification, as measured by the difference in salinity at the surface and at the bottom at stations in the hypoxic reach of the Bay, and the difference in dissolved-oxygen concentrations observed at the surface and the bottom at these same stations. Note that in the center of the hypoxic reach of the Bay, the vertical stability (as measured by the density increase from just above to just below the pycnocline, the zone of rapid change in density) in 1958 was approximately twice that observed in 1965. Combining these observations with the apparently strongly supported relationship between the volume of fresh water that enters the Bay in the spring and the subsequent summer hypoxic water volume, and with our knowledge of the physical, chemical, and biological processes active in the Bay, we can tell a story about how these various processes lead to the development of hypoxia and the subsequent reaeration.

The spring freshets bring to the upper Bay a buoyant pool of low-salinity water that first pushes the saltier water seaward at all depths and then overruns the denser lower layers, continuing to move seaward. There is a delayed response in the lower layers, with higher salinity water advancing up the estuary. This produces a highly stratified condition in the middle reaches of the Bay, one that significantly reduces vertical mixing. The freshet also brings with it a large supply of nutrients. In particular, most nitrates are brought into the Bay with the spring freshet.

The increasing solar radiation in spring and early summer stimulates biomass production of phytoplankton, which have a rapid turnover rate. Decaying organic matter sinks into the lower layers and onto the bottom, exerting a large oxygen demand. Oxygen is supplied to the surface layers of the middle Bay directly from the atmosphere and from the amount of photosynthetic oxygen production that exceeds the oxygen used by respiration. High vertical stratification severely limits the mixing of the oxygenated waters of the upper layer into the oxygen-deficient deeper layers. There is also an advective (horizontally transported) supply of dissolved oxygen into the hypoxic region in the up-estuary–directed lower-layer flow. This flow originates in the more oxygenated southern reaches of the Bay which in turn are supplied by a flux of even more highly oxygenated lower-layer waters from the adjacent continental shelf. The fact that the dissolved oxygen in the lower layers of the mid-Bay continues to decrease from midspring to early August indicates that the advective transport of oxygen from the southern Bay is insufficient to overcome the biological and chemical oxygen demand of the hypoxic waters.

In early August surface temperatures begin to drop and the decreased freshwater inflows lead to an increase in the salinities of the upper layers. Both these processes—a drop in temperature and an increase in salinity—increase the density of the water in the upper layer, causing the vertical stability to weaken, particularly in the Virginia portion of the Bay. With increased vertical mixing, the dissolved oxygen in the lower layers of the southern Bay increases, as does the advective transport of oxygen from the lower Bay into the mid-Bay hypoxic zone. With the weakening of vertical stratification in the hypoxic reach, more oxygen is mixed downward from the aerated surface layers. The volume of hypoxic water decreases rapidly.

By about the time of the autumnal equinox each year, the surface layers have been cooled to the point that the temperatures of the upper layer are lower than those of the lower layer. This results in a temperature effect on the vertical-density gradient opposite to that of the salinity difference, which further weakens the vertical stability such that the next large wind event results in an overturn of the Bay waters in Virginia. The advective transport of the higher-oxygen lower-layer waters from Virginia into the hypoxic reaches effectively terminates hypoxia for the year. The continuing existence of a longitudinal salinity gradient results in a rapid return of stratification, and a short interval of hypoxia may also return to the deepest waters in the mid-Bay trench. By mid-October, hypoxia has completely disappeared and will not be observed again until the following spring.

It remains unclear whether increased summertime stratification or increased supply of nutrients, both of which increase with freshwater inflow, is the more important factor in leading to an increased volume of hypoxic water. This is an important question for management, since the strategy to reduce the hypoxic volume is to reduce the input of nutrients from sewage treatment plants and from agricultural runoff.

Contaminants

Most contaminants are relatively insoluble in water and have a high affinity for fine particulate matter. Because of this, the motions that control the transportation, distribution, and accumulation of fine particles also control these processes for most contaminants. The most important of these motions in the Bay proper and its major tributaries is probably the long-term average estuarine circulation, as modified locally by topographically induced eddies. Departures from this mode are driven primarily by near-field and far-field winds, at time scales of from 2 to 10 days, and can affect the transportation, distribution, and accumulation of contaminants at these same time scales. However, following return to average conditions, sediments and associated contaminants are redistributed to reflect the effects of the long-term estuarine circulation.

The significance of the relationship between the variability of motions in estuaries on the one hand and the transportation, distribution, and accumulation of sediments and sediment-bound contaminants on the other has not been established. The effects of these fluctuations depend strongly on when they occur, how long they persist, and the size and geometry of the particular subestuary. Within the Bay proper these wind-driven forced fluctuations in the residual currents probably act to increase the effective horizontal dispersion but have little effect on the average advective transport of pollutants and other waterborne materials. However, at the mouth of the Bay these wind-driven motions probably contribute significantly to the exchange of dissolved and suspended constituents (including plankton) between the Bay and the adjacent open continental-shelf waters.

Tidal currents play an important role in the dispersal and accumulation of contaminants. Their effects are primarily through the alternate resuspension and redeposition of bottom sediments associated with the waxing and waning of tidal currents. On average, resuspension of sediments in the near-bottom-water layers leads to a net transport of sediment up-estuary because of the long-term net upstream flow of the lower layer of partially mixed estuaries. Resuspension coupled with the net upstream motion in the lower layer leads to a redistribution of sediments that may have been carried downstream during conditions of net seaward flow at all depths.

The alternate resuspension and redeposition of bottom sediments plays another role in contaminant dispersal. As particles are resuspended, they are transferred into an environment with different physical-chemical characteristics. These changes in environmental conditions may contribute to either a release of surface-bound contaminants into the water column or a scavenging of contaminants from the water column by the resuspended particles. The mixing of interstitial waters into the overlying water column that accompanies sediment reworking by tidal currents may also lead to a transfer of contaminants from the bottom to the overlying waters.

Extreme events such as Tropical Storm Agnes in June 1972 can result in major inputs of contaminants, both in solution and bound to fine particles. With the exception of overenrichment by enhanced nutrient input, particularly of nitrogen compounds, it is unlikely that the largest and most persistent impacts of such events are associated with the input of contaminants. They reside in the water column a relatively short time before being removed by particles and deposited on the Bay floor.

Closing Observations

Human activities since the beginning of European colonization have had limited effect on the physical characteristics of the Chesapeake Bay. Human activities have had a greater impact on the interaction between physical and biochemical processes in the Bay, as particularly illustrated by the changes in the physical and biochemical interactions controlling the intensity and extent of hypoxia.

Although rapid shoaling of the upper Bay north of Swan Point prior to the start of the twentieth century probably resulted in a decrease in the northward intrusion of brackish water, the following two factors tend to mitigate this process. Because the most saline water moves up the Bay in the deep channel, artificial dredging of the navigation channel from the vicinity of Swan Point all the way to the head of the Bay has, over the last 50 years or so, partly reversed the effects of increased sedimentation. In addition, the continued rise of sea level has approximately kept pace with the rate of sedimentation.

Shoreline and island erosion has led to a small increase in the surface area of the Bay, but the added water area is a shallow shelf along the eroding shoreline and around the eroding islands such that this increased water area has very little impact on the dynamics of circulation and mixing in the Bay.

With one possible exception, the limited changes in the topography described earlier have not resulted in any significant effect on the circulation and mixing, and the distribution of physical properties, of the Chesapeake Bay estuarine system. The exception concerns the upthrusting oyster reefs. The existence of upthrusting oyster reefs at the time of European colonization, and continuing up to the early 1800s, produced a rough, highly irregular shelf platform along the edges of the Bay and its tributaries. This irregular topography probably extended out to the "shelf break," which over large reaches of the Bay is found at a depth of about 6 m. This depth is coincident with the depth of the layer of rapid increase in the salinity, and hence in the density, of the water column in spring and summer. Although data are not available to prove the case, it is quite possible that the removal of the upthrusting oyster reefs from the shelves bordering the Bay and its tributaries may have resulted in a significant change in the degree of vertical mixing during the critical spring and summer periods of vertical stratification. If the intensity of vertical mixing were larger prior to the removal of the upthrusting reefs, then the Bay would have been less stratified during the spring and summer. The vertical variation in salinity would have been smaller than occurs today, and the transfer of dissolved oxygen from the surface layers to the bottom layers would have been greater. Consequently, oxygen depletion would have been less severe, suggesting that the removal of the upthrusting oyster reefs had profound consequences for the biological well-being of the Bay.

Removal of the upthrusting oyster reefs may have had a further profound effect on the biological control mechanisms of the Bay. It has been suggested that the overexploitation of oysters changed the Bay from a system dominated by filter-feeding organisms attached to the bottom to a system dominated by plankton. If this hypothesis is correct, the most important consequence was the dramatic increase in the time it takes for the biological community to filter suspended particles from the Bay. Before European exploitation of the oyster reefs, oysters filtered a volume of water equivalent to the volume of the Bay in a few days. The same volume now takes a year to filter. Such a change would have made the Bay far more susceptible to the undesirable effects of overfertilization by nutrients in runoff. Such feedback loops are not well understood, but they can be powerful driving forces in natural systems.

The very complexity of the Bay—the complexity of the mixing and the motions of its waters—and its responses to human perturbations tell us that we must have a constancy of commitment to improve our understanding. We need to better understand the natural processes that characterize the Bay and how human actions—past, present, and prospective—affect those processes and their consequences. Fortunately, there is strong evidence of this commitment. No estuary in the world receives as much dedicated funding as the Chesapeake Bay. Understanding is not enough: we must also manage the Bay and other natural resources based on these advances in knowledge. Such management strategies are called adaptive management—they adapt as understanding and knowledge change. This is management for the long haul, the only form of management that makes sense for as important and vulnerable a resource as the Chesapeake Bay.

BIBLIOGRAPHY

Beaven, G. F. 1960. Temperature and salinity of surface waters at Solomons, Maryland. *Chesapeake Sci.* 1: 2–11.

Boicourt, W. C. 1969. *A Numerical Model of the Salinity Distribution in Upper Chesapeake Bay.* Technical Report No. 54, Ref. 69-7 (Baltimore: Chesapeake Bay Institute, The Johns Hopkins University).

———. 1983. *The Detection and Analysis of the Lateral Circulation in the Potomac River Estuary.* Report PPRP-66 (Baltimore: Maryland Power Plant Siting Program and the Chesapeake Bay Institute, The Johns Hopkins University), 209.

Cameron, W. M., and D. W. Pritchard. 1963. Estuaries. In *The Sea,* v. 2, *Ideas and Observations on Progress in the Study of the Seas* (Interscience Pub.), 306–24.

Goodrich, D. 1985. On stratification and wind-induced mixing in the Chesapeake Bay. Ph.D. diss., Marine Sciences Research Center, State University of New York, Stony Brook.

Hayward, D., C. S. Welch, and L. W. Haas. 1982. York River destratification: An estuary-subestuary interaction. *Science* 216: 1413–14.

Hicks, S. D. 1964. Tidal wave characteristics of Chesapeake Bay. *Chesapeake Science* 5(3): 103–13.

Pritchard, D. W. 1968. Chemical and physical oceanography of the Bay. In *Proceedings of the Governor's Conference on Chesapeake Bay, Sept. 12–13* (State of Maryland).

Pritchard, D. W., and M. E. C. Vieira. 1984. Vertical variations in residual current response to meteorological forcing in the mid-Chesapeake Bay. In *Estuary as a Filter,* ed. V. S. Kennedy (New York: Academic Press), 27–65.

Schubel, J. R. 1968. The turbidity maximum of the Chesapeake Bay. *Science* 161: 1013–15.

———. 1972. The physical and chemical conditions of the Chesapeake Bay. *Journal of Washington Academy of Sciences* 2(56).

Seitz, R. C. 1971. *Temperature and Salinity Distributions in Vertical Sections along the Longitudinal Axis and across the Entrance of the Chesapeake Bay (April 1968 to March 1969).* Graphical Summary Report No. 5, Ref. 71-7 (Baltimore: Chesapeake Bay Institute, The Johns Hopkins University), 99.

CHAPTER FIVE

A Long-Term History of Terrestrial Birds and Mammals in the Chesapeake-Susquehanna Watershed

DAVID W. STEADMAN

The fossil record of birds and mammals shows relatively little extinction in the nearly two million years preceding the arrival of the first humans, when ground sloths, giant beavers, mammoths, mastodons, and other megafauna disappeared. Post-Columbian losses in the Chesapeake region include the gray wolf, wapiti, bison, and eight species of birds. Nowadays, habitat changes caused by human activities have had the greatest effect on bird and mammal communities.

Changes in vertebrate communities can be evaluated on a variety of time scales. To most ecologists, long-term studies are measured in decades, but I will use centuries and millennia as the benchmarks to describe some of the long-term changes that have occurred in the bird and mammal communities of the Chesapeake-Susquehanna watershed of New York, Pennsylvania, West Virginia, Maryland, the District of Columbia, and Virginia. Although these three-, four-, and five-digit chunks of time may seem large to biologists, they are small to most geologists and paleontologists. The faunal story I am about to tell is based on a blend of paleontological, archaeological, ethnographic, and zoological data.

During the latest Pleistocene (25,000 to 10,000 years ago), most of the northern quarter of the modern Chesapeake drainage was covered by the Laurentide ice sheet. The area was ice-free, but still much cooler than today, when the first Amerindians arrived about 11,000 years ago. The relatively mild climates of the Holocene (the postglacial interval that began about 10,000 years ago and continues today) have sustained and nurtured all but the first millennium of human occupation in the region. Enormous cultural and biotic changes began with the European and African discovery of North America (see, for example, Chaps. 3, 7, and 8). Therefore, it is useful to divide the Holocene into the prehistoric, or Amerindian (= pre-Columbian), period versus historic (= post-Columbian) period. For birds, the focus will be more on prehistoric than historic times (see Chap. 15 for a description of the past 350 years of changing bird life in the Chesapeake region).

Two other qualifiers are necessary. First, my information targets terrestrial and aquatic rather than marine faunas, simply because so much more prehistoric information is available for the terrestrial species. A second qualification is that the geographic coverage of this chapter (Fig. 5.1) extends a little beyond the actual limits of the modern Chesapeake-Susquehanna drainage. This allows inclusion of many of the most important late Pleistocene and Holocene vertebrate sites of the Mid-Atlantic region. From the standpoint of birds and mammals, extending the geographical coverage 100 km or so to the east, west, or north of the Chesapeake watershed adds greatly to the prehistoric data set and does not seem to violate any biogeographic principles or add any major habitat types that do not already occur within the Chesapeake watershed. While the Chesapeake Bay region itself "is a little too warm for tamarack and a little too cool for live oak," as Sharrer notes (Chap. 14), the headwaters of the Susquehanna River in New York and Pennsylvania drain forests of white pine, hemlock, beech, and sugar maple. These forests are very different from coastal plain forests along the Chesapeake Bay which feature various southern pines, oaks, hickory, gum, cedar, bald cypress, holly, and so on.

Not surprisingly, therefore, the greater Chesapeake region represents the approximate southern limit of the breeding range for a number of northern species of birds and mammals, such as many ducks, golden eagle, yellow-bellied flycatcher, gray-cheeked thrush, many warblers, white-throated sparrow, swamp sparrow, masked shrew, water shrew, snowshoe hare, porcupine, red squirrel, marten, fisher, and others. Similarly, various southern species reach their approximate northern or northeastern limit in this region, such as the black vulture, red-cockaded woodpecker, Carolina chickadee, brown-headed nuthatch, prothonotary warbler, yellow-throated warbler, summer tanager, Bachman's sparrow, southeastern shrew, western big-eared bat, golden mouse, rice rat, eastern harvest mouse, spotted skunk, and many others.

Much of this chapter will describe faunal losses, which fall into three

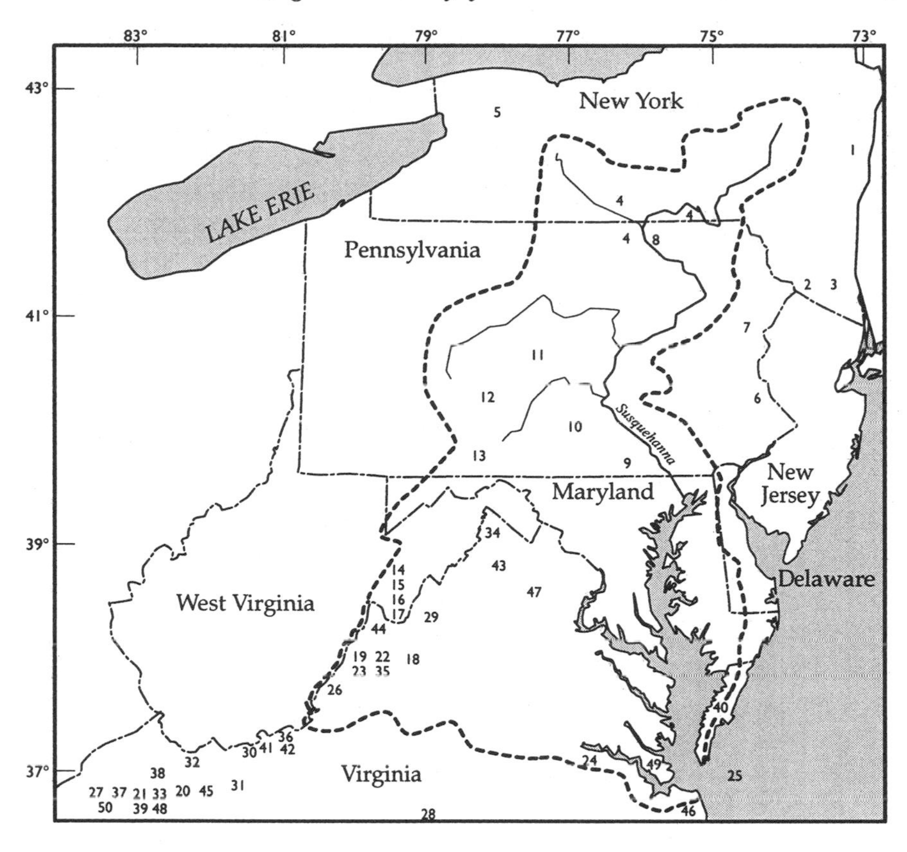

Fig. 5.1. The Mid-Atlantic region, with numbered paleontological and archaeological sites listed in Tables 5.1 and 5.2. *Note:* Dashed line indicates the limits of the Chesapeake-Susquehanna drainage.

categories. First is *extinction,* which is the loss of an entire species, such as the passenger pigeon, Carolina parakeet, giant beaver, American mastodon, or stag-moose. The second category is *extirpation,* which is the loss throughout the Chesapeake region of geographically or taxonomically distinct populations or of subspecies of extant species, such as the heath hen (an eastern race of the greater prairie chicken), gray wolf, mountain lion (= panther or cougar), and elk (= wapiti). Finally, there is *range contraction,* which is the local loss or scarcity of a species formerly more common and widespread, such as the red-cockaded woodpecker, Allegheny woodrat, or black bear. The differences between the last two categories can be subtle and subject to debate.

Because of extirpation and especially extinction, we never can regain the bird and mammal communities that existed in the past. Nonetheless, a good understanding of the paleoecology and past distribution of extinct and extirpated species is important for conservation biology. Only through such knowledge can we assess the secondary impacts of faunal losses, which may include imbalances in seed dispersal, forest composition, predator-prey relationships, or transmission of pathogens.

The Late Pleistocene

The past nearly two million years (the Pleistocene) have been characterized by a series of glacial-interglacial cycles, with glacial intervals averaging 80,000–120,000 years and interglacial intervals averaging only 10,000–20,000 years. The temperature of our current interglacial has been unusually stable, at least in the northern hemisphere. Worldwide sea-level fluctuations between glacial and interglacial intervals have been on the order of 100 m. This means that much of what we know as the Chesapeake Bay was exposed land for most of the past two million years, drained by the mighty Susquehanna River. Until about 13,000 years ago, the northern headwaters of this river were fed at the margin of the Laurentide ice sheet.

The Pleistocene vertebrate fossil record of northeastern North America shows very little extinction until about 11,000 years ago (as measured by radiocarbon dating), when the first humans arrived. During the same time, an overall warming trend (the Pleistocene-Holocene transition) was causing major changes in the distribution of most plant species (for details on the Chesapeake Bay region, see Chap. 3). Unlike in earlier geological times, the extinction event 11,000 years ago most seriously affected large mammals, those weighing more than 100 lbs. By contrast, although their geographic ranges changed, nearly all species of amphibians, reptiles, small birds, and small mammals survived. The collapse of the large mammal communities led to the demise of some dependent species, such as scavenging birds, which consumed the abundant variety of carrion provided by herds of large animals, much as in modern African game parks. When ground sloths, giant beavers, mammoths, mastodons, horses, tapir, peccaries, and so on still inhabited eastern North America, the nearly extinct California condor lived as far from California as Florida and New York.

Each glacial-interglacial transition of the past two million years was accompanied by climatically induced range changes in plants and animals. The Pleistocene-Holocene transition differed from earlier glacial-interglacial transitions because it was at this time that people from Beringia and Alaska dispersed rapidly across North America. Within mere centuries, descendants of these big-game hunters had colonized Central and South America as well. Their spears were tipped with distinctively "fluted" projectile points, which have been found with the bones of extinct mammoths and mastodons at sites radiocarbon dated to within a century or two of 11,000 years ago.

The "Pleistocene overkill theory," developed over the past four decades by Paul Martin, contends that the early big-game hunters were the primary (if not the only) cause of the collapse of American large mammal communities. Some other scientists favor changing climates and habitats as the sole or main cause of the late Pleistocene extinctions. The climate-habitat hypotheses, however, do not explain why large mammal communities that had survived so many previous glacial-interglacial cycles would suddenly collapse, just when humans first arrived. Still other scientists favor a combination of climate-habitat stress and human predation as responsible for the late Pleistocene extinctions. Regardless of cause (see various chapters in Martin and Klein, 1984), this extinction event must have seriously affected all North American terrestrial ecosystems, which once included an impressive variety of grazers and browsers, not to mention their predators and commensals.

The birds and mammals that have been recorded from late Pleistocene deposits in the greater Chesapeake-Susquehanna region are summarized in Tables 5.1 and 5.2. Like any fossil record, this is a sample, but by no means a complete set, of the species that were living in the region at the time. All but 8 of the 91 species of birds listed in Table 5.2 can still be found in the region as migrants or as summer, winter, or permanent residents. Thus, while the Pleistocene-Holocene transition resulted in range changes (especially northward shifts) for most if not all species, most of the birds that occupied eastern North America during the late Pleistocene still live there today. For migratory species, seasonality cannot be demonstrated unless the fossil record includes bones of juveniles (indicating nestlings or fledglings), medullary bone (indicating egg-laying females), or feathers (indicating stage of molt cycle). We cannot, for example, be certain whether breeding or wintering populations are represented by Virginia's late Pleistocene records of fox sparrow or pine grosbeak. The previous existence of breeding populations of these species in Virginia would be reasonable, given that other boreal species of birds and mammals are recorded from the same localities.

Table 5.1 Selected late Pleistocene (LP) nonmarine birds and mammals from within or near the Chesapeake drainage

	Fauna	Depositional location	Environment	Age	References
1	Joraleman's Cave	Albany Co., N.Y.	Cave	LP	Steadman, Craig, and Engel, 1993
2	Dutchess Quarry, lower level	Orange Co., N.Y.	Cave	LP/EH	Lundelius et al., 1983; Steadman and Funk, 1987; Funk and Steadman, 1994
3	Black Dirt sites	Orange Co., N.Y.	Lacustrine/peatland	LP	Hartnagel and Bishop, 1922; Fisher, 1955; Drumm, 1963; Funk, Fisher, and Reilly, 1970; Funk and Steadman, 1994
4	Upper Susquehanna sites	Broome & Chemung Cos., N.Y., Bradford Co., Pa.	Fluvial/peatland	LP	Fisher, 1955; Drumm, 1963; Barnosky et al., 1988
5	Hiscock	Genesee Co., N.Y.	Lacustrine/peatland	LP	Laub et al., 1988; Steadman, 1988b; Steadman and Miller, 1987
6	Durham Cave	Bucks Co., Pa.	Cave	EH	Lundelius et al., 1983
7	Hartman's Cave	Monroe Co., Pa.	Cave	LP/EH	Lundelius et al., 1983
8	Mosherville	Bradford Co., Pa.	Glacial deposit	LP	Lundelius et al., 1983
9	Bootlegger Sink	York Co., Pa.	Cave	LP	Lundelius et al., 1983
10	Carlisle Cave	Cumberland Co., Pa.	Cave	LP	Lundelius et al., 1983
11	Hosterman's Pit	Centre Co., Pa.	Cave	LP/EH	Lundelius et al., 1983
12	Frankstown Cave	Blair Co., Pa.	Cave	LP	Lundelius et al., 1983
13	New Paris No. 4	Bedford Co., Pa.	Cave; fissure fill	LP/EH	Lundelius et al., 1983
14	Eagle Cave	Pendleton Co., W.V.	Cave	LP/EH	Lundelius et al., 1983
15	Hoffman School Cave	Pendleton Co., W.V.	Cave	LP/EH	Lundelius et al., 1983
16	Mandy Walters Cave	Pendleton Co., W.V.	Cave	LP/EH	Lundelius et al., 1983
17	New Trout Cave	Pendleton Co., W.V.	Cave	LP	Holman and Grady, 1987
18	Natural Chimneys	Augusta Co., Va.	Cave	LP	Lundelius et al., 1983
19	Clark's Cave	Bath Co., Va.	Cave	LP	Lundelius et al., 1983; Eshelman and Grady, 1986
20	Saltville	Smyth Co., Va.	Salt spring	LP	Lundelius et al., 1983
21	Abingdon	Washington Co., Va.	Salt spring?	LP	Eshelman and Grady, 1986
22	Back Creek Cave #1 (= Cook Cave)	Bath Co., Va.	Rockshelter owl roost	LP/EH	Eshelman and Grady, 1986

23	Back Creek Cave #2 (= Sheets Cave)	Bath Co., Va.	Rockshelter owl roost	LP/EH	Eshelman and Grady, 1986
24	City Point	Prince George Co., Va.	—	LP	Eshelman and Grady, 1986
25	Continental shelf	Offshore Va.	Ocean bottom	LP	Eshelman and Grady, 1986
26	Covington	Alleghany Co., Va.	—	LP	Eshelman and Grady, 1986
27	Darty Cave	Scott Co., Va.	Cave	LP/EH	Eshelman and Grady, 1986
28	Denniston	Halifax Co., Va.	—	LP	Eshelman and Grady, 1986
29	Edom	Rockingham Co., Va.	—	LP	Eshelman and Grady, 1986
30	Eggleston Fissure	Giles Co., Va.	Fissure	LP	Eshelman and Grady, 1986
31	Gardner's Cave	Wythe Co., Va.	Cave	LP	Eshelman and Grady, 1986
32	Gillespie's Cliff Cave	Tazewell Co., Va.	Cave	LP/EH	Eshelman and Grady, 1986
33	Holston Vista Cave	Washington Co., Va.	Cave	LP/EH	Eshelman and Grady, 1986
34	Hot Run	Frederick Co., Va.	Spring/alluvium	LP	Eshelman and Grady, 1986
35	Hot Springs	Bath Co., Va.	Spring	LP	Eshelman and Grady, 1986
36	Klotz Quarry Cave #5	Giles Co., Va.	Cave	LP	Eshelman and Grady, 1986
37	Lane Cave	Scott Co., Va.	Cave	LP	Eshelman and Grady, 1986
38	Loop Creek Quarry Cave	Russell Co., Va.	Cave	LP/EH	Eshelman and Grady, 1986
39	Meadowview Cave	Washington Co., Va.	Cave	LP	Eshelman and Grady, 1986
40	Metomkin and other barrier islands	Accomack Co., Va.	Ocean beach	LP	Eshelman and Grady, 1986
41	Pembroke Railroad Cave #1	Giles Co., Va.	Cave	LP	Eshelman and Grady, 1986
42	Ripplemead Quarry	Giles Co., Va.	Fissure	LP/EH	Eshelman and Grady, 1986
43	Skyline Caverns	Warren Co., Va.	Cave	LP	Eshelman and Grady, 1986
44	Strait Canyon	Highland Co., Va.	Fissure	LP	Eshelman and Grady, 1986
45	Unnamed cave	Smyth Co., Va.	Cave	LP/EH	Eshelman and Grady, 1986
46	Virginia Beach (near)	Exact provenance in Va. unknown	—	LP	Eshelman and Grady, 1986
47	Warrenton	Fauquier Co., Va.	—	LP	Eshelman and Grady, 1986
48	Will Farleys Cave	Washington Co., Va.	Cave	LP?	Eshelman and Grady, 1986
49	Williamsburg (near)	York Co., Va.	—	LP	Eshelman and Grady, 1986
50	Winding Stairs Cave	Scott Co., Va.	Cave	LP	Eshelman and Grady, 1986

Note: Some of these faunas include a mixture of early Holocene (EH) material. Purely Holocene components of these faunas are not included. In Table 5.2, site numbers are linked to specific species. To save space, original publications are not cited for faunas listed by Lundelius et al. (1983) or Eshelman and Grady (1986).

Table 5.2 Nonmarine late Pleistocene (sometimes mixed with early Holocene) birds and mammals recorded within or near the Chesapeake drainage

Species	Site numbers
Birds	
Podilymbus podiceps	5,19,44
Pied-billed grebe (BMW)	
Botaurus lentiginosus	19
American bittern (BMW)	
Olor sp.	10
Swan (MW)	
Anas crecca	19
Green-winged teal (MW)	
Anas rubripes/platyrhynchos	19
American black duck/mallard (BMW)	
Anas discors/cyanoptera	18,19
Blue-winged teal/cinnamon teal (BMW)	
Bucephala albeola	18,44
Bufflehead (MW)	
Lophodytes cucullatus	19,44
Hooded merganser (BMW)	
Mergus sp.	19
Merganser (BMW)	
Oxyura jamaicensis	18,19
Ruddy duck (BMW)	
Gymnogyps californianus	5
California condor (E)	
Accipiter striatus	18,19,44
Sharp-shinned hawk (BMW)	
Accipiter cooperii	22
Cooper's hawk (BMW)	
Buteo lineatus	18,44
Red-shouldered hawk (BMPW)	
Buteo platypterus	18,19
Broad-winged hawk (BM)	
Buteo jamaicensis	18,44
Red-tailed hawk (BMPW)	
Falco sparverius	19
American kestrel (BMPW)	
Dendragapus canadensis	18,19,23,44
Spruce grouse (E)	
Lagopus mutus	19,23
Rock ptarmigan (E)	
Bonasa umbellus	12,13,16,18,19,22,23,44
Ruffed grouse (P)	
Tympanuchus phasianellus	13,18,19,22,23
Sharp-tailed grouse (E)	

Meleagris gallopavo Wild turkey (P)	6,10,12,13,18,19,33
Colinus virginianus Northern bobwhite (P)	18,19,23,44
Rallus limicola Virginia rail (BMW)	19,44
Porzana carolina Sora (BMW)	19
Gallinula chloropus Common moorhen (BMW)	19
Grus americana Whooping crane (E)	18
Pluvialis dominica Lesser golden plover (M)	19
Charadrius vociferus Killdeer (BMW)	18
Limosa sp. Godwit (M)	19
Tringa solitaria Solitary sandpiper (M)	19
Catoptrophorus semipalmatus Willet (BM)	18
Actitis macularia Spotted sandpiper (BM)	19,44
Bartramia longicauda Upland sandpiper (BM)	18
Calidris minutilla Least sandpiper (M)	18
Gallinago gallinago Common snipe (BMW)	19
Scolopax minor American woodcock (BMW)	18,19,44
Ectopistes migratorius+ Passenger pigeon (E)	16,18,19,22,23,44
Coccyzus sp. Cuckoo (BM)	19
Otus asio Eastern screech owl (P)	19,23
Bubo virginianus Great horned owl (P)	19,44
Asio otus/flammeus Long-eared/short-eared owl (BMPW)	19,44
Aegolius acadicus Northern saw-whet owl (BMW)	19,23
Chordeiles minor Common nighthawk (BM)	19
Chaetura pelagica Chimney swift (BM)	19

(continued)

Table 5.2 *(continued)*

Species	Site numbers
Ceryle alcyon Belted kingfisher (BMPW)	18,19
Melanerpes erythrocephalus Red-headed woodpecker (BMPW)	18,23
Melanerpes carolinus Red-bellied woodpecker (P)	18,19
Sphyrapicus varius Yellow-bellied sapsucker (BM)	19
Colaptes auratus Northern flicker (BMPW)	13,18,19,23,44
Dryocopus pileatus Pileated woodpecker (P)	13,19
Dendrocopos pubescens Downy woodpecker (P)	18,19
Dendrocopos villosus Hairy woodpecker (P)	18,19,23
Contopus virens Eastern wood pewee (BM)	18
Empidonax sp. Flycatcher (BM)	19
Sayornis phoebe Eastern phoebe (BMW)	18
Eremophila alpestris Horned lark (BMW)	19
Progne subis Purple martin (BM)	16
Hirundo pyrrhonota Cliff swallow (BM)	18,19
Perisoreus canadensis Gray jay (E)	18,19
Cyanocitta cristata Blue jay (P)	18,19,23
Pica pica Black-billed magpie (E)	18,44
Corvus brachyrhynchos American crow (P)	19,22,44
Parus bicolor Tufted titmouse (P)	19
Sitta canadensis Red-breasted nuthatch (BMW)	18,19
Certhia familiaris Brown creeper (BMW)	19
Cistothorus platensis Sedge wren (BM)	19
Sialia sialis Eastern bluebird (BMW)	19

Catharus sp. Thrush (BM)	18,19
Turdus migratorius American robin (BMW)	18,19
Toxostoma rufum Brown thrasher (BM)	13,18,19
Anthus spinoletta Water pipit (MW)	19
Bombycilla cedrorum Cedar waxwing (BMPW)	19
Vermivora chrysoptera Golden-winged warbler (BM)	23
Dendroica coronata Yellow-rumped warbler (BMW)	19
Seiurus sp. Ovenbird/waterthrush (BM)	19
Piranga sp. Tanager (BM)	19
Pooecetes gramineus Vesper sparrow (BMW)	19
Passerella iliaca Fox sparrow (MW)	13,18
Melospiza melodia Song sparrow (P)	18
Zonotrichia albicollis White-throated sparrow (BMW)	18
Junco hyemalis Dark-eyed junco (BMW)	19,23
Dolichonyx oryzivorus Bobolink (BM)	19
Agelaius phoeniceus Red-winged blackbird (BMW)	18,19,44
Sturnella sp. Meadowlark (BMW)	19
Molothrus ater Brown-headed cowbird (BMW)	18
Icterus galbula Northern (Baltimore) oriole (BM)	5
Icterus spurius Orchard oriole (BM)	19
Pinicola enucleator Pine grosbeak (W)	19,23
Carpodacus purpureus Purple finch (BMW)	22
Loxia sp. Crossbill (BMW)	19,23

(continued)

Table 5.2 *(continued)*

Species	Site numbers
Mammals	
Sorex arcticus Arctic shrew	9,13,14,17,18,19,23,33
Sorex cinereus Masked shrew	9,12,13,18,19,22,23,33,44
Sorex dispar Rock shrew	13,19,23,33,44
Sorex fumeus Smoky shrew	13,14,18,19,22,23,33,44
Sorex palustris Water shrew	13,18,19,33
Microsorex hoyi Pygmy shrew	9,13,18,19,33,36,41
Blarina brevicauda Northern short-tailed shrew	9,12,13,14,15,16,18,19,22,23,30, 33,36,39,41,43,44,48
Cryptotis parva Least shrew	9,14,18
Parascalops breweri Hairy-tailed mole	9,12,13,14,15,16,18,19,22,23,27, 33,44,48
Scalopus aquaticus Eastern mole	7,18,19,23
Condylura cristata Star-nosed mole	9,13,18,19,22,23,33,38,44
Myotis grisescens Gray bat	15,19,23
Myotis keenii Keen's myotis	9,11,13,18,19,23,33
Myotis leibii Eastern small-footed myotis	19,23
Myotis lucifugus Little brown myotis	11,13,18,19,23,33
Pipistrellus subflavus Eastern pipistrelle	9,13,15,18,19,23,44
Eptesicus fuscus Big brown bat	2,7,9,13,15,18,19,22,23,33,36
Plecotus townsendii Western big-eared bat	19,23
Lasionycteris noctivagans Silver-haired bat	42
Lasiurus borealis Red bat	18,19
Megalonyx jeffersonii+ Jefferson's ground sloth	3,12,20,37

Sylvilagus floridanus Eastern cottontail	9,11
Sylvilagus transitionalis New England cottontail	16
Lepus americanus Snowshoe hare	12,13,18,19,23,33
Tamias striatus Eastern chipmunk	2,7,9,13,15,16,18,19,22,23,33
Tamias minimus Least chipmunk	16,19,23
Marmota monax Woodchuck	6,7,9,13,14,15,16,18,19,22,23, 27,30,32,33,36,39,50
Spermophilus tridecemlineatus Thirteen-lined ground squirrel	9,13,14,15,16,18,19,22,23,33
Sciurus carolinensis Gray squirrel	6,7,15,18,19,22,25,27,33
Tamiasciurus hudsonicus Red squirrel	9,13,14,15,16,18,19,22,23
Glaucomys sabrinus Northern flying squirrel	9,13,14,15,16,18,19,22,23,33
Glaucomys volans Southern flying squirrel	2,9,11,13,15,18,19,22,23,27,33
Castor canadensis Beaver	6,7,14,18,22,27,42
Castoroides ohioensis+ Giant beaver	2,7,18,40
Peromyscus leucopus White-footed mouse	9,13,18,19,22,23,33
Peromyscus maniculatus Deer mouse	9,13,16,18,19,22,23,44
Neotoma magister Allegheny woodrat	2,6,7,12,13,14,15,16,18,19,22, 23,33,44
Clethrionomys gapperi Southern red-backed vole	9,12,13,14,15,16,18,19,22,23, 27,30,33,36,38,39,41,42,44,48
Phenacomys intermedius Heather vole	14,15,16,18,19,22,23,44
Microtus chrotorrhinus Rock vole	9,13,14,15,16,18,19,22,23,27,30, 33,36,39,44
Microtus pennsylvanicus Meadow vole	7,8,9,12,13,14,15,16,18,19,22, 23,30,33,39,44
Microtus pinetorum Pine vole	8,9,11,13,14,15,16,18,19,22,23, 27,30,33,38,39,41,44
Microtus xanthognathus Yellow-cheeked vole	9,13,14,17,18,19,22,23,32,33
Ondatra zibethicus Muskrat	5,6,13,14,15,16,18,19,22,27,44
Synaptomys borealis Northern bog lemming	9,13,14,15,16,18,19,22,23,30, 33,44

(continued)

Table 5.2 *(continued)*

Species	Site numbers
Synaptomys cooperi Southern bog lemming	3,11,12,13,14,15,16,18,19,22,27, 30,33,36,38,39,41,44
Dicrostonyx hudsonius Labrador collared lemming	13,17
Zapus hudsonius Meadow jumping mouse	12,13,14,18,19,44
Napaeozapus insignis Woodland jumping mouse	9,13,18,19,30,44
Erethizon dorsatum Porcupine	6,7,9,11,12,13,14,15,16,18,19, 27,33,44
Canis dirus+ Dire wolf	12,19
Canis lupus Gray wolf	7,18,42
Canis latrans Coyote	12
Vulpes vulpes Red fox	9,18,23
Urocyon cinereoargenteus Gray fox	6,7,16,42
Ursus americanus Black bear	1,6,12,18,19,22,31
Procyon lotor Raccoon	2,6,7,14,18,19,22,27,33,48
Martes americana Marten	13,14,18,19,27,38
Martes pennanti Fisher	13,18,22,27,42
Mustela erminea Ermine (short-tailed weasel)	7,19
Mustela frenata Long-tailed weasel	17
Mustela nivalis Least weasel	13,18,19,22,23
Mustela vison Mink	16,18,19,23,27
Brachyprotoma obtusata+ Short-faced skunk	12
Mephitis mephitis Striped skunk	6,7,9,12,16,18,19
Spilogale putorius Spotted skunk	42
Lutra canadensis River otter	7

Mammut americanum+ American mastodon	3,4,5,10,12,20,21,24,25,26,29, 34,35,44,45,49
Mammuthus primigenius+ Woolly mammoth	4,20,25,47
Mammuthus columbi+ Columbian mammoth	4
Equus complicatus+ Complex-toothed horse	21,44
Equus fraternus+ Brother horse	34
Equus sp.+ Horse	3?,7,12,28
Tapirus veroensis+ Vero tapir	12,42,44,48,50
Mylohyus nasutus+ Long-nosed peccary	7,12,13,18,22,41,42,44,48
Platygonus compressus+ Flat-headed peccary	2,8,34
Cervus elaphus Elk	2,6,7,11,15,19,44
Odocoileus virginianus White-tailed deer	2,6,7,9,11,12,15,16,18,19,23, 34,44
Cervalces scotti+ Stag-moose	3,12,20
Alces alces Moose	6,44
Rangifer tarandus Caribou	2,5,6,7,9,17,20,27,30,36
Bison sp. Bison	7,20,25,34,46
Bootherium bombifrons+ Woodland muskox	12,20,44
Ovibos moschatus Muskox	4

Note: See Table 5.1 for explanation of site numbers. The modern status of birds, given in parentheses following the common name, reflects regular rather than accidental occurrences: B = breeding summer resident; M = migrant (spring and/or fall); P = permanent resident; W = winter resident; E = extinct or extralocal. In the late Pleistocene, virtually all the birds listed here would have been classified as B or P. + = extinct species.

Various species of grouse are excellent habitat indicators today, and therefore are useful in reconstructing late Pleistocene habitats. Only the ruffed grouse now lives in the greater Chesapeake region, where its populations have declined historically in the lowlands. Three other species are recorded in the late Pleistocene. The spruce grouse is characteristic of boreal forests, especially those dominated by spruce. The nearest extant population is in the Adirondack Mountains of New York. The rock ptarmigan inhabits tundra, and occasionally taiga, and occurs today no nearer than northern Quebec, more than 1,000 km north of the Chesapeake region. The sharp-tailed grouse inhabits brushy grasslands and hedgerows of the upper Midwest and West, no nearer than Wisconsin or perhaps Michigan.

The scavenging black-billed magpie resembles the sharp-tailed grouse in its modern habitat requirements and geographic range, whereas the gray jay is similar in these regards to the spruce grouse. Taken together, the spruce grouse, rock ptarmigan, sharp-tailed grouse, black-billed magpie, and gray jay indicate that a late Pleistocene patchwork of cool habitats likely existed in the region, ranging from tundra and grasslands to taiga, shrublands, and boreal forests. The presence of the whooping crane (as well as various waterfowl, rails, and shorebirds) suggests that wetlands were an important part of this mosaic.

The California condor's presence in the greater Chesapeake area is related to feeding opportunities provided by mammalian megafauna. The passenger pigeon survived the Pleistocene-Holocene transition (11,000 to 9,000 radiocarbon years ago), flourished throughout the prehistoric Holocene, and remained abundant until its precipitous decline in the 1800s from deforestation and overhunting. Its extinction early in the twentieth century probably has changed the biotic character of northeastern North America more than the loss of any other single species in the past 10,000 years. Great flocks of passenger pigeons consumed and dispersed, in prodigious quantities, seeds of trees such as elm, maple, basswood, oak, beech, and chestnut. Furthermore, these prolific pigeons were ideal prey for northern goshawks, peregrines, and other predatory species. Only the much earlier losses of the American mastodon and other megafauna about 11,000 years ago had a greater influence.

The late Pleistocene mammal fauna has a number of biogeographic and paleoecological similarities with the avifauna but has withstood more losses of species. Of the 83 species of mammals recorded, only 53 have maintained populations in the region throughout the past 11,000 years. The remaining 30 species consist of 15 extirpated and 15 extinct species.

Post-Columbian overexploitation accounts for the extirpation of three large species (gray wolf, elk, and bison) that survived the Pleistocene-Holocene transition and ten millennia of Amerindian predation, only to die out in the region because of hunting and habitat losses over the past five centuries. The marten, fisher, and moose lived in northern and higher elevation areas at the time of European contact, suffered major range contractions because of hunting, trapping, and deforestation, and only in the past few decades have begun to reoccupy much of their former range in the northeast. During the sixteenth through nineteenth centuries, the impact of unregulated hunting and trapping was severe. In 1760, for example, a single, large, well-organized hunting party in Snyder County, Pennsylvania (on the west bank of the Susquehanna River), killed, stacked, and burned 114 bobcats, 111 bison, 112 foxes (both red and gray?), 109 gray wolves, 98 white-tailed deer, 41 mountain lions, 18 black bears, 3 fishers, 3 beavers, 2 elk, and 500 unidentified smaller mammals.

The remaining extirpated species in the region (arctic shrew, thirteen-lined ground squirrel, least chipmunk, heather vole, yellow-cheeked vole, labrador collared lemming, caribou, and muskox) are characteristic of habitats found today to the north or west, such as boreal forests, taiga, tundra, or prairies. The current absence of these eight species in the Chesapeake-Susquehanna drainage is due to natural changes in climate and habitat rather than to human influences. During the late Pleistocene, these species lived alongside the same species of mammals that are found in the region today. The enriched late Pleistocene small mammal communities have been called "disharmonious" because they have no modern analog. In other words, there is no region of North America today where the same set of boreal and temperate species can be found. Modern species may have occupied narrower niches during the late Pleistocene than today, thereby rendering tenuous the assumption that the modern ecology of living species can be extrapolated accurately into the geological past. Disharmonious assemblages of animals or plants have been used to argue that late Pleistocene climates were more "equable" (i.e., less seasonal, especially because of milder winters) than modern climates. This idea was refuted (I believe correctly) by Kenneth Cole in 1995.

The extinct megafauna of the Chesapeake-Susquehanna region featured Jefferson's ground sloth, giant beaver, dire wolf, American mastodon, two mammoths, two horses, vero tapir, two peccaries, stag-moose, and woodland muskox. What an amazingly different place this would be if these species still existed.

The Prehistoric Holocene

Because species of plants responded individually rather than as entire communities, the plant communities assembled during the Holocene were not identical to those that existed farther south or at lower elevations during the late Pleistocene. The distributions of surviving species of mammals also changed during the Pleistocene-Holocene transition (c. 11,000 to 9,000 years ago), again somewhat independently rather than as intact communities.

With most of the large mammals extinct, eastern Amerindians shifted to a more generalized diet. Bones from Holocene archaeological and paleontological sites may represent as many as 50 species of vertebrates, which range in size from frogs and songbirds to black bear, white-tailed deer, and elk. These sites often record species from localities outside of their known post-Columbian range, such as trumpeter swan, golden eagle, ivory-billed woodpecker, common raven, porcupine, northern flying squirrel, thirteen-lined ground squirrel, rice rat, Allegheny woodrat, varying hare, fisher, and mountain lion. Intertribal trading of animals may account for some of these records, but many if not most zooarchaeological records probably represent formerly indigenous populations. Prehistoric hunting may be involved in some of these Holocene range contractions, while natural changes in climate and habitat are probably responsible for the others.

West and south of the Chesapeake, evidence has been found of domesticated plants such as the bottle gourd, chili pepper, beans, squash, and maize as early as 7,000 years ago, perhaps even earlier. The onset of agriculture in the Chesapeake region probably occurred two or three millennia later than in the Southeast or the Mississippi Valley. Agriculture affects natural ecosystems through the reduction or loss of certain species of plants and animals, the propagation of others, and modifications of landscape, soils, and water supply through deforestation, erosion, deposition, channeling, flooding, draining, and so on. In spite of these and other processes that created clearings of various sizes, the Chesapeake region was mostly forested throughout the Holocene, including the time of European contact.

The introduction of agriculture (see Chap. 6) may have led to less hunting and less gathering of wild plants, but such a transition probably was gradual and never was complete. Hunting, fishing, and gathering remained a significant part of subsistence in prehistoric North America long after the domestication of plants and animals. At the time of European contact, North American Indians (whose total population has been estimated at anywhere from 1 to 18 million) still hunted wild turkey, ruffed grouse, passenger pigeon, black bear, white-tailed deer, elk, bison, and many other species of birds and mammals.

Post-Columbian Times

Even though Amerindians had occupied the Chesapeake Bay region for about 11,000 years, newly arriving Europeans in the sixteenth through eighteenth centuries regarded eastern North America as a wilderness teeming with wildlife. This was an appropriate and unavoidable assessment; the European landscape that the early colonists had left behind was modified extensively by millennia of human activity that, unlike in North America, had included the use of iron tools. Game was scarce in Europe, where hunting often was impossible for nonprivileged persons.

The vertebrate faunal changes that occurred after the arrival and establishment of Europeans and Africans in the Chesapeake drainage are mainly due to human influences and are related to four factors. First was the loss or alteration of natural habitats. The Chesapeake watershed was more than 90% forested when the first Europeans arrived. By the late 1800s, forest cover did not exceed 25%. Second was direct predation through hunting and trapping. Third, non-native vertebrates were introduced that either preyed upon or competed with indigenous species. Finally, non-native wildlife pathogens possibly were introduced.

The first two of these categories probably have had the greatest impact on bird and mammal communities. It is unlikely that a single indigenous species has been unaffected by the past three or four centuries of intensive agricultural, residential, and industrial activities in the Chesapeake region. To overgeneralize, species that tolerate or prefer disturbed habitats have persisted or even increased their range and population. However, many of the species that require relatively undisturbed habitats, be it forest or wetland, have undergone declines in range and population.

Among the birds of our region, the labrador duck, passenger pigeon, Carolina parakeet, and ivory-billed woodpecker have become extinct in the past two centuries. Gone also, although not representing the loss of a full species, are the greater prairie chicken (also known as the heath hen) and whooping crane. Two others, Bachman's warbler and the transient Eskimo curlew, are virtually if not actually extinct, while other species are rare or endangered. Unfortunately, reliable estimates of bird populations do not exist for the sixteenth through nineteenth centuries. Thus, for most extant species we are unable to quantify population trends that are more than a few decades old. If ornithology remains as well organized as it is today, in a century or two we will begin to understand long-term changes in bird populations.

As with the birds, details of the mammal communities in early colonial times are poorly known. Faunal descriptions from this period are biased toward the more obvious species, especially those with commercial value as food or fur. Unless cited otherwise, much of the mammal information that follows is extracted from Rhoads (1903), Doutt, Heppenstall, and Guilday (1967), Paradiso (1969), Lee (1984), and Webster, Parnell, and Biggs (1985), supplemented by the data files of the New York State Museum.

Many widespread species were lost more rapidly in the lowlands than in the piedmont, and often persisted longer in the mountains than in the piedmont. This pattern applied not only to extirpated species; much of the current distributions of black bear and bobcat reflect this differential vulnerability. Certain species were particularly vulnerable to colonial activities. The bison, for example, declined dramatically in the seventeenth and eighteenth centuries and was gone from the region by 1800. The mountain lion, gray wolf, and elk were exterminated from most of the region by 1800 or before, although these three species persisted in certain mountainous areas of New York, Pennsylvania, Maryland, West Virginia, and Virginia into the mid- to late 1800s.

A southern species, the currently endangered red wolf, once occurred as far north as southern Pennsylvania, but it has been extirpated in the Chesapeake-Susquehanna region since colonial times. The region lies within the southern range limit of several other large mammals whose range in eastern North America retreated northward as the southernmost populations were hunted or trapped during the past few centuries. These species include the snowshoe hare, lynx, wolverine, marten, fisher, and moose. With protection by game laws, habitat preservation, and restocking programs, the marten and fisher once again live in the more remote areas of New York and Pennsylvania. Very small, peripheral, perhaps transient populations of lynx and moose occur in northern New York today, whereas the wolverine seems to be gone, even from all of eastern Canada.

Some other furbearers that were widely found, such as the beaver and otter, were trapped heavily in the 1700s and 1800s, reaching population lows in the late 1800s and early 1900s. These species now reoccupy much of their former range. The beaver, in fact, is considered a pest these days in much of the Northeast.

The Future

Because we are interested in what may happen in the future, it may be informative to examine long-term trends in the extinction of native species and the introduction of non-native species. In doing so, we might begin to appreciate our bird and mammal communities in terms of time intervals of potentially evolutionary significance (millennia or more), as well as the seasons, years, or decades that most ecological studies measure. Let us consider birds, a group of organisms better understood than most. At least 20 to 40 species of North American birds were lost at the end of the Pleistocene, mainly because of the extinction of large mammals, on which they depended. Interestingly, none of these extinct species has been recorded in the Mid-Atlantic region. This is probably due, at least in part, to how poorly documented late Pleistocene birds are in the region. The states with the most records of extinct species (Florida, Texas, New Mexico, Arizona, California, Nevada, and Oregon) are those with relatively rich and well studied late Pleistocene avifaunas.

For the following 10,000 years, preceding the arrival of Europeans, we have no evidence for the extinction of birds in eastern North America. In the past 200 years, however, at least five species have been lost (great auk, labrador duck, passenger pigeon, Carolina parakeet, and ivory-billed woodpecker). Two others, the Eskimo curlew and Bachman's warbler, are probably gone, while two more (greater prairie chicken and whooping crane) are extirpated east of the Mississippi River. At least two more species persist today only in small, localized populations (red-cockaded woodpecker and Kirtland's warbler). Even if we maintain reasonable levels of political, economic, and environmental stability, some of these last six species are likely to die out in the next century or two, as may others that are endangered, threatened, or in decline today.

If we continue to lose native birds at rates exceeding one species per century, there may be more than 100 fewer species to face the changing climates and habitats to come with North America's next ice age. On a shorter time scale, many species of North American trees and shrubs, including such fundamental groups as maples, oaks, and beech, may face a northward range shift because of global warming. They must accomplish this task without the passenger pigeon, which until a century ago was the most abundant consumer and disperser of their seeds.

Similarly, we should be concerned about the introduction of non-native species that potentially influence native species. Five non–North American species of birds are established in our region today (mute swan, ring-necked pheasant, rock dove or pigeon, common starling, and house sparrow). Although most of these seem to have minimal impact on native birds, the starling does compete for nest sites with the red-headed woodpecker and eastern bluebird, and probably with other species. Modern wildlife laws and public awareness should help to prevent other exotic species from being established.

Most of this chapter has dealt with the rather depressing topic of faunal losses since the last retreat of the continental ice sheets. Many species already are gone, and some of the downward trends continue. The needs of an expanding and increasingly consumptive human population often seem incompatible with the needs of many nonhuman species. Have we lost so much that the long-term integrity of North American terrestrial ecosystems is in doubt? Have the bird and mammal communities been damaged beyond repair?

Definitive answers to these questions await the next decades and centuries of human activity. Certainly the extinctions, and probably some of the extirpations, represent situations that are beyond repair. Nevertheless, for the bird and mammal communities of the Chesapeake-Susquehanna region and the terrestrial ecosystems upon which they depend, I am not entirely pessimistic about the future. The populations of certain high-profile species, such as the osprey, bald eagle, wild turkey, and beaver, have increased in the past two decades. Already in place are a number of conservation research and management programs sponsored by federal, state, and local governments as well as by private citizens and organizations. Many of these programs now are concerned, at least in part, with the small nongame-nonraptor species that traditionally have been overlooked by wildlife biologists.

At a minimum, we are fortunate that most species of birds and mammals indigenous to the Chesapeake-Susquehanna region still exist. Whenever biologists who study continental faunas begin to despair at the declines they see, they need only glance at the situation on oceanic islands—where often 50–100% of indigenous species of birds and mammals already are gone—to realize how much they have left. Maintaining the long-term integrity of our bird and mammal communities is a difficult challenge, but it is not an impossible one.

BIBLIOGRAPHY

Barnosky, A. D., C. W. Barnosky, R. J. Nickman, A. C. Ashworth, D. P. Schwert, and S. W. Lantz. 1988. Late Quaternary paleoecology at the Newton site, Bradford Co., northeastern Pennsylvania: *Mammuthus columbi*, palynology, and fossil insects. *Bulletin of the Buffalo Society of Natural Sciences* 33: 173–84.

Brauning, D. W., ed. 1992. *Atlas of Breeding Birds in Pennsylvania* (Pittsburgh: University of Pittsburgh Press).

Brodkorb, P. 1971. Catalogue of fossil birds: part 4 (Columbiformes through Piciformes). *Bulletin of the Florida State Museum (Biological Sciences)* 15: 163–266.

———. 1978. Catalogue of fossil birds: part 5 (Passeriformes). *Bulletin of the Florida State Museum (Biological Sciences)* 23: 139–228.

Burney, D. A. 1993. Recent animal extinctions: Recipes for disaster. *American Scientist* 81: 530–41.

Clark, M. K., and D. S. Lee. 1987. Big-eared bat, *Plecotus townsendii,* in western North Carolina. *Brimleyana* 13: 137–40.

Cole, K. E. 1995. Equable climates, mixed assemblages, and the regression fallacy. In *Late Quaternary Environments and Deep History: A Tribute to Paul S. Martin,* ed. D. W. Steadman and J. I. Mead (Hot Springs, S.D.: The Mommonth Site), 131–38.

Dansgaard, W., S. J. Johnsen, H. B. Clausen, D. Dahl-Jensen, N. S. Gundestrup, C. U. Hammer, C. S. Hvidberg, J. P. Steffensen, A. E. Sveinbjörnsdottir, J. Jouzel, and G. Bond. 1993. Evidence for general instability of past climate from a 250-kyr ice-core record. *Nature* 364: 218–20.

Delcourt, H. R., and P. A. Delcourt. 1991. *Quaternary Ecology: A Paleoecological Approach* (New York: Chapman and Hall).

Diamond, J. 1992. *The Third Chimpanzee* (New York: HarperCollins).

Doutt, J. K., C. A. Heppenstall, and J. E. Guilday. 1967. *Mammals of Pennsylvania* (Harrisburg: Pennsylvania Game Commission).

Drumm, J. 1963. Mammoths and mastodons: Ice Age elephants of New York. *New York State Museum and Science Service, Educational Leaflet* no. 13.

Eshelman, R., and F. Grady. 1986. Quaternary vertebrate localities of Virginia and their avian and mammalian fauna. In *The Quaternary of Virginia—A Symposium Volume,* ed. J. N. McDonald and S. O. Bird. *Virginia Division of Mineral Resources Publication* 75: 43–70.

Fisher, D. W. 1955. Prehistoric mammals of New York. *New York State Conservationist* (Feb.–Mar.): 18–22.

Fritz, G. J. 1990. Multiple pathways to farming in precontact eastern North America. *Journal of World Prehistory* 4: 387–435.

Funk, R. E., D. W. Fisher, and E. M. Reilly Jr. 1970. Caribou and Paleo-Indian in New York State: A presumed association. *American Journal of Science* 268: 181–86.

Funk, R. E., and D. W. Steadman. 1994. *Archaeological and Paleontological Inves-*

tigations of the Dutchess Quarry Caves, Orange County, New York (Buffalo, N.Y.: Persimmon Press).

Graham, R. W., and E. L. Lundelius Jr. 1984. Coevolutionary disequilibrium and Pleistocene extinctions. In *Quaternary Extinctions,* ed. P. S. Martin and R. G. Klein (Tucson: University of Arizona Press), 223–49.

Hartnagel, C. A., and S. E. Bishop. 1922. The mastodons, mammoths, and other Pleistocene mammals of New York State. *New York State Museum Bulletin* v. 241–42.

Haynes, C. V., Jr. 1987. Clovis origin update. *Kiva* 52: 83–93.

Holman, J. A., and F. Grady. 1987. Herpetofauna of New Trout Cave. *National Geographic Research* 3: 305–17.

Laub, R. S., M. F. DeRemer, C. A. Dufort, and W. L. Parsons. 1988. The Hiscock site: A rich late Quaternary locality in western New York State. *Bulletin of the Buffalo Society of Natural Sciences* 33: 67–81.

Lee, D. S. 1984. Maryland's vanished birds and mammals: Reflections of ethics past. In *Threatened and Endangered Plants and Animals of Maryland* (Annapolis: Maryland Department of Natural Resources), 454–71.

Lee, D. S., and W. R. Spofford, 1990. Nesting of golden eagles in the central and southern Appalachians. In *Bull's Birds of New York State* (Ithaca: Cornell University Press).

Long, A., B. F. Benz, D. J. Donahue, A. J. T. Jull, and L. J. Toolin. 1989. First direct AMS dates on early maize from Tehuacán, Mexico. *Radiocarbon* 3: 1035–40.

Lundelius, E. J., Jr., R. W. Graham, E. Anderson, J. Guilday, J. A. Holman, D. W. Steadman, and S. D. Webb. 1983. Terrestrial vertebrate faunas. In *Late-Quaternary Environments of the United States,* v. 1, *The Late Pleistocene,* ed. S. C. Porter (Minneapolis: University of Minnesota Press), 311–53.

Martin, P. S. 1984. Prehistoric overkill: The global model. In *Quaternary Extinctions,* ed. P. S. Martin and R. G. Klein (Tucson: University of Arizona Press), 354–403.

———. 1990. 40,000 years of extinctions on the "planet of doom." *Palaeogeography, Palaeoclimatology, Palaeoecology* 82: 187–201.

Martin, P. S., and R. G. Klein, eds. 1984. *Quaternary Extinctions* (Tucson: University of Arizona Press).

Owen-Smith, N. 1987. Pleistocene extinctions: The pivotal role of megaherbivores. *Paleobiology* 13: 351–62.

Paradiso, J. L. 1969. Mammals of Maryland. *North American Fauna* no. 66.

Pielou, E. C. 1991. *After the Ice Age* (Chicago: University of Chicago Press).

Pirazzoli, P. A. 1993. Global sea-level changes and their measurement. *Global and Planetary Change* 8: 135–48.

Rampino, M. R., and S. Self. 1992. Volcanic winter and accelerated glaciation following the Toba super-eruption. *Nature* 359: 50–52.

Rhoads, S. N. 1903. *The Mammals of Pennsylvania and New Jersey. A Biographic, Historic, and Descriptive Account of the Furred Animals of Land and Sea, Both Living and Extinct, Known to Have Existed in These States* (Lancaster, Pa.: Wickersham Printing Co.).

Robbins, C. S., and D. Bystrak. 1977. Field list of the birds of Maryland. *Maryland Avifauna* no. 2.

Santner, S. J., D. W. Brauning, G. P. Schwalbe, and P. W. Schwalbe. 1992. Annotated list of the birds of Pennsylvania. *Pennsylvania Biological Survey Contribution* no. 4.

Schorger, A. W. 1955. *The Passenger Pigeon: Its Natural History and Extinction* (Madison: University of Wisconsin Press).

Semken, H. A., Jr. 1983. Holocene mammalian biogeography and climatic change in the eastern and central United States. In *Late-Quaternary Environments of the United States,* v. 2, *The Holocene,* ed. H. E. Wright Jr. (Minneapolis: University of Minnesota Press).

———. 1988. Environmental interpretations of the "disharmonious" late Wisconsinan biome of southeastern North America. *Bulletin of the Buffalo Society of Natural Sciences* 33: 185–94.

Shackleton, N. J. 1987. Oxygen isotopes, ice volume and sea level. *Quaternary Science Reviews* 6: 183–90.

Smith, B. D. 1992. Prehistoric plant husbandry in eastern North America. In *The Origins of Agriculture,* ed. C. W. Cowan and P. J. Watson (Washington, D.C.: Smithsonian Institution Press), 101–19.

Steadman, D. W. 1988a. Prehistoric birds in New York State. In *The Atlas of Breeding Birds in New York State,* ed. R. F. Andrle and J. R. Carroll (Ithaca: Cornell University Press), 19–24.

———. 1988b. Vertebrates from the late Quaternary Hiscock site, Genesee County, New York. *Bulletin of the Buffalo Society of Natural Sciences* 33: 95–113.

———. 1995. Ecosystem integrity: The impact of traditional peoples. *Encyclopedia of Environmental Biology* (San Diego: Academic Press), 1: 633–47.

———. 1997. Human-caused extinction of birds. In *Biodiversity II,* ed. M. L. Reaka-Kundla, D. E. Wilson, and E. O. Wilson (Washington, D.C.: Joseph Henry Press), 139–61.

Steadman, D. W., L. J. Craig, and T. Engel. 1993. Late Pleistocene and Holocene vertebrates from Joralemon's (Fish Club) Cave, Albany County, New York. *The Bulletin, Journal of the New York State Archaeological Association* 105: 9–15.

Steadman, D. W., and R. E. Funk. 1987. New paleontological and archaeological investigations at Dutchess Quarry Cave No. 8, Orange County, New York. *Current Research in the Pleistocene* 4: 118–20.

Steadman, D. W., and P. S. Martin. 1984. Extinction of birds in the late Pleistocene of North America. In *Quaternary Extinctions,* ed. P. S. Martin and R. G. Klein (Tucson: University of Arizona Press), 466–77.

Steadman, D. W., and N. G. Miller. 1987. California condor associated with spruce-jack pine woodland in the late Pleistocene of New York. *Quaternary Research* 28: 415–26.

Stewart, R. E., and C. S. Robbins. 1958. Birds of Maryland and the District of Columbia. *North American Fauna* no. 62.

Terborgh, J. 1989. *Where Have All the Birds Gone?* (Princeton: Princeton University Press).

Virginia Society of Ornithology (V.S.O.). 1979. Virginia's birdlife: An annotated checklist. *Virginia Avifauna* no. 2.

Webb, T., III, P. J. Bartlein, S. Harrison, and K. H. Anderson. 1993. Vegetation, lake levels, and climate in eastern United States since 18,000 yr B.P. In *Global Climates Since the Last Glacial Maximum,* ed. H. E. Wright, T. Webb III, and J. E. Kutzbach (Minneapolis: University of Minnesota Press).

Webster, W. D., J. F. Parnell, and W. C. Biggs Jr. 1985. *Mammals of the Carolinas, Virginia, and Maryland* (Chapel Hill: University of North Carolina Press).

Wilson, E. O. 1992. *The Diversity of Life* (Cambridge: Belknap Press, of Harvard University Press).

Zubrow, E. 1990. The depopulation of native America. *Antiquity* 64: 754–65.

CHAPTER SIX

Living along the "Great Shellfish Bay"

The Relationship between Prehistoric Peoples and the Chesapeake

HENRY M. MILLER

The relationship between American Indians and the Chesapeake Bay ecology was dynamic. The Indians demonstrated remarkable resilience to environmental changes. In the early Paleo-Indian period, they hunted large game and collected plants. Chesapeake Indians focused on white-tailed deer and smaller game in the subsequent Archaic period. They harvested shellfish increasingly in the Late Archaic and Woodland periods. During the Late Woodland, cultivation of corn and other crops began. These new patterns and Chesapeake ecology encouraged significant social changes accompanied by the rise of new political systems by the time Europeans arrived.

The magnificent estuary we call the Chesapeake Bay is intimately connected with human history and prehistory. Before there was a Chesapeake Bay, people lived on the land that now forms its channels, shores, and marshes. As the Chesapeake developed, the people adapted to the many new possibilities it offered. These people are the American Indians, and their lives were intricately bound with this largest of the estuaries in North America. In this chapter, one aspect of the human relationship with the Chesapeake Bay is presented. It is a story assembled by archaeologists from the often faint and always fragmentary traces left by Native Americans over a span of 12,000 years, since the end of the Pleistocene Ice Age.

A Changing Environment

Perhaps the most salient factor influencing the experience of Native Americans in this region is the dramatic environmental changes over the last 12,000 years. The Chesapeake region was transformed by climatic forces from a cold, glacier-influenced landscape covered by spruce and fir forests to today's temperate climate and landscape dominated by oak, hickory, and pine forests. The transition was not smooth, for numerous periods of wet or dry conditions and higher or lower temperatures were common and represented challenges for the people living in the region. At the end of the Pleistocene epoch, melting glaciers released huge quantities of water which produced very rapid sea-level rise for thousands of years. Geologists estimate that about 15,000 years ago sea level was 325 feet lower than it is at present. At that time, the Atlantic shoreline was more than 60 miles east of its present location, at the edge of the continental shelf. These lower water levels allowed the Susquehanna River and its tributaries to cut deep channels as they flowed toward the sea. By 7000 years B.P., the ocean had filled these channels, now flooded, rising to within 40 feet of present sea level, and by 4000 B.P. to within 16 feet. Estimates are that the sea rose 3 feet or more per century from 11,000 to about 8,000 years ago but only increased at a rate of one-half a foot per century at 4000 B.P. These dramatic changes had significant consequences for people living in the region.

Prehistoric Peoples of the Chesapeake

Archaeologists divide the span of human prehistory in eastern North America into three segments: the Paleo-Indian period (c. 13,000–10,000 B.P.), the Archaic period (10,000–3000 B.P.), and the Woodland period (3000–400 B.P.). These are defined by general cultural traits, key artifacts, and the prevailing environmental conditions. The Archaic and Woodland periods are subdivided into three segments, which are believed to reflect changes in development of native culture. These same divisions will be employed here for organizational purposes.

The Paleo-Indian Period

Evidence indicates that people reached America by the end of the last glaciation, if not earlier. Called Paleo-Indians, these people have long been portrayed as hunters of woolly mammoth, mastodon, and other now extinct Pleistocene megafauna. New research suggests that this is not an accurate picture. Some megafauna were probably taken when the opportunity arose, but hunters in the Middle Atlantic region more likely specialized on the less exotic white-tailed deer, elk, and caribou. Evidence from the Shawnee-Minisink site along the upper Delaware river valley shows that they hunted smaller game, collected a variety of plants, and caught some fish. Available data suggest they traveled over large areas in bands, following food sources and visiting quarry sites to obtain stone for tool manufacture. Paleo-Indians hunted and camped in some areas now inundated by the Chesapeake Bay and coastal waters of the Atlantic. Evidence for this comes from fishermen who have found the distinctive spearpoints used by Paleo-Indians and the remains of extinct animals in these areas. Because the Chesapeake Bay had not yet formed, it is unlikely that these people had access to estuarine resources.

The Archaic Period

As temperatures rose and the glaciers continued to melt, the Pleistocene megafauna became extinct and the ecology of the Chesapeake region became more diverse. The boreal forest of spruce and fir trees gradually gave way to a hemlock, oak, and spruce forest, which lasted until about 4,000 years ago. During the early part of the Archaic (10,000–8500 B.P.), the seminomadic hunting and collecting life of the preceding period continued, although it probably became more seasonally patterned. In the middle segment of this period (8500–5000 B.P.), there was an expansion in the types of tools being used, suggesting the harvesting of new food sources. People at this time apparently focused their activities on inland resources, not on the rivers and newly forming Chesapeake estuary. Early and Middle Archaic sites are not found in large numbers within this region, so the human population likely was small.

Middle Archaic people tended to concentrate their settlements around large inland marshes, such as the Dismal Swamp in Virginia and Zekiah Swamp in Maryland. Short-term camps are also found in the uplands. Sea-level rise also raised groundwater levels, resulting in the creation or expansion of these marshes in poorly drained inland areas. These inland swamps offered an abundance of plant foods and fresh water and attracted game animals. By this time, white-tailed deer was the most significant land animal being hunted. It is unlikely that these Middle Archaic people made much use of coastal areas or resources. Rapid sea-level rise and the shallow slope of much of the lands being inundated resulted in pronounced temperature, salinity, and turbidity changes in the rivers. Such an unstable environment is not conducive to the development of a rich estuarine biota. There is little evidence for major coastal settlements or significant use of estuarine resources during this time, although it must be noted that the data are incomplete since sea-level rise has inundated many of the areas that were coastal at that time. Some coastal sites from the Middle Archaic are found along the lower Bay, however, where estuarine conditions first developed. It is significant that these are small sites that contain a limited array of tools and lack shellfish remains. According to William Gardner (1987), these were temporary camps for a people whose main efforts focused upon the interior areas.

In the Late Archaic (5000–3000 B.P.), several environmental shifts occurred which had profound consequences for the native people. One was a change to somewhat drier climatic conditions. In the forest, oak, hickory, chestnut, and pine trees came to dominate, providing acorns and nuts as reliable food sources. Equally significant was a major decrease in the rate of sea-level rise. Most of the lower river valleys were flooded by that time, and the slower rate of inundation permitted conditions within the estuary to begin stabilizing. Habitats conducive to the growth of estuarine fish and shellfish were formed and their populations expanded. Among the evidence for this is that no shell middens are known in the Chesapeake region dating earlier than c. 4,500 years ago. Support for this comes from intensive coring on the adjacent Delaware Bay, where the earliest extensive oyster shell deposits date after 4700 B.P. Given that sea level was rising at approximately the same rate in both estuaries, it is improbable that shellfish were well established in the Chesapeake Bay any earlier. Development of stable estuarine conditions meant that aquatic food sources became abundant, widespread, and dependable for the first time during the Late Archaic.

Not surprisingly, the Late Archaic peoples responded by beginning to shift their settlements away from the inland swamps and toward the riverine and estuarine areas. Coastal shell deposits are the best evidence for this change. Along the lower Potomac River, excavation of the White Oak Point site by Gregory Waselkov (1982) revealed a long sequence of oyster use by Native Americans. The earliest deposit dates to about 4,000 years ago and contained a variety of shellfish, including oysters, softshell clams, marsh periwinkle, ribbed mussels, and stout tagellus. It is of interest to note that the other shellfish contributed more meat to the diet than the oysters did, suggesting that the people were trying a variety of resources and had not yet developed a preference for oysters. Fish bones were also recovered along with the remains of snakes and white-tailed deer.

Further up the Potomac River, just south of Washington, D.C., a series of large sites were found with initial habitations during the Late Archaic. Michael Stewart and William Gardner (1978) identify these as camps for the exploitation of anadromous fish. Anadromous fish are those that spend portions of their lives at sea but return to brackish-water or freshwater areas for spawning. Examples include herring, shad, striped bass, and sturgeon. These upriver sites, dating to 3,500 years ago, are believed to provide the earliest evidence for use of this important estuarine resource. Another aspect of the new settlement pattern at this time was a greater preference for living near the freshwater-saltwater transition zone. By placing base camps at these locations within the estuary, Chesapeake Indians achieved maximum access to a wide variety of food resources.

The Woodland Period

After the Archaic, there was a period of intensified use of coastal resources. Along the Potomac, a major emphasis on oyster exploitation is seen at sites such as Pope's Creek, where a massive shell midden up to 26 feet thick and covering many acres of land was created, beginning about 3,000 years ago. At the White Oak Point site on the lower Potomac, oysters, softshell clam, and ribbed mussel continued to be utilized, although oysters were now the most important. Excavations there also recovered remains of blue crab and fish dating to c. 3100 B.P. The first shell middens on the Patuxent River and Eastern Shore date to the Early Woodland. Evidence suggests that the Early and Middle Woodland Indians (3,000 to 1,000 years ago) became more sedentary and experienced population growth. Base camps continued to be located near the freshwater–brackish water interface, where the greatest variety of resources were accessible. As in earlier millennia, most of the people in the Chesapeake region during the Early Woodland lived in societies that were kin based and relatively egalitarian. However, archaeologist Richard Dent (1995) suggests that societies run by a "Big Man," who achieved the leadership position, first appeared during this period.

Seasonal mobility and exploitation of diverse resources was part of a well-developed annual schedule of food-getting activities. Coastal resources were especially significant during the late winter, spring, and summer. Hunting and collecting provided an array of animals and plants for the diet at other times during the year. White-tailed deer remained the most important mammal, with estimates from some sites that it contributed up to 80% of the meat. Besides venison, Native Americans ate animals such as beaver, ducks, opossum, rabbit, raccoon, snakes, squirrel, turtles, and turkeys. Plant foods, however, probably accounted for most calories consumed, and these included nuts, acorns, the tubers of freshwater marsh plants, the wild grains of amaranth and *Chenopodium*, and a variety of fruits and berries. Bruce Smith (1992) suggests that by this time, some of these wild grains were actually domesticated and planted, along with species such as gourds and sunflowers. In adapting to the new possibilities offered by the resources of the Chesapeake Bay, the natives established a pattern of life by the Early Woodland which endured for several thousand years. One indication is that the White Oak Point site was known and used specifically for oystering over 3,500 years, a remarkable consistency in site usage and subsistence patterning.

From the Middle Woodland (1800–1000 B.P.), there is abundant evidence for intensive use of the Chesapeake Bay's resources. People on the lower Bay continued relying on fish and shellfish during the spring and summer. One of the best sources of information about estuarine resource use during the Middle Woodland is the Addington site, located near the mouth of the Chesapeake on Lynnhaven Bay. Its excavation by Thomas Whyte (1988) yielded a remarkable sample of unusually well preserved fish, shellfish, and other animal remains. Analysis indicates that the site was consistently occupied between April and September each year. At least 31 types of fish are represented in the collection. Marine species include sea trout, croaker, spot, menhaden, white perch, sharks, skates, burrfish, needlefish, and stargazer. A range of freshwater to brackish-water fish are represented by longnosed gar, golden shiner, white catfish, yellow bullhead, and pumpkinseed. The Addington site fishermen apparently caught these using nets or weirs. Site occupants also ate shellfish (including ribbed mussels, marsh periwinkle, and oysters) and three species of clams, as well as consumed blue crabs in considerable quantities. There is no evidence of habitation at this site during the winter, indicating that it was only one part of their settlement pattern. During the autumn, the inhabitants may have moved inland, where various types of acorns and nuts were gathered and deer and smaller game hunted.

People living further up the tributaries of the Chesapeake, in the brackish-water to freshwater zone, focused on somewhat different resources, reflecting the local biota. At the Maycock Point site on the James River below Richmond, Tony Opperman found freshwater mussels were eaten so intensively that a shell midden up to 5 feet thick developed. Site residents also exploited the anadromous fish, such as sturgeon, and ate freshwater gar and catfish. Deer, turkeys, and nuts provided additional food. Also of importance in these upriver areas were the freshwater swamps, which yielded tubers and other edible plants. These varied resources were sufficiently dependable and abundant for some year-round habitation to occur at places like the Maycock Point site.

The most detailed study of coastal adaptation at a specific locality in the Chesapeake region was conducted by Stephen Potter (1993) and Gregory Waselkov (1982). They concentrated their efforts along the lower Potomac River in the Northern Neck of Virginia. Known as the Coan River, this area yielded diverse archaeological sites, including large villages, smaller hamlets, and various special-purpose sites. Potter's study of the settlement pattern revealed that sites dating around 1800 B.P. included small to medium-size coastal habitations that were semisedentary in use, along with upland hunting and collecting sites and seasonally occupied oyster-collecting camps (of which the White Oak Point site is a good example). Around 1300 B.P., the first large villages appeared, and they were inhabited for much of the year. These villages were linked to the seasonal oystering and fishing sites and to the inland hunting and collecting camps by a carefully timed seasonal round of activities. The larger size of the settlements suggests that the human population continued to increase during the Middle Woodland.

These large villages did not continue into the Late Woodland period. After 1100 B.P., settlement in the Coan River area became dispersed, with small clusters of houses being most common. About this same time in the Piscataway and Pope's Creek areas of the upper Potomac, the settlement pattern changed focus from on the rich coastal settings to the more interior floodplains. This dispersion into small and medium-sized settlements located on the lowlands is common over much of the Chesapeake region during this period. There are probably several factors involved in the widespread change. One is climate change. The Medieval Warm Period, around 1100 B.P., seems to have been a time of higher temperatures and somewhat drier conditions. On the Nanticoke River, palynologist Grace Brush found that the deposition of charcoal and heavy metals in sediments increased dramatically between 1100 and 1000 B.P. She attributes this to the onset of very dry conditions and the occurrence of numerous forest fires (see Chap. 3).

During the same period, the Chesapeake Indians initiated an important change in their economy by beginning to cultivate corn, beans, and squash. The earliest evidence for the raising of these plants in the Chesapeake region dates between 1,100 and 1,000 years ago. Human population had apparently increased substantially in size during the preceding centuries, and Stephen Potter (1993) suggests that the onset of dry conditions could have reduced food resources sufficiently to encourage the move to cultivation. Dispersal into smaller hamlets or homesteads may have been initiated by a decrease in food resources. By more broadly distributing the human population, harvesting pressure on natural resources in any single locality would be lessened. It seems that agriculture was slowly integrated into the subsistence system and became a crucial food resource.

About 700 B.P., people again started concentrating into larger villages. Along the outer coastal plain, these villages tended to be internally spread out, with houses placed adjacent to the horticultural fields. In some locations, however, especially on the inner coastal plain near the Fall Line, where the potential for conflict with Piedmont Indians was high, villages were often compact and fortified. Hostilities and warfare became more common throughout the region about 600 years ago, as indicated by the presence of fortified sites. Reasons for increased conflict include population growth, efforts to control surplus foods, the rise of new social and political institutions, limited resources, and, eventually, unequal access to European trade goods.

Greater reliance on domesticated crops seriously affected the overall cycle of life for the Chesapeake Indians and altered their relationship with one another and with nature. Mobility as a strategy to resolve social conflicts or food shortages became less viable because horticulture demanded greater residential permanence. It required people to input large amounts of labor in the springtime to clear fields and plant crops and continued effort to tend the growing plants. This upset the carefully timed sequence of seasonal food getting and population movements. One consequence may have been the significant change in oyster utilization which occurred at the White Oak Point site about 700 years ago. The intensity of oyster harvesting increased substantially and roasting pits were used for the first time. This probably indicates oysters were being dried for storage instead of consumed at the site, which Waselkov (1982) believes was one of the responses to scheduling conflicts caused by the adoption of agriculture. Entire groups could not move to the oyster-collection site in the spring because of the need to prepare the fields and plant crops. The Indians apparently devised a division of labor so that some members of a group would collect, prepare, and transport this and other important springtime foods for those engaged in agriculture. Managing this more complex approach with larger human populations may have played an important role in the development of chiefdoms in the Chesapeake area.

Native Life at the Time of European Contact

The best evidence regarding Native American life after their adoption of cultivation comes from the writings of English explorers such as John Smith and William Strachey in the early seventeenth century. These accounts, which have been thoroughly studied and synthesized by Helen Rountree (1989), provide many valuable insights regarding the Indian's relationship to the Chesapeake environment and food-getting practices at the end of the Late Woodland period.

In the early seventeenth century, we know that the Chesapeake Indians spoke Algonquin and lived in semipermanent settlements scattered along the shores of the region's many creeks and rivers. Greater cultural and political complexity had arisen and a new level of social organization, the chiefdom, had emerged by the Late Woodland. Our best evidence for their settlement pattern comes from a remarkable map compiled by Captain John Smith during his exploration of the Bay in 1608. Inspection of the map clearly reveals the coastal orientation of the villages. Most people apparently lived in hamlets ranging in size from 2 to 50 houses. A few larger villages existed which were the residences of the chiefs. Early settlers' accounts indicate that these settlements were moved as the need arose, prompted by access to fertile soils, more firewood, and other factors. Within these villages, people placed their houses with little apparent order, choosing to spread the buildings among the cultivated fields. These historical observations regarding settlements correspond well with archaeological data from Late Woodland sites.

All the English observers agreed that the Indians relied both upon domestic plants and a variety of wild food sources for their sustenance. Food-getting activities were a prominent element of their lives and were expressed in many aspects of their culture, including gender roles, social organization, and beliefs. The Chesapeake Indians divided the many food-related tasks strictly along gender lines. Women raised the crops and foraged for various wild plant foods. They may also have harvested oysters and clams. Although hunting is often viewed as the most important activity, the women's cultivation and collecting efforts actually provided most of the food consumed by the Chesapeake Indians. Men contributed meat through hunting and fishing, taking many different types of animals. They considered venison the most prestigious meat, as it was frequently consumed by chiefs and their guests. Deer yielded hides for clothing and essential materials for tool manufacture such as bone, sinew, and glue.

From the colonial accounts, it appears that the Chesapeake Indian's division of labor worked well and yielded ample food. Despite European perceptions that native peoples were engaged in a constant struggle for survival, modern anthropological research indicates that band, tribal, and chiefdom level societies usually acquire sufficient foods without demanding excessive labor from their members. Even native peoples in harsh environments such as the Kalahari desert of Africa manage to find enough food while also enjoying abundant leisure time. Residents of much richer environmental settings such as the Chesapeake would have done likewise.

In food acquisition, the Woodland peoples and their predecessors employed a strategy based upon the idea of using many different ecological zones. As particular resources within each zone became available during the year, efforts focused there. The zones included rivers and creeks, tidal marshes, the fertile lowland forests bordering the rivers, inland swamps, and upland forests. Efficient harvesting of foods in each zone required careful timing and a mobile lifestyle. The Chesapeake Indians accomplished this through an annual subsistence cycle, which they refined over the centuries. To be successful, this strategy required use of a variety of foods, mobility, and an intimate knowledge of the locations, seasonal availability, nature, and uses of many different animals and plants. Because of the close relationship between native subsistence practices and the cycles of nature, native foodways can be best described by reference to their annual pattern of food acquisition.

In the winter, stored foods such as corn, beans, acorns, and nuts, along with both preserved and fresh meats from deer, raccoon, turkey, and other wild animals were consumed. The Chesapeake Indians possessed no domesticated animals except the dog. As the late winter gave way to early spring and stored food stocks dwindled, attention turned to the migratory waterfowl that made their biannual visits to the Chesapeake. The abundant and easily collected oysters and clams became very important during this period of limited food resources, as confirmed by data from numerous archaeological sites. Especially crucial to the spring diet were anadromous fish such as shad and herring, which appeared in large schools during March and April. Fishermen caught them most effectively in the narrow upper reaches of the rivers and streams using weirs and nets.

Preparation and planting of the fields also occurred during the spring. Lacking metal axes and confronted with often huge trees, the natives employed an efficient means of land clearance. Called slash-and-burn cultivation, this method involved slashing a ring of bark from trees, which killed them. The men probably left the largest trees standing but cut down smaller ones with stone axes or by a combination of burning and cutting. Fires that would be started later among the ground litter of fallen limbs and leaves not only killed any undergrowth and cleared the ground but released potassium and other nutrients into the soil. Women then worked the rich, sandy loam with hoes or sticks and planted crops in April and May. Men helped with the strenuous task of field clearance, but women conducted most of the agrarian activities. Principal crops were maize, or "Indian Corn," as the colonists labeled it, and several varieties of beans. Secondary crops included squash, Jerusalem artichokes, gourds, melons, and sunflowers. Corn was undeniably the major and most prestigious food crop. Studies suggest that 50% or more of the calories consumed derived from corn.

During the summer, while the corn and other crops matured, efforts concentrated on the extraordinary abundance of seafood that filled the Chesapeake Bay. Colonial accounts tell us that the native people ate many different fish along with crabs and turtles. Women collected wild strawberries, blackberries, mulberries, greens, seeds, and fruits such as grapes and plums during this season. To augment the rapidly diminishing supplies of stored corn, the natives collected a wild plant called the tuckahoe. Common in freshwater marshes and low-lying coastal areas, the starchy tuber of this plant could be made into bread. By midsummer, garden vegetables were available and the first corn eaten.

Chesapeake Indians experienced the greatest abundance of food in the autumn, when they held their principal ceremonies and feasts. Corn and other crops were harvested, the rivers still contained many fish, and the forests yielded acorns, walnuts, chestnuts, and hickory nuts. They made bread using acorns of the white oak and extracted oil from walnuts and hickory nuts. In late autumn, efforts turned to the acquisition of large numbers of deer for both meat and hides. Hunting groups traveled toward the Fall Line of the Chesapeake where they conducted communal deer drives. Movement to the Fall Line for hunting was apparently a response to the depletion of deer populations on the outer coastal plain. According to the English, up to a dozen deer were often taken in a single drive. A portion of this meat was dried for later consumption. Afterward, people returned to their coastal villages for the winter.

Thus, the lifestyle of the natives conformed closely to the rhythms of nature. Activities concentrated upon fishing, hunting, planting, or collecting, depending upon the time of year. The physical appearance of the people also apparently changed according to the season. As one description by an early English settler notes: "It is strange to see how their bodies alter with their diet, even as the deare and wilde beastes they seem fat and leane, strong and weake" (Strachey, 1953: 80). Hence, the native people's physical condition varied with the changing availability of resources. This is perhaps the most powerful testimony to the fact that the Chesapeake Indians lived in union with the cycles of nature, not apart from them.

Significant changes in Chesapeake Indian society were also under way at the end of the Late Woodland period. Instead of the egalitarian bands or tribes that characterized much of Chesapeake prehistory, social and political centralization and the rise of complex societies occurred in a few locations. The English colonists found hereditary paramount chiefs who ruled many villages in a political system known as a chiefdom. These were new and still-evolving political institutions unlike anything which had previously existed in the region. On the upper coastal plain of the James and York Rivers was the Powhatan chiefdom, while the Piscataway, or Conoy, chiefdom was established on the Potomac River, below the Fall Line. Chiefdoms did not emerge on the lower coastal plain.

A major research question is why these complex societies arose on the upper coastal plain. Potential explanations for the development of this political institution have been summarized by Stephen Potter (1993) and include conflict and trade with the Piedmont peoples, stress induced by environmental change, sporadic European contact, growing populations, restricted food resources, and soil distributions. The issue remains unresolved, but it is certain that a complex interplay of factors caused the emergence of these ranked, chiefdom societies in the Chesapeake region. One factor is the nature of the estuary's ecology. Lewis Binford (1964) has argued that intergroup competition for anadromous fish was an important variable. The spring fish runs were a key element in the subsistence system, and the spawning fish were most readily obtained at certain locations on the upper coastal plain near the Fall Line. Given the larger Late Woodland populations, this resource may have become even more valuable. Binford proposes that the people living on the upper coastal plain had a monopoly on this critical resource and were able to use it to their political advantage. Although scholars agree that this was not the only or even the most important factor in the rise of these chiefdoms, the unequal distributions of crucial resources certainly had a role. Thus, Chesapeake ecology not only helped shape the economic life of the native people but also seems to have had an influence on Chesapeake Indian social and political evolution.

Environmental Impact and the Chesapeake Indians

We know the native inhabitants of the Chesapeake region used many of the Chesapeake's resources, but the potential impact of their activities on the environment is more difficult to assess. The sheer abundance of fish, migratory waterfowl, and shellfish along with the relatively small size of the Native American population makes it unlikely they had any major effect on the fauna of the Chesapeake. Population estimates are very uncertain, but there were perhaps 30,000 to 45,000 people living along or near the Chesapeake Bay in 1600. The only evidence of serious resource depletion involves the white-tailed deer. According to colonial accounts, deer populations were significantly reduced in some areas by the time of European settlement. Archaeological data show that Indians had intensively hunted this animal since at least the middle Archaic, and it is suggested by Randolph Turner (1976) that overhunting of deer in the Tidewater area of Virginia began as early as the Late Archaic period. It seems likely that intense hunting pressure over thousands of years also caused some changes in the nature of the deer population. Smaller animals and a variety of plants were probably severely reduced in numbers or even annihilated in the vicinity of major settlements, but these were almost certainly temporary and spatially restricted depletions.

It is worth noting that the Chesapeake Indians at the time of European settlement did not practice any conservation measures to protect natural resources. Evidence for this comes from a number of the early settlers such as William Strachey, who wrote that "at all tymes and seasons they destroy . . . hares, partridges, Turkeys, fatt or leane, young or old in eggs in breeding tyme, or however, they devowre, at no tyme sparing any that they can katch in their power" (1953: 82). The native people were not conservation oriented for one important reason—they had no need to be. A small human population and the abundant food resources of the Chesapeake region made such efforts unnecessary. At the same time, it is clear that they did not needlessly kill or waste resources. They consumed most animals and used non-meat portions such as hides, feathers, bones, and sinew for clothing, tools, and other purposes.

The greatest ecological impact from prehistoric human actions in the Chesapeake probably came from burning of the forests and grass-covered barrens. Evidence indicates that the forests were being regularly burned during the Late Woodland period. Accounts of the early European colonists say that little undergrowth existed on the forest floor in many areas. Maryland settler Andrew White S.J. reported that there was "great variety of woode, not choaked up with undershrubs, but commonly so farre distant from each other as a coach and fower horses may travale without molestation" (Hall, 1910: 40). Burning was used to clear areas for settlement, it allowed management of vegetation and some animal resources over large areas, it encouraged the growth of certain food plants, and it was a valuable strategy for hunting success. Native people regularly burned forests in many portions of North America, but we do not know how early the practice began in the Chesapeake. Sediment analysis by Grace Brush (see Chap. 3) has identified periods of high charcoal deposition believed to indicate large forest fires, but these could have resulted from natural causes. Given that burning was so widely used across North America, however, it must have been of considerable antiquity and was probably long used in the Chesapeake region. Archaeologist Sanderson Stevens (1991) suggests that burning came into use around 4,000 years ago in the region. Further analysis of sediment cores will yield more insights about this issue.

Burning not only encouraged the growth of seed plants such as *Chenopodium* and amaranth but also created a greater diversity of plant life, especially browse plants, upon which deer feed. The result was an expansion of the amount of browse, thereby providing the opportunity for growth in the deer population. Given the importance of deer in the native economy, the benefits of burning to the Chesapeake Indians is obvious. Use of this practice also reveals the error in the European assumption that the America they found was a pristine wilderness untouched by humans.

The native peoples altered important elements of the terrestrial environment, but what direct impact did they have upon the Chesapeake Bay and its estuarine ecology? Burning of the forest may have had some effect on the estuary. Forest fires produce nutrient-rich ash and charcoal and create some land surfaces more exposed to rainfall. However, evidence suggests that regular burning does not result in major deforestation but tends to remove the smaller understory plants, leaving the forest canopy largely intact. Nevertheless, burning probably resulted in additional materials washing into the Bay tributaries, as suggested by pollen core samples. If correct, the practice of periodic burning over a span of centuries would have increased the amount of nutrients entering the estuary. In a drainage area covered by forests and with very low rates of average sedimentation, the frequent input of even small quantities of additional nutrients would encourage greater biological activity within the estuarine system and make a rich habitat even more productive. Whether burning also resulted in some change in Chesapeake water quality is uncertain.

Human actions probably did affect estuarine resources in highly localized settings. Intensive harvesting of shellfish, for example, may have depleted specific places but prehistoric oyster shells do not normally display the significant size reduction that indicates overharvesting. Oystering sites such as White Oak Point were used repeatedly over a span of several thousand years, suggesting that any depletions were temporary and oysters remained plentiful. In this context, it is of interest to note that the Algonquin word *Chesapeake* can be translated as "great shellfish bay," offering linguistic support for the sheer abundance of these animals.

When cultivation of corn was adopted by the Chesapeake peoples in the Late Woodland, the potential for greater soil erosion and a decline in soil fertility increased. It nevertheless appears that native technology and agrarian methods did not result in any serious alteration of soil quality. Indian cultivation only required the removal of small areas of forest, with the fields reportedly ranging between 2 and 200 acres. Crops were raised until soil fertility declined, then the Indians cleared new fields and abandoned the old ones, allowing them to undergo reforestation. In so doing, the Chesapeake Indians relied upon natural processes to replenish fertility of the soil over several decades. As a result, no long-term degradation of soil productivity occurred and fertilizers were unnecessary. Because these fields were surrounded by forested tracts and second growth, the potential for severe soil erosion was also quite low. Such a method of shifting-field, hoe cultivation was an efficient, sustainable system of agriculture that had little if any negative environmental effects. Overall, evidence indicates that the Native American practices had only a slight impact on the ecology and resources of the Chesapeake Bay itself.

Conclusions

The story of the Chesapeake Bay is also in important ways the story of the Chesapeake Indians. They have been a part of the region's ecosystem since well before the estuary was created. Archaeology indicates that these people successfully coped with dramatic changes in climate and ecology over thousands of years. They achieved this through an adaptation based primarily on knowledge of the ecology and on behavioral flexibility, not technology. There were certainly notable technological advances such as pottery, weirs, and the bow and arrow, but overall the Indians solved problems created by major and minor climatic shifts over twelve millennia through mobility, environmental knowledge, and flexibility.

Study of the archaeological record left by the Chesapeake Indians provides a unique perspective on the evolution of the estuary and its resources. For over half the time it took the Chesapeake to reach its current size, it offered very limited aquatic foods to humans. Such insights are valuable for understanding how estuarine resources develop and respond to changing conditions. Only in the latter half of its existence did the Chesapeake start providing a rich bounty of dependable foods for people. Availability of these abundant foods had effects on human cultures and on the people themselves. In a review of Chesapeake Indian health, Douglas Ubelaker and Philip Curtin (see Chap. 7) show that the Indians experienced high infant mortality and had numerous health problems. Life expectancy for individuals who reached the age of 15 was only 35 years. Despite these relatively short life spans, data suggest that people living in the Tidewater area had better health than those in the Piedmont. It is likely that the range and richness of food resources offered by the Chesapeake Bay was a key factor in creating this difference in health.

The seventeenth-century Chesapeake Indians inherited a culture that was well adapted to the Bay ecology and capable of impressive resilience in the face of repeated changes in the natural environment. It was to be a massive shift in the cultural environment, however, to which these successful societies would be less capable of resisting. In less than two centuries after European settlers arrived, a way of life was forever lost and cultures which had evolved over thousands of years were largely destroyed; although their human descendants do remain in the Chesapeake region today. That destruction took with it an intricate and refined body of ecological and animal behavioral knowledge we are unable to fully reproduce. With this loss, the fragile traces remaining from those 12,000 years of human life take on added significance. They allow us to recover clues to the estuary's evolution and to learn from the remarkable experiences of a people who actually witnessed the birth of the Chesapeake and were the first to live along its majestic shores.

BIBLIOGRAPHY

Barse, M. F. 1988. *A Preliminary Archaeological Reconnaissance of the Middle Portion of the Patuxent River, Charles, Calvert, Prince Georges, and Anne Arundel Counties.* Maryland Geological Survey, Division of Archaeology, File Report #219 (Baltimore).

Binford, L. R. 1964. Archaeological and ethnohistorical investigations of cultural diversity and progressive development among aboriginal cultures of coastal Virginia and North Carolina. Ph.D. diss., University of Michigan (Ann Arbor: University Microfilms).

Custer, J. 1988. Coastal adaptations in the Middle Atlantic region. *Archaeology of Eastern North America* 16: 121–35.

Dent, R. 1995. *Chesapeake Prehistory: Old Traditions, New Directions* (New York: Plenum Press).

Fletcher, C. H., III. 1988. Holocene sea-level history and neotectonics of the United States Mid-Atlantic region: Applications and corrections. *Journal of Geology* 96: 323–37.

Gardner, W. M. 1982. Early and Middle Woodland in the Middle Atlantic: An overview. In *Practicing Environmental Archaeology: Methods and Interpretations,* ed. R. Moeller (Washington, Conn.: American Indian Archaeological Institute), 53–87.

———. 1987. Comparison of Ridge and Valley, Blue Ridge, Piedmont, and Coastal Plain Archaic period site distribution: An idealized transect (preliminary model). *Journal of Middle Atlantic Archaeology* 3: 49–80.

Hall, C. C., ed. 1910. *Narratives of Early Maryland* (New York: Charles Scribner's Sons).

Kraft, J. C. 1976. *Radiocarbon Dates in the Delaware Coastal Zone (Eastern Atlantic Coast of North America).* Delaware Sea Grant Technical Report No. 14 (Newark: University of Delaware).

McNett, C. W., ed. 1985. *Shawnee Minisink: A Stratified Paleoindian-Archaic Site in the Upper Delaware Valley of Pennsylvania* (New York: Academic Press).

Potter, S. R. 1993. *Commoners, Tribute, and Chiefs: The Development of Algonquian Culture in the Potomac Valley* (Charlottesville: University Press of Virginia).

Reinhart, T. R., and M. E. N. Hodges, eds. 1990. *Early and Middle Archaic Research in Virginia: A Synthesis.* Archaeologial Society of Virginia, Special Publication No. 22 (Richmond: Dietz Press).

———. 1991. *Late Archaic and Early Woodland Research in Virginia: A Synthesis.* Archaeological Society of Virginia, Special Publication No. 23 (Richmond: Dietz Press).

———. 1992. *Middle and Late Woodland Research in Virginia: A Synthesis.* Archaeological Society of Virginia, Special Publication No. 29 (Richmond: Dietz Press).

Rountree, H. C. 1989. *The Powhatan Indians of Virginia: Their Traditional Culture* (Norman: University of Oklahoma Press).

Rountree, H. C., and T. E. Davidson. 1997. *Eastern Shore Indians of Virginia and Maryland* (Charlottesville: University Press of Virginia).

Smith, B. 1992. *Rivers of Change: Essays on Early Agriculture in Eastern North America* (Washington, D.C.: Smithsonian Institution Press).

Stevens, J. S. 1991. A story of plants, fire, and people: The paleoecology and subsistence of the Late Archaic and Early Woodland in Virginia. In *Late Archaic and Early Woodland Research in Virginia,* ed. T. R. Reinhart. Archaeological Society of Virginia, Special Publication No. 23 (Richmond: Dietz Press), 185–220.

Stewart, R. M., and W. M. Gardner. 1978. Phase II Archaeological Investigations near Sam Rice Manor, Montgomery County, Md., and at 18PR166 and 18PR172 near Accokeek, Prince George's County, Md. Report submitted to the Washington Suburban Sanitary Commission.

Strachey, W. [1612] 1953. *The Historie of Travell into Virginia Britania,* ed. L. B. Wright and V. Freund. Hakluyt Society, 2nd ser., v. 103 (London).

Turner, E. R. 1976. An archaeological and ethnohistorical study of the evolution of rank societies in the Virginia Coastal Plain. Ph.D. diss., Pennsylvania State University.

———. 1992. The Virginia Coastal Plain during the Late Woodland Period. In *Middle and Late Woodland Research in Virginia: A Synthesis,* ed. T. R. Reinhart and M. N. Hodges. Archaeological Society of Virginia, Special Publication No. 29 (Richmond: Dietz Press), 97–136.

Waselkov, G. 1978. Evolution of deer hunting in the Eastern Woodland. *Mid-Continental Journal of Archaeology* 3: 15–34.

———. 1982. Shellfish gathering and shell midden archaeology. Ph.D. diss., University of North Carolina (Ann Arbor: University Microfilms).

Whyte, T. R. 1988. Fish and shellfish use in the Woodland Period on the Virginia coast. *Journal of Middle Atlantic Archaeology* 4: 102–20.

Wittkofski, J. M., and T. R. Reinhart, eds. 1989. *Paleoindian Research in Virginia: A Synthesis.* Archaeological Society of Virginia, Special Publication No. 19 (Richmond: Dietz Press).

CHAPTER SEVEN

Human Biology of Populations in the Chesapeake Watershed

DOUGLAS H. UBELAKER AND PHILIP D. CURTIN

Some Old World pathogens stayed behind when humans migrated across Beringia, while others such as tuberculosis and other infectious diseases were transported to the New World. Human skeletal evidence shows that the health of the Indian population varied, depending on availability of food sources such as calcium-rich oysters. Increasing population density encouraged the spread of disease, especially new pathogens that were brought by the Europeans and Africans and depended on a nucleus of non-immune population.

Relations between humans and their parasites in North America can be seen as a drama in three acts. The first act began with the arrival of humans and their pathogens during the Ice Age, up to about 10,000 years ago. The second act began in the late fifteenth century with the coming of Africans and Europeans and a new group of pathogens. The third act began in the second half of the nineteenth century, when humans began to conquer decisively the most serious infectious diseases that had troubled them.

The initial mode of arrival over the land bridge from Asia helped to determine the kinds of pathogens humans could bring with them. Many Old World parasites, for example, had evolved in ways that required not only a human host but a secondary host as well—often an insect or a snail. Many that required a secondary host were left behind.

Still another group of diseases failed to cross to the Americas because they had not yet evolved. Some pathogens had begun to parasitize human hosts before the agricultural revolution brought humans into close contact with domestic animals like dogs, cats, pigs, mice, and rats. But it was the new, postagricultural contact between humans and other animals that made it possible for diseases such as cowpox to mutate into smallpox.

All known immigrants to the Americas before the 1500s were preagricultural, including the Inuit, who began to arrive about 2000 B.C.E. All these early immigrants came as small bands of hunters and foragers. Many diseases can maintain themselves only in a human community of a certain minimum size. This is especially the case for those that pass directly from one human host to another and for those that either kill quickly or leave the victim with a substantial immunity to further attack. In small communities, these pathogens can easily immunize such a large part of the population that the pathogens die out for lack of new hosts, though they may be reintroduced when a subsequent generation of nonimmunes appears.

The community size required to maintain diseases of this type varies considerably from one pathogen to another. Measles is a classic example: recent estimates suggest that measles in isolated communities will die out if the population at risk is much less than about a half million. Other similar diseases, sometimes called crowd diseases, include mumps, smallpox, chickenpox, rubella, common cold, cholera, falciparum malaria, and perhaps influenza. The urban civilizations that developed in the Americas became dense enough to harbor these pathogens, but only when they were introduced from Europe and Africa after Columbus's voyages.

Ecologists have paid little attention to the role of human pathogens in the ecosystem, though medical specialists have taken an ecological approach to disease for some time. Pathogens behave in much the same way as other parasites and predators, and they follow a similar evolutionary process. The species that survive are those that have adapted and readapted to the changing ecological niches in which they find themselves. Microscopic parasites, like larger predators, have a symbiotic relationship to the host community. New and more virulent strains can appear and probably have appeared many times in the past, but increased virulence is viable only within limits. A superparasite that kills off the host community will die out for lack of prey. On the other hand, a pathogen that mutates toward less infectiveness may also disappear from its inability to infect new victims. Successful evolution for the parasite thus tends toward a balance—enough virulence to pass readily to new hosts, but not so much that it wipes out the host community.

The hosts also adapt through evolution, but large primates such as humans evolve quite slowly. A human generation takes about twenty years.

By contrast, some microorganisms pass through a complete generation in less than twenty minutes. Humans and other large animals respond more rapidly to pathogens through their immune system than through selective evolution. The immune system can detect most parasitic attacks and create antibodies against the attackers. If the victim survives, it acquires some degree of immunity against further attack.

Some immunities last a lifetime, while others are limited to a few months, like immunity to the common cold. Many immunities can be passed by a mother to her offspring, but most of these immunities last only briefly. Although a few genetic immunities are lifelong, like the immunity of some West Africans to vivax malaria, most are more limited, enough to make the infection less serious rather than preventing it altogether.

Some viral infections, like poliomyelitis and yellow fever, are also far less serious in childhood than they are later in life, for reasons that are not well understood. Other diseases, often identified as childhood diseases, are more common in children because after most children in a large community have been affected, they grow to adulthood with an acquired immunity. Some immunities are complete and lifelong, like the protection that follows survival of an attack of yellow fever or smallpox.

Over the millennia before the land bridge disappeared, human diseases and their countervailing immunities took one direction in the intercommunicating regions of Afro-Eurasia and another in the Americas. Agriculture, which made dense human populations possible, was discovered in both areas independently, but most of our present crowd diseases appeared only in Afro-Eurasia. Some new diseases, such as Rocky Mountain spotted fever (and, more recently, Lyme disease) did appear in the Americas, but the ancient Americans escaped many of the diseases still evolving in the Old World. Their level of immunity to these diseases was therefore negligible, and the pathogens' attack took an especially devastating form known as virgin-soil epidemics.

The Americas before 1492 were not a disease-free paradise. Pre-Columbian human skeletons indicate the presence of tuberculosis, intestinal parasites, and some form of treponemal disease, among others. Across the continent, infant mortality was high and life expectancy at birth relatively low.

It is common to think of the Columbian encounter as involving only Europe and the Americas, but tropical Africa also played a part as European mariners on the Atlantic began to reach south along the Saharan coast of Africa, establishing a routine pattern of voyages by the 1440s.

Regular communication among tropical Africa, Europe, and the Americas brought together three disease environments that had been at least partly isolated. Most European diseases were known south of the Sahara, though Africans had weaker immunities to tuberculosis and pneumonia than they had to most other European diseases. Some important African diseases, however, were confined to the tropical zone, mainly because their insect vectors could not survive cold weather. The European disease environment thus lacked the yellow fever and falciparum malaria that were common in Africa. This meant that newly arrived Europeans in tropical Africa died in great numbers, with death rates up to 500 per thousand in the first year of residence.

At least one of Columbus's crewmen on the first voyage was of African descent, and many other Africans followed—some, but not all, as slaves. This meant that the Americas received new pathogens from both Old World disease environments. From Europe came the crowd diseases the Americas had so far avoided. The list included influenza, plague, and the whole range of children's diseases, such as measles, whooping cough, scarlet fever, and mumps. Smallpox first reached the Caribbean from Africa, but later it came from Europe as well.

Malaria came from both Europe and Africa, but as a different species from each. *Plasmodium vivax* was common in the Mediterranean and existed as far north as Great Britain. The usual European vectors were species of anopheles mosquito confined to Europe, but the parasite found that it could shift to other anopheline species in the Americas. Vivax malaria had been a serious health problem in the Mediterranean basin, and it became serious in the Americas, but it was a debilitating illness that caused "ague and fever," rarely an early death.

Africa's contribution was *P. falciparum,* which is fatal far more often. The consequences for the Americas were disastrous. The Indian population of the tropical lowlands was virtually wiped out by the combination of smallpox, falciparum malaria, and yellow fever. Some individuals survived and contributed to the gene pool of later populations, but separate Indian communities in the Caribbean virtually disappeared by the end of the 1500s.

African and European immigrants to the Americas also suffered from contact with unfamiliar diseases, but in time all the populations of the Americas adjusted to the new disease environment. By the mid-nineteenth century, African-derived populations in the Americas had achieved net natural increase everywhere. The American Indian populations that had encountered European diseases in earlier centuries not only stopped their decline but began to grow again. Europeans in the humid tropics were slow to attain a net natural increase, but many did so by the nineteenth century. In North America they had one of the highest rates of net natural increase in the nineteenth century.

The third act, sometimes called the mortality revolution, began about the middle of the nineteenth century and continued into the twentieth. By the 1820s, life expectancy at birth had reached about 40 years in the United States and the major countries of western Europe. By 1914, it was more than 50. And by 1990, it was more than 75. We will not deal with these changes in detail or in the Chesapeake context, though they were a major cause of increasing human populations in the watershed which in turn brought increasing pressure to bear on other species within the ecosystem. Some aspects of that pressure are discussed in almost every chapter of this volume.

The Chesapeake Watershed before Columbus

When Europeans arrived on the Chesapeake Bay in the early seventeenth century, they found the land already occupied by peoples with several languages and cultures. Algonquian-speaking tribes like the Nanticoke were present in Maryland and Virginia; Delawares were located to the northeast; and Iroquoian speakers were north and northwest of the Bay.

Although scientists agree that the American Indians lived in and adapted to all the diverse environments of the Americas, they disagree about population size. Estimates of the Indian population for the Americas as a whole before Columbus range from about 8.5 million to 100 million, and for America north of Mexico from 900,000 to 18 million. The widely divergent estimates reflect both the fragmentary evidence and different interpretations of the biological condition of the American Indian population. Those who estimate a relatively low population size rely heavily on early ethnographic records that suggest considerable pre-1492 morbidity and mortality. Advocates for large population size argue that before 1492 the American Indian population had few disease problems, which, coupled with high fertility, allowed rapid population growth. Some of the higher estimates assume that epidemics or pandemics had already reduced the numbers between the time of Columbus's first voyage and the first reliable ethnohistorical accounts.

Although many of the diseases that decimated the American Indian population in the sixteenth and seventeenth centuries and later were new and clearly of European origin, skeletal analysis indicates that even before 1492 morbidity was increasing over time throughout the Americas, apparently because of the increasing population density that accompanied the shift from seminomadic hunting and foraging to more sedentary agricultural occupations.

Various measures of morbidity are used in skeletal analysis. Cavities in the teeth reflect dental hygiene, diet, and constitutional factors. Sticky, starchy food residue on the teeth allow bacteria to colonize and to release chemicals that break down the surface of the tooth. When the tooth surface is sufficiently altered, it collapses, creating a cavity. Throughout the Americas, the pre-Columbian frequency of dental caries increases through time in association with the shift to agriculture.

Other indicators of morbidity include dental hypoplasia (a particular kind of underdevelopment of the teeth), the rate of growth in stature, evidence of infectious disease, evidence of anemia, lines of increased density in long bones, and trauma. These problems are easily detected in skeletal remains and are commonly registered in skeletal analysis. Hypoplasia in teeth and lines of increased density in long bones are formed when normal growth and dental formation are temporarily slowed from physiological stress, such as periods of malnutrition and high fever.

These skeletal indicators of stress vary somewhat from one region to another, but most of them, throughout the Americas, increase from the agricultural revolution onward. Diet is clearly the causal factor in dental caries and probably is significant for the other problems as well. Increasing population density, sanitary problems, and the shift from hunting to sedentary farming contributed to the pattern of increasing morbidity, but life expectancy was low in agricultural and nonagricultural populations alike, with life expectancy at birth in the upper teens and twenties, reflecting high infant mortality.

Health of American Indians

The only source of information on human health in the Chesapeake Bay region before European contact is archeologically recovered human remains. Such remains at the Tollifero and Clarksville sites in southern Virginia, just south of the Chesapeake watershed, are valuable because they come from the same area at different time periods. They show not only health conditions before the European arrival but also how those conditions probably changed over time. The Tollifero sample includes 37 skeletons and dates from the Early Woodland. The Clarksville site dates from the Middle to Late Woodland and is represented by 80 human skeletons.

Significant infant mortality was present at both sites, with 16–17% of all individuals in the samples under the age of three. Higher juvenile mortality at the later, Clarksville site, meant that life expectancy was greater at the earlier, Tollifero site. The remains from the two sites suggest that over time the populations had less dental attrition (wear on the teeth), probably reflecting changes in diet and methods of food preparation. Frequencies of dental cavities (caries), infection (periosteal lesions), and trauma all increased through time. The increase in dental cavities likely represents more dietary reliance upon agricultural products during the Late and Middle Woodland periods. The increased frequency of periosteal lesions probably reflects increases in population density stimulated by the shift toward agriculture and related changes in settlement pattern. Greater evidence of trauma in the later sample likely represents an increase over time in intragroup and intergroup conflict, still another probable result of increasing population and political complexity.

In the Chesapeake watershed, variation among different locations shows the value of the Bay's resources for the human condition. Joan W. Chase (1988) studied pathological differences between human remains from the Piedmont and those from the Coastal Plain. Her Piedmont sample included nine inland sites from Virginia and Maryland, and she selected one site from the Coastal Plain. Chase looked specifically at those aspects of the skeleton which mark physiological stress of various types. She found that the Coastal Plain sample from southern Maryland showed higher frequencies of cribra orbitalia and porotic hyperostosis (abnormal porosity in the upper orbits and in the cranial vault of individuals). These two conditions are thought to represent a skeletal response to anemia.

Most other measures of morbidity were more apparent in the Piedmont samples. These samples displayed higher frequencies of lines of arrested growth in the long bones, of cribra orbitalia and porotic hyperostosis in adults, of alveolar resorption, dental caries, and enamel hypoplasia. As discussed earlier, the lines of arrested growth and enamel hypoplasia mark disruption in the normal growth pattern of bones and teeth, respectively, probably from nutritional deficiency or other forms of physiological stress. The alveolar resorption represents loss of bone around the teeth, resulting from periodontal gum disease, perhaps coupled with caries or other forms of dental disease.

In general, the Piedmont-Coastal comparison suggests that, although the coastal populations depended more on agriculture and lived in greater population densities than the Piedmont folk, they had a lower disease experience. Chase suggests that the diet, which included products of the Bay and its tributaries, offset the disadvantages of an agricultural settlement pattern and population density observed elsewhere. The calcium-rich oyster may have offered Bay-area Indians a nutritional advantage over those of the Piedmont, especially by minimizing the impact of growth disruptions.

The trend toward increasing morbidity through time continues into the decades just before and after Columbus's voyages. Data are available from two samples, the Shannon site remains from Montgomery County, Virginia, and the Juhle site in Charles County, Maryland. The Shannon site dates from between A.D. 1550 and 1650. Mecklenburg's study of 106 skeletons from the Shannon site showed that 20–25% died before the age of 5. The individuals in the sample had minimal attrition but high infant mortality and abundant dental disease.

More detailed information about these periods is available from an analysis of two of the ossuaries from the tidewater Potomac region in Charles County, Maryland. The ossuaries in this area date from the fifteenth or sixteenth centuries and resulted from a mortuary procedure in which the dead apparently were initially placed in death houses above ground. Only after several years were the remains gathered and buried in a large pit. Two of these ossuaries were excavated and found to contain the remains of at least 131 and 161 individuals, respectively.

Ages at death were estimated from radiographic study of dental formation (in immature individuals) and microscopic assessment of femoral cortical remodeling (in adults). The resulting ages were then grouped to establish demographic profiles. Reconstructed life tables for the two populations represented by the ossuaries suggest life expectancies at birth of 21 and 23 years (Table 7.1). Thirty percent of the population apparently died before the age of 5. Life expectancy at age 15 was about 20 years for the two ossuaries, suggesting that an individual in that population at age 15 could expect to live on average to be 35 years old.

The crude mortality rate (number of persons dying per year per thousand in the population) was 44 and 48 for the two ossuaries. These figures suggest high mortality because, in comparison, the rate was only about 30 for fifteenth-century ruling families in Europe and for nineteenth-century U.S. whites.

Table 7.1 Life table reconstructed from Ossuary II age distribution, Juhle site, Maryland

Age interval (x)	No. of deaths (Dx)	% of deaths (dx)	Survivors (lx)	Probability of death (qx)	Total no. years lived between x and x + 5 (Lx)	Total no. years lived after lifetime (Tx)	Life expectancy ($e^o x$)
0	56	29.787	100.000	0.2979	425.533	2,297.930	22.98
5	12	6.383	70.213	0.0909	335.108	1,872.397	26.67
10	7	3.723	63.830	0.5830	309.843	1,537.289	24.08
15	14	7.447	60.107	0.1239	281.918	1,227.446	20.42
20	9	4.787	52.660	0.0909	251.333	945.528	17.96
25	12	6.383	47.873	0.1333	223.408	694.195	14.50
30	21	11.170	41.490	0.2692	179.525	470.787	11.35
35	20	10.638	30.320	0.3509	125.005	291.262	9.61
40	13	6.915	19.682	0.3513	81.123	166.257	8.45
45	11	5.851	12.767	0.4583	49.208	85.134	6.67
50	8	4.255	6.916	0.6152	23.943	35.926	5.19
55	4	2.128	2.661	0.7997	7.985	11.983	4.50
60	0	0.000	0.533	0.0000	2.665	3.998	7.50
65	1	0.592	0.533	1.0000	1.333	1.333	2.50

Source: Modified from Ubelaker (1974: 63).

Ethnohistoric Evidence

Seventeenth- and eighteenth-century accounts offer somewhat conflicting opinions about the health of Indians living in the Chesapeake region. Several early observers described the Indians in Virginia and North Carolina as tall and healthy. For example, in about 1708, John Lawson wrote of North Carolina Indians: "As there are found very few, or scarce any, Deformed, or Cripples, amongst them, so neither did I ever see but one blind Man; and then they would give me no Account of how his Blindness came."

In contrast, Gabriel Archer comments in 1607 after visiting Indians along the James River in Virginia: "The great diseaze reignes in the men generally, full fraught with noodes botches and pulpable appearances in their forheades, we found aboue a hundred." Writing six years later, also after having visited James River tribes, William Strachey adds: "And vncredible yt is, with what heat both Sexes of them are given over to those Intemperances, and the men to preposterous Venus, for which they are full of their owne country-disease (the Pox) very young."

Population Size

Research for the *Handbook of North American Indians,* organized by the National Museum of Natural History of the Smithsonian Institution, included assessment of population size for all North American tribes. In 1988, this information and other data were used to produce estimates of population size for each of the North American groups. Combining individual tribal estimates gave regional totals of 357,700 (based on a range from 205,000 to 503,200) for the northeast United States and 204,400 (based on a range from 155,800 to 286,000) for the southeast. This implies a population density of about 19 persons per 100 km^2 in the northeast (likely range 11–26) and about 22 per 100 km^2 (range 17–31) in the southeast. A total of estimates for North America north of the urban areas of central Mexico was 1,894,350, with a likely range of 1,213,475 to 2,638,900. North American population density was about 11 per 100 km^2 with a possible range from 7 to 15.

This same research generated new estimates for groups living in the Chesapeake region at the time of first European contact. Tribal groupings to be discussed are those employed in the handbook project. The groups located in the immediate area of the Bay as well as in the drainage areas to the northeast and west are the Delaware, the Nanticoke and neighboring tribes, the Virginia Algonquians, and the Susquehannock, plus Iroquoians living on the upper tributaries of the Susquehanna in what is now New York state. Of these, the Virginia Algonquians were likely the most numerous at 16,000 (range 14,000–21,000), with a population density of about 81 persons per 100 km^2 (range 71–106). The total number of these groups living south of the present Pennsylvania–New York line was likely about 45,000, at a density of about 42 per 100 km^2 (range 31,000–59,000) (Table 7.2). The Bay watershed was, in short, more densely settled than either the northeastern or the southeastern United States as a whole.

Table 7.2 Estimates of population size of American Indian groups in the Chesapeake Bay area at the time of initial European contact

	Minimum		Best estimate		Maximum	
Tribal group	No.	No. per km^2	No.	No. per km^2	No.	No. per km^2
Delaware	8,000	14	11,000	19	13,000	22
Nanticoke and neighboring tribes	3,000	21	10,000	71	13,000	93
Virginia Algonquians	14,000	71	16,000	81	21,000	106
Susquehannock	6,000	41	8,000	54	12,000	81
Total	31,000	29	45,000	42	59,000	55

A New People and New Diseases

The Indians

As in other regions of the Americas, the Europeans' arrival brought new diseases to the Chesapeake watershed, but this took place nearly a century after similar events in the Caribbean and Middle America. The earliest European contacts with the Chesapeake were scattered, brief visits by passing ships. These occasional visits brought one serious but unidentifiable epidemic to the Bay region as early as 1561. Many of the initial epidemics, however, burned themselves out, as infectious disease often did in small communities, by infecting and immunizing a large part of the population.

The most serious spread of alien diseases came when Europeans settled as permanent residents. Smallpox, the worst killer of all, was relatively slow to be established in this part of North America. It had swept as an epidemic through the Caribbean, Mexico, and Andean South America beginning in 1519, with reported mortality at 35–50% of the population. It did not, however, move north from Florida or the Gulf coast at that same time, probably because the population of the pine barrens of Georgia and northern Florida was relatively sparse. The disease thus burned itself out in small communities rather than spreading in a massive epidemic, as it did among the dense populations in such places as the valley of Mexico.

Data about disease in North America in the 1600s and 1700s are hard to interpret. The Europeans who kept the records knew more about their own diseases than those of the Indians, but their reports, even of their own diseases, are clinically hazy. An outbreak of "burning ague" occurred in Rhode Island in 1723. It might have been malaria, but the case-fatality rate was higher than that of any recognized disease of the time. The most likely hypothesis is that it was a mutation of some other pathogen, one that killed so many of its victims that it ran out of hosts and eventually disappeared.

Smallpox symptoms are hard to mistake, and the earliest occurrence reported in the Chesapeake watershed occurred near its northern headwaters. Smallpox reached Albany in 1634–35, a sidelight of its deadly sweep across New York state and southern Ontario in the 1630s. It certainly struck the Iroquoian population of the upper tributaries of the Susquehanna, but with no clear record whether it moved south down the valley. The first clear cases of smallpox in the Chesapeake tidewater came only in 1667.

This relatively late arrival of smallpox came about largely because of the disease's epidemiology in Europe. Smallpox is one of the few diseases that leaves its victims with an absolute, lifelong immunity. (This is one reason it was possible to destroy it altogether in the long run.) Where a community is large and population dense, smallpox is almost unavoidable. It becomes a childhood disease, with a presumed case-fatality rate lower for children than for adults because of some immunity carried over from the mother. The point is uncertain, however, because we have so little evidence about the epidemiology of smallpox among children before Jennerian vaccination was introduced two hundred years ago. If subclinical infection was common among children, as it is, for example, with yellow fever, the possibility of childhood-acquired immunity exists.

In other, smaller and sparser communities, smallpox occurs as an occasional epidemic that sweeps through the population with a case-fatality rate of 20–40%, leaving behind immune survivors. The disease then disappears for lack of further nonhuman hosts to act as a reservoir. It could still return, when population growth has produced enough nonimmunes to make a new epidemic possible. This was the smallpox pattern in tropical Africa in recent centuries and no doubt in the sixteenth century as well.

This pattern meant that only European children were likely to transmit smallpox, the adults being both immune and noninfectious, and few European children came to the Chesapeake in the early years of settle-

Table 7.3 Variability in population reduction in the Chesapeake Bay area

Tribal group	No. at contact	No. at nadir	Approximate date of nadir	Reduction in nos.	% reduction
Delaware	11,000	1,500	1900	9,500	86
Nanticoke and neighboring tribes	10,000	100	1900	9,900	99
Virginia Algonquians	16,000	500	1900	15,500	97
Susquehannock	8,000	0	1800	8,000	100
Total	45,000	2,100	1900	42,900	95

ment. Smallpox's early appearance in the northern Chesapeake watershed reflected the larger number of children among the settlers on the New England coast, and perhaps also in Dutch New York.

In North America, the initial pattern of smallpox among the Indians was a series of epidemics, each of which may have taken as much as 35% of the non-immune population, followed by a new epidemic once the non-immune population grew large enough to sustain one. Dean R. Snow worked out a computer simulation of this pattern for the northeastern United States, and it is clear for the southern part of the Chesapeake watershed too. These patterns help to explain why the Indian population in the southern United States declined more steeply in each succeeding century. Recent population estimates show an Indian population decline in the 1500s of 23%; in the 1600s, a decline of 33%; and in the 1700s, of 43%.

The population decline in the Chesapeake Bay area was more serious than it was elsewhere in eastern North America. The 45,000 inhabitants probably present at the time of initial European contact were reduced to only about 2,000 by 1900, a reduction of 43,000, or 95% (Table 7.3). Virginia Algonquians were reduced by 97% and the Nanticoke and neighboring tribes declined by 99%, while the Susquehannock lost their identity completely. Such reductions, or apparent extinction in the case of the Susquehannock, were not necessarily caused by mortality alone. Many American Indians surely died through disease, but many joined other groups, were relocated, or otherwise became invisible to European census takers.

The Chesapeake Bay region was not typical of demographic patterns elsewhere in the southeast. South of Virginia, populations from Africa and Europe grew more slowly and Indian populations declined more slowly. In Virginia east of the mountains, and presumably for the rest of the southern Chesapeake watershed, African and European populations grew more rapidly than they did elsewhere. In 1685, Virginia's population included four-fifths of all Europeans and four-fifths of all Africans in the whole south. By 1775, Indians were still a majority south of Virginia, but there were very few in the Virginia portion of the Chesapeake watershed.

In the late eighteenth and early nineteenth century, the trends began to reverse as the Indian population first leveled off and then started to show small natural increases in some places. The mechanism for this change is not well understood, and the data are neither as detailed nor as clear as one might hope. The explanation for smallpox, however, seems reasonably clear. When smallpox shifted from the pattern of recurrent epidemics to become a childhood disease, it became less serious for the community. Part of the gain came about because infant mortality is less damaging to society as a whole than adult mortality is, even when case-fatality rates are identical. It is also possible that the childhood case-fatality rate for smallpox was lower than it was for adults.

With the passage of time, the other European childhood diseases became childhood diseases for the Indians as well. As a result, the Indian population that reached its nadir in the Chesapeake region sometime in the late eighteenth or early nineteenth century began to increase by natural growth. By the 1990s, for the United States as a whole, the Indian population may have returned to or exceeded its size at the time of contact.

No one knows exactly what happened to the descendants of the original Indian population of the Chesapeake watershed. That population had decreased sharply in the seventeenth and eighteenth centuries by a combination of high mortality and emigration. In the twentieth century it grew again by a combination of natural growth and the immigration of Indians from other parts of the country, such as the Lumbie community in Baltimore, which originated along the Lumber River in North Carolina.

The Europeans

Although the Europeans introduced the diseases that decimated the Indians, they also entered a new disease environment themselves for which they were ill prepared. The first settlement at Jamestown suffered enormous losses in its early years. Of the 108 settlers who landed in 1607, only 38 were still alive at the end of the first year—a death rate of 648 per thousand. Over the winter of 1608–9, the loss was 880 per thousand in only six months. A little later, over the years 1619–22, 70% of those who landed died during the three-year period. Plymouth Colony also lost about half its population over the first winter, though the losses at Roanoke in 1585–86 were negligible.

Historians have tried several explanations of the high initial losses at Jamestown. Some suggested yellow fever, but yellow fever was not found in the Americas until after the 1640s. Others considered malaria, but death rates from vivax malaria would have been much lower and falciparum malaria had not yet arrived from Africa. Deficiency diseases like beriberi, scurvy, and pellagra may well have contributed, but the fatality rates for their victims were usually lower than those found for the Jamestown settlers. Carville Earle worked out an ingenious explanation that stressed Jamestown's location on the James River, examining salinity at different seasons, and he concluded that the most likely cause was a combination of salt poisoning, typhoid, and dysentery. Given the lack of a good clinical description of the causes of death, his explanation seems as likely as any but still uncertain. The crisis period for the settlers passed, in any event, after they moved away from Jamestown.

For the later 1600s, we have several careful estimates of European demography in the southern part of the Chesapeake watershed. One study, based on parish death and baptismal records in Charles County, Maryland, from 1652 to 1699, shows a life expectancy at birth of about 26 years—not far different from the 21- and 23-year life expectancies yielded by the life tables reconstructed for Charles County Indians of the fifteenth and sixteenth centuries.

Wyndham Blanton worked out another reconstruction of Chesapeake demography, based on the settler population of seventeenth-century Virginia. He estimated birth and death rates from the age at death, arriving at a probable crude birth rate of 40 per thousand and a death rate of about 35 per thousand—both in the same high range as the Maryland estimates. Those rates should have led to a net natural decrease of about 0.5 per thousand. At that figure, however, immigration should have yielded a Virginia population of European descent of about 130,000 by 1707. In fact, the population of Virginia reached only about 70,000. The probable cause of the shortfall was the skewed sex ratio—before 1707, only about 20% of the immigrants from Europe were women.

Even without the skewed sex ratio, the Virginia and Maryland health picture was worse than that of contemporaneous Europe or New England—worse than many parts of the less-developed world today—and the principal cause was malaria. Vivax malaria arrived with the earliest European settlers, but the death rates from malaria were low until the 1680s, when the increasing intensity of slave imports from Africa brought in falciparum malaria. Vivax and falciparum both found a vector in the form of *Anopheles quadrimaculatus,* a mosquito that flourished throughout most of North America. In the 1600s, malaria, mainly vivax, was present as far north as New England, but it was less serious there and the malaria line shifted south about the middle of the 1700s, leaving New England malaria-free.

Far into the nineteenth century, the Chesapeake Bay lay within the northern fringe of an east-coast malaria belt, and malaria was a significant cause of death. During the Civil War, a quarter of all illness in the Union army was from malaria, and the average death rate from malaria among federal white troops was 17.38 per thousand for the whole of the war. Wartime conditions also produced a resurgence of malaria in the Bay area. The 1870 census indicated a malaria belt on either shore of the Bay, where malaria caused between 6% and 9% of all deaths. By contrast, in most of the Susquehanna valley and the upper watershed, malaria caused less than 1% of all deaths. In absolute terms, malaria then accounted for 55–90 deaths per thousand in the Bay area itself, and less than 10 per thousand in most of the watershed north of the Mason-Dixon line.

The Africans

Most of the older works on American history mention 1619 as the date when immigration from Africa began. One cargo of slaves from Africa did indeed land on the shores of the Chesapeake in that year, but substantial immigration from Africa began only in the last third of the seventeenth century. In 1700, the African-derived population in Virginia east of the mountains was less than 9% of the total; by 1790, it had become more than 40% (Table 7.4).

The immigrants from Africa brought at least one medical practice that was to be important for everyone living in the Chesapeake region. That was variolation, a technique for immunizing against smallpox by inserting material from a smallpox pustule under the skin of the person seeking protection. The result was often a light case of the disease followed by lifelong immunity. A serious case was also possible, but the case-fatality rate was much lower than it was when the disease was transmitted in the usual way, by droplets from an infected person's lungs entering the lungs of a new victim. It is not known where variolation was

Table 7.4 The population of Virginia east of the mountains, by ancestry, 1685–1790

	Indian	European	African	Total
1685	2,900	38,100	2,600	43,600
1700	1,900	56,100	5,500	63,500
1715	1,300	74,100	20,900	96,300
1730	900	103,300	49,700	153,900
1745	600	148,300	85,300	234,200
1760	400	196,300	130,900	327,600
1775	300	279,500	186,400	466,200
1790	200	442,100	305,500	747,800

Source: Wood (1989: 38).

first discovered, but it was practiced in Africa and Asia long before it turned up in Europe.

In 1706, Cotton Mather learned about variolation from a slave. He later reported the practice to the Royal Society in Britain, and knowledge of the technique also reached Britain from the Ottoman Empire in 1721. From the 1720s onward, variolation was widely practiced in America and by slave traders shipping slaves from Africa. Inoculated slaves commanded a higher price on the American market. It is impossible to know how widely the practice spread beyond the African American community, especially among the Indians, but the close relations between the two communities in much of the American South makes this spread extremely likely—one more step in Native Americans' adjustment to the new diseases from overseas.

The African disease environment was different from that of Europe, and the Africans had a different set of immunities. Africans had known smallpox, but not smallpox as a childhood disease. They had known yellow fever and even had a degree of genetic immunity to it; people born and raised in Africa often had an acquired total immunity as well. Some Africans and African Americans inherited the sickle-cell trait, which could cause sickle-cell anemia later in life but also protected the individual to some degree against falciparum malaria. Most West Africans had a genetic immunity to vivax malaria, but they had a weaker immunity to tuberculosis and pneumonia than Europeans had.

Once Africans reached the New World, their immunities were especially beneficial in tropical regions like the Caribbean, where the falciparum malaria and yellow fever they brought with them soon became the leading cause of death. Quantitative data are not available to show how far this immunological superiority of Africans over Europeans might extend into North America. It certainly reached as far as the Carolina low country in the seventeenth and eighteenth centuries. It probably also extended into the southern part of the Chesapeake watershed, where falciparum malaria was endemic and yellow fever was an occasional summer visitor.

With time, the African American immunity pattern changed. Sickle-cell trait declined, because the intense falciparum malaria found in the African continent was not there to reinforce it, and African Americans gained a new level of immunity against diseases of the lungs. By the early 1700s, African American populations in the Chesapeake began to show a net natural increase, more than a century earlier than the similar shift in the Caribbean.

DURING THE PAST CENTURIES the Chesapeake region was one setting for a worldwide process of integrating human communities from at least three different disease environments and with three different sets of pathogens. Integrating diseases and immune systems carried a painful price in mortality for all concerned, but in the end the populations adjusted to the new disease environment.

In the past century and half, the rise of scientific medicine was an additional cause of the mortality revolution, which contributed in turn to the enormous increase in human population. Human success in meeting the challenge of disease had a major impact on all other biological communities. Adjusting to the burgeoning human presence was by far the most serious challenge for other biological communities that share the ecosystem—in the Chesapeake watershed and worldwide.

BIBLIOGRAPHY

Archer, G. 1969. Relatyon of the description of our river. In *The Jamestown Voyages under the First Charter,* ed. P. L. Barbour. Hakluyt Society, 2nd ser., v. 136 (London), 80–98.

Axtel, J. 1988. *After Columbus: Essays in the Ethnohistory of Colonial North America* (New York: Oxford University Press).

Blanton, W. B. 1961. Epidemics, real and imaginary, and other factors influencing seventeenth century Virginia's population. *Bulletin of the History of Medicine* 31: 454–62.

Boyd, M. F. 1941. An historical sketch of the prevalence of malaria in North America. *American Journal of Tropical Medicine* 21: 223–44.

Burnet, M., and D. O. White. 1972. *Natural History of Infectious Disease,* 4th ed. (Cambridge: Cambridge University Press).

Chase, J. W. 1988. A comparison of signs of nutritional stress in prehistoric populations of the Potomac Piedmont and Coastal Plain. Ph.D. diss., The American University, Washington, D.C.

Cockburn, A. 1963. *The Evolution and Eradication of Infectious Diseases* (Baltimore: Johns Hopkins Press).

———. 1971. Infectious disease in ancient populations. *Current Anthropology* 12: 45–62.

Cohen, M. N., and G. J. Armelagos. 1984. *Paleopathology at the Origins of Agriculture* (Orlando: Academic Press).

Curtin, P. D. 1961. The white man's grave: Image and reality, 1780–1850. *Journal of British Studies* 1: 94–110.

———. 1968. Epidemiology and the slave trade. *Political Science Quarterly* 83: 191–216.

———. 1978. African health at home and abroad. *Social Science History* 10: 369–98.

———. 1990. The end of 'white man's grave'? Nineteenth-century mortality in West Africa. *Journal of Interdisciplinary History* 21: 63–88.

———. 1993. Disease exchange across the tropical Atlantic. *History and Philosophy of the Life Sciences* 15: 169–96.

Demeny, P. 1990. Population. In *The Earth as Transformed by Human Action: Global and Regional Changes in the Biosphere over the Past 300 Years,* ed. R. L. Turner II et al. (New York: Cambridge University Press).

Dobson, M. J. 1989. Mortality gradients and disease exchanges: Comparisons from Old England and Colonial America. *Social History of Medicine* 2: 259–97.

Dobyns, H. F. 1966. Estimating aboriginal American population: An appraisal of techniques with a new hemispheric estimate. *Current Anthropology* 7: 395–416.

———. 1983. *Their Number Become Thinned: Native Population Dynamics in Eastern North America* (Knoxville: University of Tennessee Press).

Duffy, J. 1953. *Epidemics of Colonial America* (Baton Rouge: Louisiana State University Press).

Earle, C. V. 1979. Environment, disease, and mortality in Early Virginia. In *The Chesapeake in the Seventeenth Century,* ed. T. W. Tate and D. L. Ammerman (Chapel Hill: University of North Carolina Press).

Feest, C. F. 1973. Seventeenth-century Virginia Algonquian population estimates. *Quarterly Bulletin of the Archaeological Society of Virginia* 28: 66–79.

———. 1978. Virginia Algonquians. In *Handbook of North American Indians,* v. 15, ed. B. G. Trigger (Washington, D.C.: Smithsonian Institution Press), 253–70.

Fenner, F., D. A. Henderson, I. Arita, Z. Zezek, and I. D. Ladnyi. 1988. *Smallpox and Its Eradication* (Geneva: World Health Organization).

Goodman, A. H., and J. C. Rose. 1991. Dental enamel hypoplasias as indicators of nutritional stress. In *Advances in Dental Anthropology,* ed. M. A. Kelley and C. S. Larsen (New York: Wiley-Liss), 279–93.

Henige, D. 1998. *Numbers from Nowhere: The American Indian Contact Population Debate* (Norman: University of Oklahoma Press).

Hopkins, D. 1983. *Princes and Peasants: Smallpox in History* (Chicago: University of Chicago Press).

Hoyme, L. E., and W. M. Bass. 1962. Human skeletal remains from the Tollifero (Ha6) and Clarksville (Mc14) sites, John H. Kerr Reservoir Basin, Virginia. In *Archeology of the John H. Kerr Reservoir Basin, Roanoke River Virginia–North Carolina,* ed. C. F. Miller. Smithsonian Institution, Bureau of American Ethnology Bulletin 182 (Washington, D.C.), appendix, 329–400.

Innis, F. C. 1993. Disease ecologies of North America. In *The Cambridge World History of Human Disease,* ed. K. F. Kiple (New York: Cambridge University Press).

Jaffe, A. J. 1992. *The First Immigrants from Asia: A Population History of the North American Indians* (New York: Plenum Press).

Kiple, K. F., and V. H. King. 1981. *Another Dimension of the Black Diaspora* (New York: Cambridge University Press).

Kroeber, A. L. 1939. *Cultural and Natural Areas of Native North America.* University of California Publications in American Archaeology and Ethnology 38 (Berkeley).

Kunitz, S. J. 1993. Disease and mortality in the Americas since 1700. In *The Cambridge World History of Human Disease,* ed. K. F. Kiple (New York: Cambridge University Press), 328–34.

Kupperman, K. O. 1979. Apathy and death in Early Jamestown. *Journal of American History* 66: 24–40.

Larsen, C. S., R. Shavit, and M. C. Griffin. 1991. Dental caries evidence for dietary change: An archaeological context. In *Advances in Dental Anthropology,* ed. M. A. Kelley and C. S. Larsen (New York: Wiley-Liss), 179–202.

Lawson, J. 1937. *Lawson's History of North Carolina* (Richmond: Garrett and Massie).

Mecklenburg, C. W. 1969. Human skeletal remains from the Shannon Site, Montgomery County, Virginia. M.A. thesis, University of Washington.

Pielou, E. C. 1991. *After the Ice Age: The Return of Life to Glaciated North America* (Chicago: University of Chicago Press).

Rouse, I. 1986. *Migrations in Prehistory: Inferring Population Movements from Cultural Remains* (New Haven: Yale University Press).

Rutman, D. B., and A. H. Rutman. 1976. Of agues and fevers: Malaria in the early Chesapeake. *William and Mary Quarterly* 33: 31–60.

Smith, D. B. 1978. Mortality and the family in the colonial Chesapeake. *Journal of Interdisciplinary History* 8: 403–27.

Snow, D. R. 1992. Disease and population decline in the Northeast. In *Disease and Demography in the Americas,* ed. J. W. Verano and D. H. Ubelaker (Washington, D.C.: Smithsonian Institution Press), 177–86.

Snow, D. R., and K. M. Lanphear. 1988. European contact and Indian depopulation in the Northeast: The timing of the first epidemics. *Ethnohistory* 35: 15–29.

Stead, W. W. 1992. Genetics and resistance to tuberculosis. *Annals of Internal Medicine* 116: 937–41.

Strachey, W. [1612] (1953). *The Historie of Travell into Virginia Britania,* ed. L. B. Wright and V. Freund. Hakluyt Society, 2nd ser., v. 103 (London).

Thornton, R., J. Warren, and T. Miller. 1992. Depopulation in the Southeast after 1492. In *Disease and Demography in the Americas,* ed. J. W. Verano and D. H. Ubelaker (Washington, D.C.: Smithsonian Institution Press), 187–95.

Ubelaker, D. H. 1974. *Reconstruction of Demographic Profiles from Ossuary Skeletal Samples: A Case Study from the Tidewater Potomac.* Smithsonian Contributions to Anthropology No. 18 (Washington, D.C.: U.S. Government Printing Office).

———. 1988. North American Indian population size, A.D. 1500–1985. *American Journal of Physical Anthropology* 77: 289–94.

———. 1989. Patterns of demographic change in the Americas. *Human Biology* 64: 361–79.

Verano, J. W., and D. H. Ubelaker. 1991. Health and disease in the Pre-Columbian world. In *Seeds of Change,* ed. H. J. Viola and C. Margolis (Washington, D.C.: Smithsonian Institution Press), 209–24.

———, eds. 1992. *Disease and Demography in the Americas* (Washington, D.C.: Smithsonian Institution Press).

Walsh, L. S., and R. R. Menard. 1974. Death in the Chesapeake: Two life tables for men in early Colonial Maryland. *Maryland Historical Magazine* 69: 211–27.

Winslow, C.-E.A., W. G. Smillie, J. A. Doull, and J. E. Gordon. 1952. *The History of American Epidemiology* (St. Louis: C. V. Mosby).

Wood, P. H. 1974. *Black Majority: Negroes in Colonial South Carolina from 1670 through the Stono Rebellion* (New York: Knopf).

———. 1987. The impact of smallpox on the native population of the eighteenth-century South. *New York State Journal of Medicine* 87: 30–36.

———. 1989. Changing population of the Colonial South: An overview by race and region, 1685–1790. In *Powhatan's Mantle: Indians in the Colonial Southeast,* ed. P. H. Wood, G. A. Waselkov, and M. T. Hatley (Lincoln: University of Nebraska Press), 35–103.

CHAPTER EIGHT

A Useful Arcadia

European Colonists as Biotic Factors in Chesapeake Forests

TIMOTHY SILVER

The arriving European settlers called the new world of abundant flora and fauna an Arcadia, but they set to clearing the savage forests to establish civility, plant their crops, and use the resources as commodities. Their activities altered the interaction between plants, animals, and the physical environment and changed all the forest processes, from tree assemblage and drainage patterns to the wildlife populations.

In early spring 1524, a single ship flying a French flag weighed anchor in turbulent seas off the Carolina coast. Pointing his lone craft northeast, the vessel's commander, Giovanni da Verrazzano, began quartering slowly up the Atlantic seaboard. A stiff breeze blew from the shore. With the wind came a soothing, almost narcotic, fragrance that captain and crew likened to some "aromatic liquor." The sailors soon realized that the intoxicating bouquet came not from the land itself but from its trees. Glimpsed from the ship, the forests were "so beautiful and delightful that they def[ied] description." A bit farther north, Verrazzano and his men went ashore to explore the sweet-scented woods. The trees so impressed the mariners that they called the country "Arcadia," a name used by a sixteenth-century poet to describe an imaginary land of splendid forests. Verrazzano's Arcadia, however, was real enough. It lay on the eastern shore of Chesapeake Bay, perhaps in what is now Worcester County, Maryland, or Accomac County, Virginia. (Wroth, 1970: 82–83, 136–37)

English colonists who followed Verrazzano and other explorers to this Arcadia on the Bay found that its forests were as appealing to the eye as to the nose. Europe had only about 25 tree species suitable for construction and ship timber; North America had some 525, many of which could be observed along the Bay and its tributaries. Moreover, by the mid-sixteenth century, England's forests had already been stripped of firewood and building material, leaving primarily the smaller timber of second-growth woods. Along the Chesapeake and the rivers that fed it, the newcomers found a seemingly infinite variety of trees that were astonishing in their "bulk and antiquity" (Illick, 1924: 150). As William Strachey, secretary of the Jamestown colony, observed, "we sometimes meet with pleasant Plaines, small rising hills, and fertile valleys" all "overgrown with trees and woods, being a plain wilderness, as God first ordained it" (1953: 39). (Cronon, 1983: 20–21)

But if settlers admired the Chesapeake wilderness, they also feared it. From their youth, most Europeans had learned that dark forests were dangerous places. Those who had heard medieval fairy tales knew that a peasant who wandered too far into such woods could easily get lost and face cruel death at the hands of wild animals or wild men. Even the word *savage,* used by the English to describe both the Irish and Indians, derived from the Latin term *silva,* meaning wood. This Chesapeake Arcadia might delight the senses, but, like the Garden of Eden, it was also a place where unspeakable evil waited to destroy a civilized people or, worse yet, tempt them into a life of sin and savagery. If colonists were to maintain their deportment, trees had to be felled, fields planted, and woodland darkness had to give way to an open, airy, rural landscape that more closely resembled the European countryside.* (Thomas, 1983: 194–96; Williams, 1989: 10–12; Nash, 1982: 14–18; Cronon, 1996: 70–71; Silver, 1990: 104)

This struggle with the Chesapeake forest would not be easy, but colonists believed it would yield splendid dividends. Dense natural vegetation suggested that choice crop land might be found "where fine timber or grapevines grow" (Jones, 1956: 77). The trees from such ground could provide housing and warmth for settlers, as well as wood and other

*Many of the ideas regarding European perceptions and transformation of the landscape contained here and in the following paragraphs are explored in more detail in Timothy Silver, *A New Face on the Countryside: Indians, Colonists, and Slaves in South Atlantic Forests, 1500–1800* (New York: Cambridge University Press, 1990).

commodities to sell on the international market. As they ventured farther inland up the Bay's tributaries, explorers compiled elaborate lists of the various trees and their potential uses. "The timber of these parts is very good, and in aboundance," wrote a visitor to Maryland in 1635. Sometimes unsure of his nomenclature, the writer continued, "The White Oake is good for Pipe-staves, the red Oake for wainescot. There is also Walnut, Cedar, Pine, and Cipresse, Chesnut, Elme, Ashe, and Popler, all of which are for Building, and Husbandry" (Hall, 1910: 79). Though they shuddered at its wildness, the earliest colonists knew that this was a useful Arcadia, one that must be tamed and improved by their presence—and made to turn a profit. (Jones, 1956: 77)

What settlers could not know was that their Chesapeake wilderness, which seemed so ancient and undefiled, had an intriguing life of its own. The trees that excited Europeans were part of eastern North America's vast deciduous forest, a general term for the band of vegetation that extended from the Great Plains to the Atlantic and from the Gulf of Mexico to northern Maine. It is a humid forest in which abundant rainfall encourages the growth of large broad-leaved trees, including the various oaks seen by early visitors to the Bay. Each year sap rises and falls in the heartwood of such trees, turning the dark green tangle of the summer forest canopy into the seemingly lifeless open woods of winter. (Vankat, 1979: 132; Greller, 1988: 288; Cronon, 1983: 37–39)

Changing seasons are not the only natural forces at work in these woodlands. Ecologists generally do not speak of one Chesapeake forest, but instead prefer to define several prominent zones of forest associations throughout the Bay's drainage. The two most important influences, or "environmental factors," that determine such regions are climate and topography. Along the headwaters of the Susquehanna River, oaks mingled with sugar maples and eastern hemlocks to create a transition forest that separated more southerly deciduous vegetation from the conifer and boreal forests to the north. On the eastern slopes of the Appalachians, one of the most striking hardwoods was the American chestnut, a monumental tree (some estimates suggest that one mature chestnut could shade 40 yards of forest floor) that helped give the mountain forest its ecological definition as an oak-chestnut region. Across the broad expanse of the Piedmont Plateau from Pennsylvania to southern Virginia, white, red, post, and black oaks combined with yellow poplars and mockernut and shagbark hickories to form a narrow band of vegetation usually classified as oak-hickory. Along the narrow Atlantic Coastal Plain where the various rivers find their way into the Bay, English people first encountered a southern mixed hardwoods forest. There the telltale oaks and hickories claimed the higher ground, while red maples, various gums, Atlantic white cedars, and bald cypresses dominated the lower swampy environs at the extreme southern end of the Chesapeake. (Sutton and Sutton, 1986: 32; Vankat, 1979: 142–47; Hudson, 1997: 190–91; Christensen, 1988: 320–26; Rountree, 1989: 24–26; Chap. 3, in this volume)

Climate and topography were important long-term determinants of forest composition, but other environmental factors had more immediate effects on the appearance of the woods. Among the most dramatic of those natural forces were the storms that blew across the Bay at all seasons. During spring and summer, southwesterly winds spawned violent thunderstorms and tornadoes. In autumn, hurricanes might stall along the Chesapeake, pounding the region with heavy rains and wind. Winter brought blizzards or, perhaps more commonly, sleet and ice storms. (Middleton, 1984: 54–55)

The huge trees in Chesapeake forests worked to minimize the impact of such violent weather on the surrounding landscape. The spreading branches of the oaks and hickories helped shelter ground vegetation from the fierce gales of winter. The standing trees also broke up and scattered precipitation, reducing surface runoff and controlling water levels in Chesapeake streams. Most Chesapeake rivers ran clear, even during times of high water. (Vankat, 1979: 26–27; Trimble, 1974: 32, 22–25)

As the storms spent themselves in the forest, wind, rain, and ice inevitably shattered or uprooted some of the trees, creating open spaces in areas that otherwise would have remained densely wooded. Farther inland, on the slopes of the Appalachians, high winds and landslides started by torrential rains tore away vegetation and topsoil. On rare occasions, lightning ignited fires that smoldered for days or weeks in the Coastal Plain and upland Piedmont forests. Although such blazes usually burned relatively small patches of ground and rarely consumed larger trees, they, too, could work as clearing agents, removing forest litter and woody plants that grew beneath the canopy of oaks and hickories. (Moehring, 1968: 13–14; Barden and Woods, 1973: 356–57; Spurr and Barnes, 1973: 353)

Once partially cleared by storm or fire, smaller open patches might be seeded from neighboring trees. But in larger cleared areas, other plants and trees began a long process of reclaiming lost terrain. Depending on drainage patterns and topography, sunny clearings might first become covered with various weeds, broom sedge, blackberries, and wild strawberries. Sassafras, a tree Europeans valued as a curative, also sprouted prolifically after fires in some parts of the Piedmont and Coastal Plain. But for Europeans the most recognizable trees on naturally cleared ground were pines: shortleaf and loblolly in lower-lying areas; pitch, Virginia, and Eastern white in the higher latitudes and elevations. (Billings, 1978: 105–6; Silver, 1990: 20–23; Chap. 3, in this volume)

Owing to the influences of climate, topography, weather, and fire, the Chesapeake woods took on a somewhat jumbled, almost hodgepodge appearance. In their scramble to list every tree that might be sold, European observers sometimes unintentionally documented the ways in which various environmental factors had influenced forest patterns. Rather than a dense, unbroken stand of mature timber, colonists probably saw a patchwork of forest communities, marked by trees of different sizes and ages. When asked to comment on eastern Virginia's pine forests, colonist John Pory (although he knew nothing of conifers invading cleared ground) noted in 1620 that among the hardwoods, pines grew only "skatteringe here one and there one" (Kingsbury, 1906–35: v. 3, 303). (Silver, 1990: 24; Chap. 3, in this volume)

The various forest communities were home to an equally diverse wildlife population. In the bottom lands along river floodplains, early explorers found bear, deer, and beaver, which promised plenty of meat and a lucrative market in skins and furs. Traveling through the oak and hickory forests, Europeans noted the abundance of wild turkeys and gawked in amazement at the autumn flights of passenger pigeons. The low-lying areas at the mouths of rivers provided habitats for a wide variety of ducks, geese, swans, and other migratory wildfowl. Gray and red wolves were prominent in Chesapeake forests and swamps. Sometimes traveling in packs, the wolves hunted deer and smaller mammals at night, frequently raising their voices in what English settlers regarded as a menacing serenade. (Strachey, 1953: 126; Byrd, 1967: 196–98; Silver, 1990: 24–31; Chaps. 5 and 15, in this volume)

Colonists, however, knew little of the intricate relationship between these animals and the forests they inhabited. Chesapeake forests, like all such natural communities, depended on energy to sustain them. That energy moved through the woods by way of various food webs. Forest food webs were driven by the energy in sunlight, which plants and trees converted into starch and protein through photosynthesis. Herbivores, such as deer, squirrels, and beaver, got the energy they needed by consuming plants. Carnivores, such as wolves, mountain lions, and hawks, in turn fed on the herbivores or other carnivores. Organisms known as decomposers consumed dead plant and animal remains, recycling energy and nutrients that would otherwise be lost. Consequently, new plant growth helped determine the number of deer, and deer the number of wolves—a fundamental dynamic that gave the forests life. (Dassman, 1964: 29; Silver, 1990: 31)

Under certain circumstances, wildlife could profoundly affect forest composition. When beavers dammed creeks, silt accumulating in ponds might eventually reach the water's surface, creating a marsh and, finally, a meadow in an area that might otherwise have been forested. Migrating passenger pigeons could clear an oak forest of acorns or distribute the seeds elsewhere, thereby disrupting patterns of new tree growth. Insects and fungi killed older trees and opened new clearings in the forests. Ecologists classify the changes wrought by wildlife and other organisms as "biotic factors." Such forces are as crucial to forest composition and change as the environmental influences of climate, topography, weather, and fire. (Spurr and Barnes, 1973: 373–74; Campbell, 1968: 64–65; Schorger, 1955: 268–69; Silver, 1990: 32)

When humans take up residence in a forest, they, too, become important biotic influences. As the first Europeans knew all too well, the Chesapeake forests already supported a large human population, namely, the Indians who lived along the Bay. Wherever settlers ventured, they saw evidence of Indian habitation. Agricultural fields cleared by the natives sprouted grasses and weeds. Where Indians employed fire to facilitate hunting and travel, undergrowth disappeared from the forest floor, leaving open woods. In the river bottom lands, where Indians hunted intensively, European explorers found deer and other game scarce. (Chap. 6, in this volume; Rountree, 1989: 40–41; Russell, 1983: 80–83; Barbour, 1986: v. 1, 164; Strachey, 1953: 124)

In ecological terms, the presettlement Chesapeake forest was hardly the unmarred "plain wilderness" described by William Strachey and others. It was no passive entity, no Eden to be altered or improved by their presence. Nor was it a "climax forest" in which ancient trees lived unchanged for centuries on end. Instead, the Chesapeake Arcadia to which Europeans came was a complex living system, influenced by climate and weather, sporadic patterns of forest growth, and energy flow among its biota. European settlers, with their fears of dark woodlands and dreams of profits, were simply additional organisms—new biotic factors—in an already dynamic natural world. (Cronon, 1983: 11–12; Vankat, 1979: 45)

Most Europeans began life in the Chesapeake woods by clearing a patch of ground on which to plant corn and tobacco: one crop for food, one for market. Although difficult to cultivate, both crops can prosper on partially cleared land. And because both plants had been grown by Indians, seventeenth-century colonists found it convenient to adopt native methods for agricultural clearing. Not bothering to remove all the timber, settlers simply stripped bark from the largest trees (a process called "girdling") to limit leaf growth. They then set fire to the undergrowth, leaving the forest floor clear enough for planting. Like the Indians, colonists found that leafless standing trees did not inhibit growth of corn and tobacco. (Gray, 1958: v. 1, 21–24; Davies, 1974: 141–45; Adair, 1930: 434–35; Chap. 11, in this volume)

But for much of the seventeenth century, Europeans found life difficult along the Chesapeake. A paucity of women (only some 20% of immigrants during the century were female) and the prevalence of malaria among white settlers worked to curb the birthrate and limit population growth. Laborers to break new ground were in short supply, meaning that, at first, tobacco farmers were unable to remove large stands of trees. Given such difficulties, agricultural clearing by colonists in the 1600s probably had little more impact on the forested landscape than Indian farming. Studies of fossil pollen and sediment from the colonial period show that ragweed (a plant that sprouts prolifically on bare ground and often serves as an indicator of agricultural clearing) increased only slightly during the first decades of European settlement. Moreover, the absorbent sandy soils and level topography of the Coastal Plain helped limit erosion and flooding in the region. For the most part, topsoil remained intact, allowing small plots to be cultivated as long as they produced sufficient crops of tobacco or corn. (Beverley, 1947: 154; Chap. 7, in this volume; Earle, 1975: 24–30; Chaps. 3 and 11, in this volume; Silver, 1990: 164)

Over the course of the eighteenth century, however, Chesapeake agriculture changed. Bad harvests and higher grain prices in Europe encouraged colonists to plant more wheat, a crop sown broadcast in thoroughly cleared and plowed fields. In addition, a number of Chesapeake farmers and European visitors began to point out the supposed inefficiency of girdling and burning, explaining that the process wasted much wood that otherwise might serve as fuel and building material. Perhaps most important, wealthier planters began to use African slaves in their fields. African men were put to work felling the largest trees while slave women and children hacked away at the underbrush with axes and hoes. Once the heaviest growth had been removed, slaves burned the field in the accustomed manner, producing ground that could be tilled with plows and draft animals after only a few years. (Byrd, 1940: 92–93; Gill, 1978: 381–85; Cronon, 1983: 116–17; Chap. 13, in this volume)

During the mid-1700s, as the birthrate rose and life expectancy improved slightly, more settlers moved into the oak and hickory forests of the Piedmont. There the effects of colonial agriculture became more apparent. Violent storms that had once lashed at the standing forest now dumped heavy rains on plowed hillsides. Silt from the resulting erosion accumulated in the narrow Piedmont stream valleys, increasing the chances for a destructive flood. Spawned in part by the clearing of Piedmont forests, disastrous spring freshets struck the Chesapeake in 1724, 1738, and 1752. But the worst of the colonial floods came in 1771. In the Coastal Plain, swollen rivers exceeded regular tide levels by 40 feet. Sediment clogged stream channels, making some rivers unnavigable. Contemporaries estimated damages from the 1771 flood at two million pounds sterling. (Trimble, 1974: 43–45; Middleton, 1984: 57–59)

By the last quarter of the eighteenth century, colonists could note other climatic effects of agricultural clearing. Without trees to control and moderate sunlight, fields were subject to extreme variations in temperature. In summer, the plots baked to a hard, dry surface that could be broken only by the most efficient plows and hardiest animals. In winter, agricultural tracts did not retain daytime heat, and more than one planter noticed that the earliest and latest frosts occurred on his cleanest land. In clearing the landscape to suit their agricultural needs, European colonists had also altered the ways in which the forests reacted to the Chesapeake climate and weather. (Greene, 1965: 433, 462, 612, 635; Silver, 1990: 112–14; Moehring, 1968: 10; Cronon, 1983: 122–23)

For the English people who settled along the Chesapeake, cleared fields became symbols of civility, irrefutable evidence that settlers had subdued the wilderness. But taming Arcadia also meant profiting from its timber resources. Initially, however, shipping trees and logs across the Atlantic proved a costly enterprise. English merchants first insisted that valuable cargo space aboard colonial ships be devoted to more lucrative products such as tobacco and deerskins. (Force, 1836–46: v. 2, tract vii, 5; Strachey, 1953: 130; Albion, 1926: 240)

In the 1620s, however, colonists at the falls of the James River began to construct mechanical sawmills. Of German design, the mills usually consisted of one or two water-driven sash saws, which might turn out a few hundred feet of roughly finished lumber each day. By century's end, mechanical sawing was also a common activity on larger tobacco farms, where planters cut boards for their own use and sold lumber to their poorer neighbors. Using slaves to construct and run the mills, wealthy planters began to dominate the Chesapeake lumber business. During the last half of the eighteenth century, these rural timber men found new urban markets for their products. Baltimore's swift growth from a Bayside village into a bustling commercial center created a steady demand for boards and building material. Norfolk emerged as the most prominent port at the southern end of the bay and became the principal destination for wood products shipped from the James River basin. (Eisterhold, 1971: 150–53; Peterson, 1975: 66; Reynolds and Pierson, 1925: 643–46; Bardin, 1923: 35; Williams, 1989: 94)

Baltimore and Norfolk were crucial to the timber trade for another reason. Both became leading ports for shipping wood products to the West Indies. On those islands, extensive clearing of local forests for sugar plantations had created a severe shortage of wood and lumber. Baltimore merchants were especially active in the island trade, contracting with planters from Virginia and the Carolinas for lumber to meet the needs of West Indian colonists. The proximity of the Caribbean market helped reduce shipping costs, and through the Baltimore and Norfolk trade, wood from Chesapeake forests became an important commodity in the Atlantic economy. Timber from the Bay's hinterland now had to meet the needs of Europeans who lived many miles from the forests. (Eisterhold, 1971: 150–51; Kinney, 1972: 386–87; Silver, 1990: 114–18; Cronon, 1992: 38–40)

Caribbean merchants were especially interested in buying barrel staves, critical items for packaging island exports of rum, molasses, and raw sugar. As the earliest European visitors to the Chesapeake had predicted, white oak proved ideal for staves. Moreover, the small hickories that grew in the same woods could be turned into the pliable hoops that held barrel staves in place. By the early 1770s, products for barrel making became regional specialties along the northern reaches of the Bay, with stave exports far outstripping the trade in sawn boards. (Michaux, 1865: v. 1, 138; Williams, 1989: 102–3)

At the southern end of the Bay, merchants also bought and sold white oak. But at the mouth of the Chesapeake, lumbermen were more inclined to harvest Atlantic white cedar and bald cypress. Both trees grew in swampy habitats and produced water-resistant wood perfect for making shingles and siding. South of the York River, planters with access to swamps taught their slaves to rive shingles in winter, when tobacco and other crops required less work. By the last quarter of the eighteenth century, cedar and cypress shingles were the most important wood products traded to and from Norfolk. (Williams, 1989: 102–3; Silver, 1990: 119)

Whether they lived in the burgeoning towns or in the countryside, Chesapeake colonists also needed much wood to meet their everyday needs. Stone was scarce along the Bay, and virtually every dwelling, tobacco house, and outbuilding had to be constructed of oak or some other durable wood. In addition, colonists used wooden fences to keep wandering livestock away from crops. Oak, chestnut, cedar, and cypress were the preferred materials because they resisted moisture and decay. (Earle, 1975: 34; Cowdrey, 1983: 54; Willson, 1948: 11; Michaux, 1865: v. 1, 67–69, v. 2, 85)

Even so, most wood cut for local use never found its way into fences, lumber, staves, or shingles. Instead, colonists used it to heat houses and cook food. Unlike Indians, who relied primarily on dead falls for firewood, European settlers took their fuel from the standing forest. And they usually burned wood in open fireplaces, an inefficient practice that sent much warmth up the chimney. Jasper Danckaerts, who traveled along the upper Chesapeake in 1679 and 1680, found settlers' houses uncomfortably drafty and cold. As he observed, "if you are not so close to the fire as to almost burn yourself, you cannot keep warm, for the wind blows through them [the dwellings] everywhere" (James and Jameson, 1913: 96). (Pryor, 1984: 17–18, 32–36; Cronon, 1983: 120)

The same trees that nourished the timber trade and sustained tobacco agriculture also fed the insatiable fireplaces. Planters preferred oak and hickory, which, as cordwood, produced "ardent heat," and left "a heavy, compact and long-lived coal" (Michaux, 1865: v. 1, 91). Residents of Baltimore, Norfolk, and other Chesapeake towns had to purchase such fuel from colonists who lived closer to the oak and hickory forests. Because the region had few roads, rivers became the primary means of transporting timber and firewood. Planters and lumbermen in the interior either shipped cordwood in boats or simply floated rafts of cut logs down Chesapeake streams to the more populous regions. (Eddis, 1969: 21; Carman and Tugwell, 1939: 189)

The timber trade eventually took a toll on the forests. By the 1770s, colonists who lived along the lower reaches of Chesapeake rivers found it difficult to acquire the most valuable timber products. William Eddis, governor of Maryland, noted in 1770 that the cost of slave labor and river shipment made firewood "an expensive article in all the considerable towns" (Eddis, 1969: 21). Planters in the countryside fared little better. That same year, along the Rappahannock River, in Virginia's Northern Neck, Landon Carter was one of many landowners who could say "We now have full 3/4 of the year in which we are obliged to keep constant fires; we must fence our ground with rails and build and repair our houses with timber and every cooking room must have its fire the year through. Add to this the natural deaths of trees and the violence of the gusts that blows them down and I must think that in a few years the lower parts of this colony will be without firewood." Carter hoped Chesapeake colonists might soon discover coal in the uplands. If not, he planned to search for some efficient method of burning the pines that flourished on his old fields. (Greene, 1965: 382)

Carter's fears notwithstanding, such observations primarily reflected local shortages, not widespread decimation of Chesapeake woods. Nevertheless, the trade in timber and firewood had a significant impact on the complex processes that determined the composition of the oak-hickory and mixed hardwood forests. Combined with agricultural clearing, selective cutting of oak and hickory likely left large tracts of Piedmont and Coastal Plain forests open to colonization by loblolly pine or red maple. Atlantic white cedar can flourish on open ground, but its regeneration process is so peculiar that, once cut, the trees are slow to replace themselves. In the swamps, removing cedar for shingles and siding usually led to an increase in bays or other hardwoods. The exact nature of the change in species composition is difficult to measure. But clearly some Europeans witnessed the process. By 1800, travelers in Virginia could observe that "amid forests of Oak, tracts of 100 or 200 acres are not infrequently seen covered with thriving young pines" (Michaux, 1865: v. 3, 123). (Buell and Cain, 1943: 91–94)

Other effects were not so obvious. To save time and overland transportation costs, slaves frequently cut trees from banks and slopes adjacent to streams. Eliminating the forest canopy caused the water to grow warmer and increased evaporation. As in cleared fields, rainwater ran off more rapidly and soil dried quicker. Rivers might rise during spring freshets, but overall, most stream levels dropped. During the height of summer, some smaller Chesapeake streams no longer carried enough water to power the settlers' sawmills. Fish populations might also have suffered as river levels fell and stream temperatures rose. At the very least, the log dams built for colonial sawmills restricted spawning runs of fish migrating upstream from the Bay. Even if a stream had no dams, ocean-going fish sometimes had to circumvent logjams created by rafts of timber lost en route to market. In selectively cutting and transporting trees for sale in the towns, European settlers had begun to affect forest hydrology and drainage patterns—subtle alterations that were practically invisible to colonists but had important long-term implications for the

Chesapeake woods. (Lee, 1980: 124–25; Cowdrey, 1983: 57; Silver, 1990: 134–35, 136)

Clearing ground and cutting timber also affected wildlife populations. When settlers removed trees from river bottom lands, they eliminated some of the habitat available to bears, pigeons, and turkeys. Just as the timber trade had turned white oak and cedar into items for the Atlantic market, an ongoing traffic in skins and pelts transformed deer and beaver into commodities valued by European tanners and furriers. By 1800, both animals were scarce along Chesapeake rivers. As those species were eliminated or driven from settled regions, other animals prospered. Weeds and grasses that grew in stump-strewn fields or along fence lines attracted farm game such as quail, rabbits, and the omnivorous opossum. Within forest communities in which the lives of plants and animals were inexorably intertwined, the slightest disturbance in vegetation patterns rippled through the labyrinthine food webs, affecting virtually every organism in the forest. (Catesby, 1977: v. 1, xxv; Richardson, 1977: 40; Matthiessen, 1959: 69)

Some of the most powerful ecological shockwaves occurred when Europeans introduced new animals into the woods. In 1611, English ships brought "one hundred Kine and other Cattell" to Jamestown. The domestic beasts adapted well to their new home. Not since the extinction of the mastodon and mammoth ten thousand years earlier had such large animals roamed the woodlands of eastern America. Cattle flourished because they moved into a long-vacant ecological niche where they faced scant competition for food and virtually no threat from indigenous disease. Pigs also prospered along the Bay. Among the most prolific of all domestic animals, pigs soon eclipsed cattle as the most visible animals on and around colonial farms. Because it was so difficult to clear fields for pasture and because corn and other livestock feed remained scarce and expensive, most Chesapeake planters fenced their crops and allowed their animals to roam the woods in search of food. By 1700 livestock served as a major source of animal protein for colonists. And like tobacco, oak staves, cedar shingles, and deerskins, beef and pork became important trade items both in eighteenth-century towns and on Caribbean islands. (Barbour, 1986: v. 2, 241; Crosby, 1986: 278–82; Chap. 5, in this volume; Laing, 1952: 1, 159; Miller, 1988: 186–87, 193)

No matter what domestic beasts meant to Europeans, once set loose along the Chesapeake, cattle and hogs were simply new organisms in the vast forest food webs. There the introduced animals sometimes became prey for the region's native carnivores, including wolves. Whether wolves took enough cattle to slow the growth of Chesapeake herds remains an open question. Modern studies suggest that wolves much prefer wild game to domestic animals and usually attack cattle when other prey is in short supply. Seventeenth-century colonists, however, still found ample reason to fret about the howling beasts. English folklore frequently portrayed wolves as savage man-eaters, and many colonists believed that killing the animals was an important community service. Moreover, when wolves did attack Chesapeake cattle, the predators tended to take calves, a practice that must have angered and frustrated struggling settlers trying to sustain a viable herd for meat and dairy products. Along the Chesapeake, as in other English colonies, wolves were quickly declared a public nuisance and local governments began to offer bounties on the animals. Wolf-killing for profit became such a popular activity in Virginia and Maryland that clerks who dispensed the bounties had to remove the ears or tongue from every wolf's head they received—lest some devious hunter try to turn in the same trophy twice. The strategy worked. By the 1720s, wolves had virtually disappeared from the settled regions, leaving livestock to prosper in peace. (Cronon, 1983: 132; Cowdrey, 1983: 49; Caras, 1967: 72–73; Bomford, 1993: 84; Hening, 1819–23: v. 2, 236, 274, v. 6, 153; *Acts of the Assembly* . . . , 1723: 68; Silver, 1990: 176–77; Kilty, 1800: chs. 4 and 8)

Delivered from their enemies, cattle and hogs became significant agents of change in the Chesapeake woods. The beasts foraged selectively on palatable grasses, woody plants, and new growth of hardwood trees. Each animal probably needed roughly 20–30 acres of such browse in order to stay healthy. As the livestock business flourished in the eighteenth century, the forests of the Coastal Plain and eastern Piedmont began to show effects of overgrazing. Along Piedmont streams and in clearings where livestock congregated, the animals trampled ground cover and compacted topsoil, increasing the odds for serious erosion. At century's end, the French naturalist François André Michaux found that upland Virginia forests had "a squalid appearance" due to "the injury they are continually sustaining from the cattle which range through them at all seasons, and which in winter are compelled by the want of herbage to subsist upon the young sprouts and the shoots of the preceding year" (1865: v. 1, 34). (Johnson, 1952: 109–13; Campbell, 1947: 262–64)

Colonists were not blind to the changes in Chesapeake forests. From the 1690s on, lawmakers pondered the potential problems associated with changing plant and animal populations. Virginia created the first closed hunting season on deer in 1699. Other restrictions followed, and by the 1790s both Virginia and Maryland had imposed complete moratoriums on commercial deer hunting. In 1692, a Maryland statute noted the need to preserve woods such as white cedar which were suitable for shingles or siding. By 1724, the colony's legislators acknowledged the growing scarcity of white oak and other trees "fit for clapboards or coopers' timber" (Kinney, 1972: 380).

For the most part, these forest laws probably had little impact on Chesapeake colonists or the woods they inhabited. The constant renewal and refinement of hunting regulations suggests that such restrictions often went unheeded. The same Maryland laws that sought to protect oak and cedar also allowed free use of all other timber to anyone constructing mills or building roads. It seems that conservation became fashionable only when profits suffered. Still, by 1800, those who lived along the Bay knew scarcity and understood that local resources were finite, notions that Verrazzano and other early European visitors could scarcely have imagined. (Cowdrey, 1983: 56; Silver, 1992: 709)

What had happened in Arcadia? In short, new organisms—Europeans—had entered and changed the forests. But like the Indians, their human predecessors along the Chesapeake, colonists were qualitatively different from other forest biota. Their relationship with the natural world was inevitably interwoven into that enigmatic tapestry of group experience known as culture. Two complementary facets of European culture worked to distinguish actions of colonists from those of other organisms and from Indians. English people saw themselves as civilizers, sent to alter and refine a savage woodland before it transformed them. Settlers also sought to turn forest resources into commodities for the Atlantic market, goods that met the needs of Europeans in other parts of the world. The appearance of Chesapeake towns, which allowed colonists to live away from the frightening forest *and* facilitated trade in its products, was perhaps the best evidence that colonists had succeeded in their plans to civilize and sell. (Dubos, 1965: 2–10; Silver, 1990: 33)

But no matter what settlers thought, they enjoyed only a small measure of control in the Chesapeake woods. For the forest itself was neither stable nor benign. Europeans or no Europeans, profits or no profits, towns or no towns, storms still rumbled across the Bay, pines invaded open ground, predators consumed their prey. Every human endeavor took place within this larger framework of interaction between plants, animals, and the physical environment. Thus, even as colonists acted on their cultural impulses to make Arcadia useful—when they cleared fields, cut timber, hunted animals, or bred livestock—they remained biotic factors within the forest community. As such they altered established processes, influencing everything from drainage patterns to wildlife populations. As Europeans became aware of their own impact, they uncovered an ironic truth that would govern human relationships with Chesapeake woodlands for years to come: Forests that had once threatened civilized people now needed their protection.

BIBLIOGRAPHY

Acts of the Assembly Passed in the Province of Maryland from 1692 to 1715. 1723 (London).

Adair, J. 1930. *A History of the North-American Indians, Their Customs &c.*, ed. S. C. Williams (Johnson City, Tenn.: Watauga Press).

Albion, R. G. 1926. *Forests and Sea Power* (Cambridge: Harvard University Press).

Archer, G. 1907. The Description of the Now-Discovered River and country of Virginia, with the Liklyhood of Ensuing Ritches by England's Ayd and Industry. *Virginia Magazine of History and Biography* 14: 375–78.

Barbour, P. L., ed. 1986. *The Complete Works of Captain John Smith (1580–1631)* (Chapel Hill: University of North Carolina Press for the Institute of Early American History and Culture).

Barden, L. S., and F. W. Woods. 1973. Characteristics of lightning fires in the southern Appalachian forests. *Proceedings of the Tall Timbers Fire Ecology Conference* 13: 345–61.

Bardin, P. C. 1923. The outline history of the sawmill. *Hardwood Record* 55(8): 35–36, 46.

Beverley, R. 1947. *The History and Present State of Virginia*, ed. L. B. Wright (Chapel Hill: University of North Carolina Press).

Billings, W. D. 1978. *Plants and the Ecosystem* (Belmont, Calif.: Wadsworth Publishing).

Bomford, L. 1993. *The Complete Wolf* (New York: St. Martin's Press).

Buell, M. F., and R. F. Cain. 1943. The successional role of southern white cedar, *Chamaecypaius thyoides*, in southeastern North Carolina. *Ecology* 24: 85–93.

Byrd, W., II. 1940. *The Natural History of Virginia, or the Newly Discovered Eden*, ed. R. C. Beatty and W. J. Malloy (Richmond, Va.: Dietz Press).

———. 1967. *Histories of the Dividing Line Betwixt Virginia and North Carolina*, ed. W. K. Boyd (Mineola, N.Y.: Dover).

Campbell, R. S. 1947. Forest grazing in the southern Coastal Plain. *Proceedings of the Society of American Foresters*, 262–70.

———. 1968. Manipulating biotic factors in the southern forest. In *The Ecology of Southern Forests 17th Forestry Symposium*, ed. N. E. Linnartz (Baton Rouge: Louisiana State University Press), 58–68.

Caras, R. A. 1967. *North-American Mammals: Fur-Bearing Animals of the United States and Canada* (New York: Gallahad Books).

Carman, H. J., and R. Tugwell. 1939. *American Husbandry* (New York: Columbia University Press).

Catesby, M. 1977. *The Natural History of Carolina, Florida, and the Bahama Islands.* 2 vols. (reprint, Ann Arbor, Mich.: University Microfilms).

Christensen, N. L. 1988. Vegetation of the southeastern Coastal Plain. In *North American Terrestrial Vegetation,* ed. M. G. Barbour and W. D. Billings (New York: Cambridge University Press), 317–63.

Cowdrey, A. E. 1983. *This Land, This South: An Environmental History* (Lexington: University Press of Kentucky).

Cronon, W. 1983. *Changes in the Land: Indians, Colonists, and the Ecology of New England* (New York: Hill & Wang).

———. 1992. Kennecott journey: The paths out of town. In *Under an Open Sky: Rethinking America's Western Past,* ed. W. Cronon, G. Miles, and J. Gitlin (New York: Norton), 28–51.

———. 1996. The trouble with wilderness; or, getting back to the wrong nature. In *Uncommon Ground: Rethinking the Human Place in Nature,* ed. W. Cronon (New York: Norton), 69–90.

Crosby, A. W. 1986. *Ecological Imperialism: The Biological Expansion of Europe, 900–1900* (New York: Cambridge University Press).

Dassman, R. F. 1964. *Wildlife Biology* (New York: Wiley).

Davies, K. G. 1974. *The North Atlantic World in the Seventeenth Century* (Minneapolis: University of Minnesota Press).

Dubos, R. 1965. *Man Adapting* (New Haven: Yale University Press).

Earle, C. V. 1975. *The Evolution of a Tidewater Settlement System: All Hallow's Parish, Maryland, 1650–1783.* University of Chicago Department of Geography Research Paper no. 170.

Eddis, W. 1969. *Letters from America,* ed. A. C. Land (Cambridge: Harvard University Press).

Eisterhold, J. A. 1971. Colonial beginnings of the South's lumber industry: 1607–1800. *Southern Lumberman* 223: 150–53.

Force, P. 1836–46. *Tracts and Other Papers, Relating Principally to the Origin, Settlement, and Progress of the Colonies in North America, from the Discovery of the Country to the year 1776.* 4 vols. (Washington, D.C.: P. Force).

Gill, H. B. 1978. Wheat culture in colonial Virginia. *Agricultural History* 52: 380–93.

Gray, L. C. 1958. *History of Agriculture in the Southern United States to 1860.* 2 vols. (Gloucester, Mass.: Peter Smith).

Greene, J. P., ed. 1965. *The Diary of Colonel Landon Carter of Sabine Hall, 1752–1778.* 2 vols. (Charlottesville: University Press of Virginia).

Greller, A. M. 1988. Deciduous forest. In *North American Terrestrial Vegetation,* ed. M. G. Barbour and W. D. Billings (New York: Cambridge University Press), 287–316.

Hall, C. C., ed. 1910. *Narratives of Early Maryland.* Original Narratives of Early American History, gen. ed. J. F. Jameson (New York: Scribner's).

Hening, W. W. 1819–23. *The Statutes at Large of Virginia: being a Collection of all the laws of Virginia from the first session of the legislature, in the year 1619.* 13 vols. (New York: R. & W. Bartow).

Hudson, C. 1997. *Knights of Spain, Warriors of the Sun: Hernando de Soto and the South's Ancient Chiefdoms* (Athens: University of Georgia Press).

Illick, J. 1924. The story of the American lumbering industry. In *A Popular History of American Invention,* ed. W. Kaempffert (New York: Scribner's), 2: 150–98.

James, B. B., and J. F. Jameson. 1913. *The Journal of Jasper Danckaerts, 1679–1680.* Original Narratives of Early American History, gen. ed. J. F. Jameson (New York: Barnes and Noble).

Johnson, E. A. 1952. Effects of farm woodland grazing on watershed values in the southern Appalachian mountains. *Journal of Forestry* 50: 109–13.

Jones, H. 1956. *The Present State of Virginia,* ed. R. L. Morton (Chapel Hill: University of North Carolina Press).

Kilty, W. 1800. *Laws of Maryland* (reprint, New Haven, Conn.: Microfilm Research Publications).

Kingsbury, S. M. 1906–35. *The Records of the Virginia Company of London* (Washington, D.C.: Government Printing Office).

Kinney, J. P. 1972. *The Development of Forest Law in America* (New York: Arno Press).

Laing, W. N. 1952. Cattle in early Virginia. Ph.D. diss., University of Virginia.

Lee, R. 1980. *Forest Hydrology* (New York: Columbia University Press).

Matthiessen, P. 1959. *Wildlife in America* (New York: Viking).

Michaux, F. A. 1865. *The North American Sylva: Or a Description of the Forest Trees of the United States, Canada, and Nova Scotia,* trans. J. J. Smith. 5 vols. (Philadelphia: Rice, Rutter, and Co.).

Middleton, A. P. 1984. *Tobacco Coast: A Maritime History of Chesapeake Bay in the Colonial Era* (reprint, Baltimore: Johns Hopkins University Press).

Miller, H. M. 1988. An archaeological perspective on the evolution of diet in the colonial Chesapeake, 1620–1745. In *Colonial Chesapeake Society,* ed. L. G. Carr, P. D. Morgan, and J. B. Russo (Chapel Hill: University of North Carolina Press for the Institute of Early American History and Culture), 176–99.

Moehring, D. M. 1968. Climatic elements in the southern forest. In *The Ecology of Southern Forests 17th Forestry Symposium,* ed. N. E. Linnartz (Baton Rouge: Louisiana State University Press), 5–17.

Nash, R. 1982. *Wilderness and the American Mind.* 3rd ed. (New Haven: Yale University Press).

Peterson, C. E. 1975. Early lumbering: A pictorial essay. In *America's Wooden Age: Aspects of Its Early Technology,* ed. B. Hindle (Tarrytown, N.Y.: Sleepy Hollow Restorations), 62–83.

Pryor, E. B. 1984. *Agricultural Implements Used by the Middle-Class Farmers in the Colonial Chesapeake.* The National Colonial Farm Research Report No. 16 (Accokeek, Md.: The Accokeek Foundation).

Reynolds, R. V., and A. H. Pierson. 1925. Tracking the sawmill westward: The story of the lumber industry in the United States as unfolded by its trail across the continent. *American Forests* 31: 643–48.

Richardson, J. L. 1977. *Dimensions of Ecology* (Baltimore: Williams and Wilkins).

Rountree, H. C. 1989. *The Powhatan Indians of Virginia: Their Traditional Culture* (Norman: University of Oklahoma Press).

Russell, E. W. B. 1983. Indian-set fires in the forests of the northeastern United States. *Ecology* 64: 78–88.

Schorger, A. W. 1955. *The Passenger Pigeon: Its Natural History and Extinction* (Madison: University of Wisconsin Press).

Silver, T. 1990. *A New Face on the Countryside: Indians, Colonists, and Slaves in South Atlantic Forests, 1500–1800* (New York: Cambridge University Press).

———. 1992. Outlaw gunners and hunting law in the English colonial South. *Transactions of the Fifty-seventh North American Wildlife and Natural Resources Conference,* 706–10.

Spurr, S. H., and B. V. Barnes. 1973. *Forest Ecology* (New York: Ronald Press).

Strachey, W. 1953. *The History of Travell into Virginia Brittania,* ed. L. B. Wright and V. Freund (London: Hakluyt Society).

Sutton, A., and M. Sutton. 1986. *Eastern Forests* (New York: Alfred A. Knopf).

Thomas, K. 1983. *Man and the Natural World: A History of the Modern Sensibility* (New York: Pantheon).

Trimble, S. W. 1974. *Man-Induced Soil Erosion on the Southern Piedmont, 1700–1970* (Ankeny, Ia.: Soil Conservation Society of America).

Vankat, J. L. 1979. *The Natural Vegetation of North America: An Introduction* (New York: Wiley).

Williams, M. 1989. *Americans and Their Forests* (New York: Cambridge University Press).

Willson, L. M. 1948. *Forest Conservation in Colonial Times* (St. Paul, Minn.: Forest Products History Foundation).

Wroth, L. C. 1970. *The Voyages of Giovanni da Verrazzano* (New Haven: Yale University Press).

CHAPTER NINE

Reconstructing the Colonial Environment of the Upper Chesapeake Watershed

ROBERT D. MITCHELL, WARREN R. HOFSTRA,
AND EDWARD F. CONNOR

Establishing an environmental baseline from which to measure landscape change in the Chesapeake since the arrival of Europeans is a crucial contribution to ecological history. This chapter provides insight into how this might be achieved using the land survey records of Frederick County, Virginia, between 1730 and 1800 to reconstruct the vegetation and land cover of the northern Shenandoah Valley during the early contact period. This information offers valuable clues to changing land use under colonial conditions of land ownership, resource use, and a developing economy.

European occupation of the Chesapeake region depended upon the transformation of what colonial settlers termed *wilderness* or *wasteland* into property. The condition of interior environments encountered by migrating European and Tidewater settlers during the early eighteenth century significantly influenced the locations and forms of colonial settlements. These environments were not pristine lands unaltered by previous native activities, although the nature and extent of the alterations remain problematic. Geographers and historians have long been interested in the first European penetrations of the mountainous upper reaches of the Chesapeake watershed in western Maryland and Virginia, but they have rarely been concerned with what is the focus of this chapter: what the process of appropriating property revealed about the relationships between society and environment.

Our treatment of Chesapeake history rests on three assumptions. We need, first, an interdisciplinary approach combining natural and social sciences because natural processes often operate independently of human action. Humans act to modify these processes, but nature responds to human activity in unexpected ways. Second, to understand this mutual interaction, we need to reach beyond the traditional documentary sources of the historian. Land-survey records in Virginia, for example, reveal important information on vegetation cover by identifying particular boundary markers, "witness trees," which are often described in some detail. When these surveys are organized geographically into continuous cadastral patterns, they provide the best data we are likely to acquire on the distribution of forested, open, and cultivated lands, as well as on the location of farmsteads, routeways, and other sites and paths of economic activity. Third, whether the intent is to test hypotheses about global processes or simply to reconstruct land cover from the bottom up, such research can take place only at local levels of inquiry and at the smallest spatial scales of individual settlement sites and land parcels.

We focus on the Shenandoah Valley of Virginia, with evidence derived particularly from Frederick County, the earliest and most effectively occupied county on colonial Virginia's far western frontiers. The Shenandoah River and the upper branches of the Potomac River form the bulk of the drainage basin of the western part of the Chesapeake watershed, the Shenandoah occupying the northern half of the Great Valley of Virginia between the Blue Ridge and the front range of the Allegheny Plateau. The Conococheague River occupies a similar position north of the Potomac in the Great Valley section of Maryland and adjacent Pennsylvania.

Settlement of the Upper Chesapeake

English authorities were unfamiliar with the area immediately west of the Blue Ridge until the second decade of the eighteenth century, when Virginia's lieutenant governor, Alexander Spotswood, and his entourage climbed the Blue Ridge in 1716 and looked over the Shenandoah Valley. Within three years, Virginia had begun to formulate a frontier policy designed to organize buffer settlements against potential French encroachments in the upper Ohio Valley and also against claims made by holders of the Northern Neck Proprietary grant to lands drained by the upper reaches of the Potomac. Maryland was slower to organize a western land policy, in part because of boundary disputes with Pennsylvania and in part because of a general perception that much of western Maryland was devoid of trees (and was thus referred to as the "Barrens"). Colonial settlement of Maryland's Piedmont and Great Valley sections was delayed until the 1740s.

In Virginia, on the other hand, authorities had begun to develop a more elaborate settlement policy under Lt.-Gov. William Gooch during the early 1730s. His administration issued orders to individuals willing to settle families west of the Blue Ridge with the stipulation that within two years the grantors would settle one family per 1,000 acres granted. The goal was rapid settlement of the Appalachian frontier rather than profitable land speculation. Thus grantors were eager to sell off these lands as freeholds to prospective families. Despite this goal, settlers were left relatively free to define their own properties and to choose sites for farmsteads. By contrast, lands under the control of the Northern Neck Proprietary in the Virginia Piedmont and along the Shenandoah River carried no stipulation requiring settlement within a particular period of time.

Settlers from western Europe and the Middle Colonies, who began to occupy the Shenandoah Valley during the late 1720s and early 1730s, found themselves in an extremely isolated situation. Settlement in the Virginia and Maryland Piedmont had barely begun, and the Great Valley section of Maryland remained unoccupied. At the same time, colonial officials knew very little about the lands west of the Blue Ridge that they were trying to control on behalf of the British Empire. They imposed no particular spatial structure on settlement in the form of counties or townships, and towns were allowed to develop as a consequence of settlement rather than as a condition for it. The English survey system then in use, called metes and bounds, imposed no preconceived order on the shape of property holdings other than the rules that settlers generated themselves in the search for settlement sites and promising natural resources. Environmental considerations, therefore, were paramount in the decisions settlers made about sites for dwellings and farms. Encoded in the pattern of settlement, consequently, was very basic information about the physical environment and the way Europeans interpreted it.

Local and regional historians have long proposed that the particular environmental conditions of the Shenandoah Valley at the time of European contact contributed greatly to rapid settlement of the area. This tradition fostered what has been termed "the prairie myth." The idea of an immense grassland prairie covering the floor of the valley became enshrined by Samuel Kercheval during the 1830s when he stated that "much of the greater part of the country between what is called the Little North Mountain and the Shenandoah River, at the first [European] settling of the valley was one vast prairie . . . with the exception of narrow fringes of timber immediately bordering on the water-courses" (1833: 52, 305). Although Kercheval's contemporary Edmund Ruffin had argued as early as 1825 that the extent of prairie cover in the northeastern United States was exaggerated, most writers who broached the subject during the nineteenth and early twentieth centuries perpetuated the prairie idea. The record of fossil pollen from the Shenandoah Valley indicates the presence of both tree and plant species characteristic of more open land, but the data so far are too sparse to establish the abundance and distribution of open land.

The most extensive review of the literature on the western Chesapeake watershed suggested that although the Shenandoah Valley was probably the most burned-over region of the central Appalachians, the consequences for the landscape were not vast areas of prairie. William Robison in 1960 stated: "In all probability the traditional belief had a sound basis, but the extent of the precolonial prairies must have been exaggerated . . . the valley was not a continuous prairie where one site was much like the next but rather it included certain limited areas that had been cleared and were therefore desirable spots for settlement" (160–62). Robert Mitchell, after reviewing the witness-tree references in 455 surveys for the southern Shenandoah Valley made between 1735 and 1755, found that white oaks, hickories, and black oaks constituted 75% of the markers. He concluded that "the early settlers would have encountered a region covered with predominantly open to relatively dense woodlands of hardwoods and pines in various phases of succession interspersed with relatively open grassland areas of varying extents" (1977: 24). Yet there has been no systematic attempt until now to re-create land cover from a detailed reconstruction of witness-tree data from land surveys. If the Shenandoah Valley at the time of European settlement was indeed extensively covered with prairie, then the natural or cultural mechanisms maintaining such an ecosystem need to be explained.

The Documentary Record

Historians who have attempted to reconstruct colonial environments in North America have relied heavily on travel accounts, natural histories, and diary notations. Colonial accounts of the Shenandoah Valley's vegetation cover suggest a mosaic of forest and open land rather than an extensive prairie. Although John Lederer viewed the northern Shenandoah Valley from the Blue Ridge in 1670 and commented on the "savannahs" of the Piedmont, he made no mention of such landscapes to the west. Louis Michel, who envisioned the creation of a Protestant colony in the region, visited the northern part of the valley between 1704 and 1707. He commented that there was "land that is dry and barren and where it is difficult to pass through the wild brush-wood. On the contrary, there is good land, where are great forest trees of oak, and where much game abounds" (quoted in Kemper, 1921: 2). Alexander Spotswood's expedition spent little more than a day west of the Blue Ridge. There was no report of a grassland or prairie, but one member, John Fontaine, remarked that members of the party ate "very good grapes, and saw a vine which bore a sort of wild cucumber, and a shrub which bore a fruit like unto a currant" (105–7). The presence of such plants suggests an environment more wooded than grass-covered.

Travel literature, however, must be used with caution. Travelers generally were not well versed in natural history and were often highly selective rather than comprehensive in their accounts. Some wrote from the perspective of the commercial potential of lands they traversed; others were more political in outlook and displayed their prejudices against colonial farm practices, frontier societies, and other social institutions; and still others indiscriminately incorporated accounts of regions unvisited into their descriptions. Given these caveats, landscape depictions were probably the least distorted elements in their travel experiences, and where travelers had some expertise in natural history their accounts become especially valuable.

Forest Patterns and Forest Transformation

Most travelers, and presumably settlers, judged land quality by its vegetation cover. Good soils were associated with hardwood forests, principally oak-hickory, and poor soils with pine woods. Lands that did not support forests and especially those covered with brush or scrub were considered infertile or barren. Open, grassy areas were noted particularly for their settlement possibilities, and comments on their appearance or extent are often extremely useful. If literate travelers had knowledge of Hugh Jones's early eighteenth-century description of the Virginia back country, they would have expected that "the whole country is a perfect forest, except where the woods are cleared for plantations, and old fields, and where have been formerly Indian towns, and poisoned fields and meadows, where the timber has been burnt down in fire-hunting or otherwise; and about the creeks and rivers are large rank morasses or marshes, and up the country are poor savannahs" (1956: 74).

The earliest travel notations made after pioneer settlers moved into the Shenandoah Valley during the early 1730s date from Moravian migrants passing through the region during the late 1740s and early 1750s. Traversing a shale zone just south of Frederick County in October 1753, one Moravian diarist noted that "the country was pretty barren, overgrown with pine trees" and, a few miles farther south, he and his companions camped and "put our horses in the woods." In the southern end of the valley a few days later he noted, "Although it is very hilly here, yet it is a fruitful country. It has few stones, but consists of the fattest, black soil. It is settled mostly by English and Irish people" (Hinke and Kemper, 1904: 144–48).

Oaks and hickories appeared most often in eighteenth-century accounts of forests in the valley. The naturalist John Bartram, traveling the short distance from Winchester to Stephensburg (Stephens City) in 1759 along a limestone-shale contact zone, "rode over very stoney ground producing great red Ceder pines . . . leaved 2 & 3 & 3 leaved broad leaved willow oak" (Bartram Papers). In the summer of the same year, Andrew Burnaby encountered "majestic woods; the whole interspersed with an infinite variety of flowering shrubs" (1798: 74, 157–59) and mentioned twenty-six species of trees including oak, hickory, maple, cedar, pine, sassafras, dogwood, locust, redbud, tulip tree, catalpa, chinkapin, persimmon, and chestnut. In Frederick County he observed that "the low grounds upon the banks of this river [Shenandoah] are very rich and fertile; . . . [the people] live in the most delightful climate, and richest soil imaginable; they are everywhere surrounded with beautiful prospects and sylvan scenes; lofty mountains, transparent streams, falls of water, rich valleys, and majestic woods; the whole interspersed with an infinite variety of flowering shrubs" (Burnaby, 1798: 73–74). The valley section of Frederick County contained the bulk of the region's more than 10,000 residents by this time, and Burnaby's landscape description is the best available after 30 years of settlement (Mitchell, 1977: 95–96).

The persistence of extensive forests and woodlands was a consistent observation in travel accounts throughout the remainder of the eighteenth century. Hardwood forests invariably covered limestone-floored valley bottom lands, while pines and scrubbier hardwoods were found on shale lands and bordering mountain slopes. Some travelers described forest openings in various stages of reversion to woodland. Philip Fithian, in 1774, saw "Glades, quite bare of Timber, and covered with Shrubs, Ground-Oak, Hazles, and c." (1934: 107). Andrew Burnaby remarked that abandoned tobacco lands when left fallow became "beautifully covered with Virginian pines" (1798: 92), an observation also supported by Isaac Weld, who wrote that "lands left waste in this manner throw up, in a very short time, a spontaneous growth of pines and cedars" (1807: v. 1, 152).

During succession, cleared land and old fields invariably produced messy areas of shrubs and thickets. John Fontaine had noted such areas on the eastern side of the Blue Ridge as early as 1716. Nicholas Cresswell and John Smyth remarked on several areas of "impenetrable thicket" on their journeys through the central Appalachians between 1774 and 1783. Other travelers, however, described the forest understory in some places as more open and devoid of heavy undergrowth. Ferdinand Bayard traveling through eastern Frederick County in 1791 passed through "forests whose hardy and tall trees did not permit bois-du-chien [dogwood] or bramble to overgrow the open spaces between the trees. A deep green turf covered those spaces and invited the tired traveler to rest" (1798: 68).

Earlier, in 1775, Philip Fithian rested near Winchester and "for the Sake of Meditation, took a Ramble into the long Bosom of a tall, dark Wood—We chose, at some distance from each other, a green shady Spot; threw ourselves carelessly on the cool Grass" (1934: 11–12). Fithian also commented on numerous flowering plants covering the valley floor, a phenomenon first noted by John Bartram in a letter to Peter Collinson describing his journey across the Blue Ridge in Frederick County and north along Opequon Creek: "Indeed beyond ye mountains in virginia & pensylvania, there is a great variety that I saw & ye inhabitants say ye ground is covered with dilicate beautiful flowers in ye spring which is not to be found after hot weather comes on" (Berkeley and Berkeley, 1992: 104). When Collinson germinated the seeds Bartram sent, he discovered "Curious Flowers" including "New Jacea with hoary rough Leaves—a very pretty Dwarf Gentian with a large Blew Flower, . . . Gratiola, . . . Dracocephalon, . . . Chrysanthemums or sun flowers, [and] Asters" (1992: 145). Several varieties of wild grape were also associated with disturbed land.

Settlers employed several methods of forest clearance, including burning, clear-cutting with ax and saw, and girdling. Only the last method is described in local accounts. Thomas Anburey wrote in 1779: "Their manner of clearing the land is, by cutting a circle round the tree through the bark quite to the wood, before the sap rises, which kills it; they then clear the small brushwood and cultivate the ground, leaving the trees to rot standing, which happens in a very few years; and after receiving the circular wound, they never more bear leaves" (1789: v. 2, 188). Nicholas Cresswell, in his 1777 account of setting up a farm in the northern Shenandoah Valley, observed that the best land in large tracts was still inexpensive "tho' perhaps one third of it is cleared from woods" (1924: 197). If Cresswell's estimate is reasonably accurate, then approximately one-third of the prime agricultural land on the valley floor would have been cleared by the time of the American Revolution, particularly in the vicinity of Winchester, the region's largest town. Twenty years later, Isaac Weld was to observe:

> In the neighborhood of Winchester it is so thickly settled, and consequently so much cleared, that wood is now beginning to be thought valuable; the farmers are obliged frequently to send ten or fifteen miles even for their fence rails. It is only, however, in this particular neighborhood that the country is so much improved; in other places there are immense tracts of woodlands still remaining, and in general the hills are all left uncleared. (1807: v. 1, 231)

Patterns and Causes of Open Land

The outcome of settlement, land clearance, and farm building throughout the eighteenth century was a variegated landscape with open bottom lands, cleared farmland, small villages and towns, and surviving wood lots between streams and on hill and mountain slopes. Part of this landscape mosaic, however, had existed before colonial settlement in the form of frequent variations in land cover that included not only hardwood forests and more open woodlands but also various forms of open land largely devoid of trees. Travelers made note of these patterns and provided several explanations for their existence.

Prior to the settlement of the Shenandoah Valley, travelers crossing the Piedmont had described various forms of landscape clearings. John Lederer, in 1670, frequently referred to these clearings as *savanae,* a descriptive term which he applied on his map to the entire eastern foothill zone of the Blue Ridge. Thomas Batts and Robert Fallam, who followed Lederer into Appalachian Virginia a year later, encountered rich meadows along stream channels on the Piedmont, occasionally with grasses as tall as a man (Alvord and Bidgood, 1912; Briceland, 1987). Yet Thomas Anburey could write of the area in the vicinity of Charlottesville in 1779,

"The country is so much covered with woods, that you travel a long time without seeing an habitation" (1789: v. 2, 196).

Several eighteenth-century travelers commented on natural meadows along stream channels in the Shenandoah Valley. Philip Fithian found along Opequon Creek near Winchester "large and rich Meadows—Many have good Grass upon the Uplands" (1934: 19), and Ferdinand Bayard "was twelve miles from Winchester, when well-kept meadows foretold the proximity of a less cursed land, and of some intelligent planter . . . On the other side [of the planter's house] was a garden which overlooked a fairly large meadow, watered by a rather large stream" (1798: 58–59). According to the Duc de la Rochefoucauld-Liancourt at the end of the century, "the banks of [Shenandoah River] are, in some instances, covered with fine natural grass" (1800: v. 3, 195), a phenomenon that Fithian had earlier noted along the lower Susquehanna River in the form of "large Plains, or as the Inhabitants call them, Glades, quite bare of Timber" (1934: 107). Few travelers estimated the dimensions of such open areas and none suggested that they were so extensive as to constitute entire landscapes of grassland. Fithian comes closest when he claimed for nearby Pennsylvania "large open Plains, cleared either by the Indians, or by accidental Fire, hundreds of Acres covered with fine grass" (91).

Fithian's tentative explanations reflect the fact that most travelers who commented on open areas assumed that these areas would have been capable of producing forests. Although fire and clearance by Indians were mentioned as principal causes of open land, fires of the intensity necessary to produce a crown burn were probably not common enough to clear extensive areas—most fires served only to reduce the forest understory and ground cover and place it in a new cycle of succession (Day, 1953; Garren, 1943; Holland, 1979; Little, 1974; Prunty, 1965; Russell, 1983). Isaac Weld made this point in 1796: "In general there is but very little brushwood in the woods of America, so that these fires chiefly run along the ground; the trees, however, are often scorched, but it is very rare for any of them to be entirely consumed" (1807: v. 1, 62). There were, moreover, no resident Indian groups in the Shenandoah Valley at the time of European contact, and few travelers made direct references to clearings called "Indian old fields." On the first map of Virginia to show the upper Chesapeake watershed in any detail, the Joshua Fry and Peter Jefferson map of 1751, the only extensive area warranting an "old field" designation was the "Shawno Fields" of the South Branch of the upper Potomac River. Travelers themselves occasionally were blamed for creating slow-burning fires. The Marquis de Chastellux, while visiting Natural Bridge at the southern end of the Shenandoah Valley in 1782, commented that "one is surprised to find everywhere in these unsettled forests the traces of several fires. These accidents are sometimes caused by the carelessness of travelers, who light a fire when they go to sleep and neglect afterwards to extinguish it" (1786: v. 2, 407–8).

Other travelers believed that periodic flooding produced most areas of open land associated with meadows and marshes along stream courses (Fontaine, 1972: 91), while a few pointed to stream damming as a result of beaver or buffalo activity (Cresswell, 1924; Fithian, 1934: 147). By far the most astute observation was made by naturalist John Bartram while visiting the Shenandoah Valley in 1759. He attributed natural meadows to stream damming from formations of calcareous (chalky) earth or marl. Marl forms as a calcium carbonate precipitate on rocks or wedged tree trunks, sticks, and twigs along streams in limestone areas. After crossing Opequon Creek on his way to Winchester, Bartram observed:

> Ye efects of ye incrusting limestone waters which is of that nature that where it runs it incrusts round brush or leaves or stones or any thing in its course frequently stoping its course & overflowing ye adjacent low grounds amongst ye leaves brush or grass or weeds which it incrusts when ye winters frost is sharp it penetrates this crust which falls in scales & enricheth ye ground exceedingly. I observed a bank 8 foot deep at a mill race all of this encrusted limestone matter converted into firm soil & many times of such a firm consistency as to make numerous dams quite cross large creeks so that ye floods frequently overflowed large quantities of low grounds & enriched them much with its calcarious matter when disolved by ye frost rain dews & sun. (Bartram Papers)

This is the first documented reference to travertine-marl deposits known in American scientific literature.

Taken together, contemporary observations concur sufficiently to lend credence to certain conclusions. The Shenandoah Valley undoubtedly was not covered by extensive prairies at contact time. Open areas certainly existed but only within a mosaic of forests, bottom lands, and clearings caused by a variety of cultural and natural processes. Periodic flooding caused by unusually heavy rains, or occasional snowmelts, or beaver or marl damming maintained meadows and marsh areas along watercourses. It is also possible that the periodic action of ice along riverbanks, soil compaction by herd animals such as buffalo at salt licks, and even intense burns emanating from lightning strikes and sulfur or saltpeter deposits in soils could help to maintain open areas. Yet fire activity, whether of natural or human origins, rarely would have been capable of extensive forest destruction. Fire served to keep forest understory down rather than to clear the forest crown.

The region's forests themselves held little value for early settlers and were commonly viewed as a hindrance to farming. The first two generations of colonial settlers might have stripped one-third of the valley floor of its forests for agricultural purposes. Even as settlers drained marshes, reclaimed bottom lands, and cleared fields and pastures on adjoining uplands, the landscape of the Shenandoah Valley remained a mosaic of open and wooded land.

The Survey Record

The availability of an extensive and continuous survey record for Frederick County affords us another perspective on the contemporary travel literature and allows reconstruction of the colonial environment in greater detail. Surveys for grants and patents in the northern Shenandoah Valley consistently employed trees as witness markers. The Crown, through the Virginia Council—and through Lord Fairfax, as proprietor of the Northern Neck (the land lying between the Potomac and the Rappahannock Rivers)—provided individuals or groups with rights to specified amounts of land. The granting authority issued warrants authorizing the survey of claimed lands. The surveyors then went out, usually during the spring or fall months, and marked the metes and bounds of tracts with the aid of two chain carriers, who measured distances with a surveyor's chain. Surveyors sketched out tracts in their field notebooks and completed the final drawings and descriptions in their offices. The surveys were next entered into the record, and prospective landowners were expected to pay any required fees. The land office then issued land grants indicating the size, location, and delineation of tracts and the method of ownership (Hughes, 1979).

Clerks entered the grants in the Crown or the Northern Neck land books. Because the grants were not always transcribed accurately, however, historians have to use the tract descriptions outlined in the original surveys, an example of which is illustrated by Fig. 9.1. Not all of the original surveys have survived for Frederick County as designated by 1779, but we were able to retrieve information from 1,000 surveys representing 70% of the original surveys recorded between 1730 and 1780.

We compiled data on ownership, location, markers or witness trees,

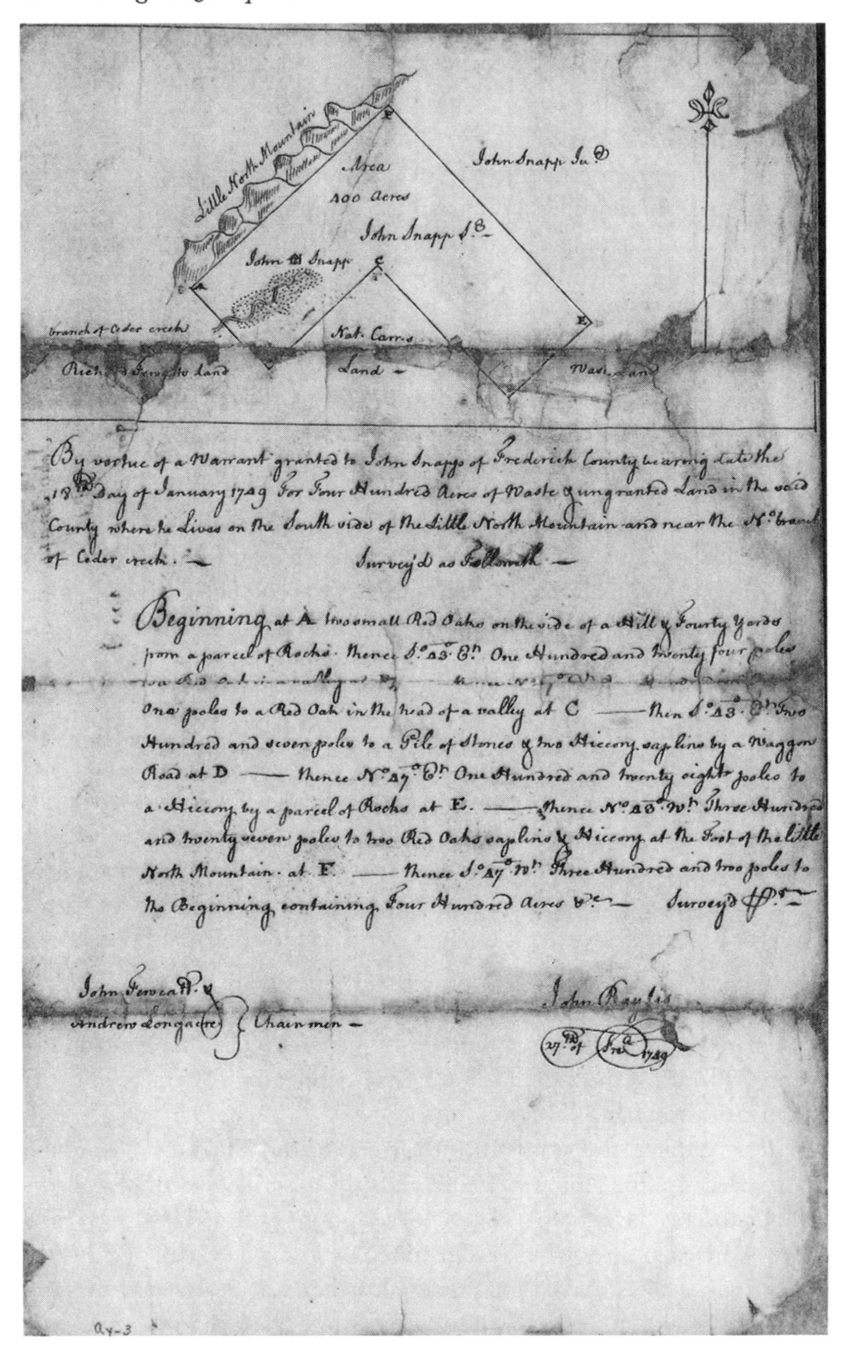

By vertue of a Warrant granted to John Snapp of Frederick County bearing date the 18th Day of January 1749 For Four Hundred Acres of Waste & ungranted Land in the said County where he Lives on the South side of the Little North Mountain and near the No. branch of Ceder creek. — Survey'd as Followeth —

Beginning at A two small Red Oaks on the side of a Hill & Fourty Yards from a parcel of Rocks. thence S° 45° E. One Hundred and Twenty four poles [illegible] One poles to a Red Oak in the head of a valley at C —— then S° 45° E. Two Hundred and seven poles to a Pile of Stones & two Hiccory saplins by a Waggon Road at D —— thence N° 47° E. One Hundred and twenty eight poles to a Hiccory by a parcel of Rocks at E. —— thence N° 45° W. Three Hundred and twenty seven poles to two Red Oaks saplins & Hiccory at the Foot of the Little North Mountain. at F. —— thence S° 47° W. Three Hundred and two poles to the Beginning containing Four Hundred Acres &c. — Surveyd ℘ —

John Fewcatt
Andrew Longacre } Chain men —

John Baylis
27th of Nov. 1749

Fig. 9.1. Survey of 1749 for John Snapp Sr., 400 acres

and other pertinent information in two databases. One was a coordinate file, consisting of the Universal Transverse Mercator coordinates for each marker in a grant. The second database, and the more important for present purposes, was the parcel information file, which contained the description of the land grant and the nature of the markers. Data about each marker included the object itself (tree, stake, or stone), the species of tree if marked by a tree, comments about the tree (young, old, sapling, forked, etc.), and any pertinent comments made by the surveyor with reference to each marker (such as "on a ridge," "in a valley") or to the area between markers (such as "into the woods," "through a meadow").

The Forest Community

We recovered information on 7,802 witness markers distributed over a 566-square-mile area on the floor of the Shenandoah Valley in Frederick County, a territory slightly more than half the land area of Rhode Island. Surveyors used terminology representing 51 different types of markers, 46 of which referred to different species and varieties of trees and the remaining five to stones, stakes, piles, posts, and poles. The proportion of marks represented by trees remained between 90% and 94% throughout the entire period from 1730 to 1780. There was no mention in any surveys of prairies or savannas, although there were numerous references to types of open land. The most commonly marked tree species was white oak (*Quercus alba* L.), followed by pine (probably *Pinus virginiana* L.), hickory (*Carya* sp.), northern red oak (*Quercus rubra* L.), black oak (*Quercus velutina* Lam.), and chestnut oak (*Quercus prinus* L.). The frequency of species identified changed noticeably during the century, however. The proportion of marked trees that were white oak declined from more than 45% during the early 1740s to 30% by 1780. A similar trend was evident for hickory, red oak, and black oak. Pine species, on the other hand, increased from less than 10% during the 1740s to almost 30% by 1780, with a distinct increase in frequency after the early 1760s. Similarly, chestnut oaks increased in frequency from less than 5% during the 1750s to more than 10% during the late 1770s. Four of every five trees marked received no further comment from surveyors. Most comments referred to the size or age of trees, with the most frequent reference being to a "sapling." Saplings, somewhat unexpectedly, were marked more frequently during the 1730s and 1740s than during the 1760s and 1770s.

A potential problem in examining data on tree species derives from the use of vernacular names. The colloquial terms used by eighteenth-century surveyors do not necessarily correspond to common names in use today, much less to scientific names. A number of authorities have worked on this problem, and the equivalents in Table 9.1 can be accepted with a high level of confidence (Spurr, 1951; McAtee, 1938; Michaux, 1865; Catesby, 1771; Ewan and Ewan, 1970).

An even more serious problem is surveyor bias. Different surveyors were likely to have had varying abilities and inclinations to distinguish and mark particular species. Species of hickory and northern red and black oak, for example, are difficult to distinguish without fruits, which appear mainly in the summer and fall. A surveyor was encouraged by law to rely on the boundaries of previously surveyed lands and not to exceed a three-to-one ratio for the length and width of the surveyed land, though this was not always followed in practice. The precise boundary was otherwise at the surveyor's discretion, as was the selection of survey points to be marked. A surveyor could thus choose to mark points using a limited range of tree species, either because of a bias in favor of certain long-lived species that showed the blazed marks of the ax well or because of the way the surveyor had been trained.

We conducted one analysis to test the existence of bias. Forty-three surveyors were involved in surveying Frederick County during the eighteenth century, but only six accounted for 667 surveys, or two-thirds of all extant surveys, covering almost 210,000 acres. There was a range among the six from 197 surveys for Robert Rigg to 64 for William Baylis. Baylis was part of two father-and-son practices, John and William Baylis and Robert and Thomas Rutherford. While Rigg and a John Mauzy demonstrated no clear biases, John and William Baylis seldom identified black oak, pine, or, together with Robert Rutherford, chestnut oak. Rutherford's son John, however, did not exhibit such a bias. These conclusions could simply indicate that the Baylis and Rutherford assignments focused on parts of the county with forests of somewhat different composition than areas covered by other surveyors. The similarities in the frequencies with which different surveyors marked other tree species might suggest either that no bias existed for those species or that the bias was shared by all surveyors. Such a common bias would be difficult to detect.

Although surveyor bias could cause certain tree species to be over- or underestimated in terms of the overall forest community, it is unlikely to distort our estimate of the abundance or extent of forested or unforested land. The small size of many of the land grants (one-quarter were for less than 200 acres), the surveyors' descriptions of the vegetation, and the lack

Table 9.1 Common names used by surveyors for marked tree species and their modern scientific and common names

Eighteenth-century common name	Twentieth-century common name	Scientific name
Pine	Pine	Pinus sp.
White pine	White pine	Pinus strobus
Cedar eastern	Red cedar	Juniperius virginianus
Yew	Yew	Taxus canadensis
Spruce pine	Hemlock	Tsuga canadensis
Spanish oak	Southern red oak	Quercus falcata
White oak	White oak	Quercus alba
Chestnut oak	Chestnut oak	Quercus prinus
Black oak	Black oak	Quercus velutina
Box oak	Post oak	Quercus stellata
Red oak	Northern red oak	Quercus rubra
Scrub oak	Scrub oak	Quercus ilicifolia
Water white oak	Swamp white oak	Quercus bicolor
Swamp white oak	Swamp white oak	Quercus bicolor
Oak	Oak	Quercus sp.
Beech	Beech	Fagus grandifolia
Chestnut	Chestnut	Castanea dentata
White hickory	White hickory	Carya tomentosa
Shagbark hickory	Shagbark hickory	Carya ovata
Hickory	Hickory	Carya sp.
Walnut	Walnut	Juglans sp.
Black walnut	Black walnut	Juglans nigra
White walnut	White walnut	Juglans cinerea
Sugar maple	Sugar maple	Acer saccharum
Maple	Maple	Acer sp.
Box elder	Box elder	Acer negundo
Locust	Locust	Robinia pseudoacacia
Sycamore	Sycamore	Platanus occidentalis

(*continued*)

Table 9.1 *(continued)*

Eighteenth-century common name	Twentieth-century common name	Scientific name
Gum	Gum	Nyssa or Liquidambar
Black gum	Black gum	Nyssa sylvatica
Sweet gum	Sweet gum	Liquidambar styraciflua
Poplar	Poplar	Populus sp.
Ash	Ash	Fraxinus sp.
Lynn	Basswood	Tilia sp.
Dogwood	Flowering dogwood	Cornus florida
Elm	Elm	Ulmus sp.
Hackberry	Hackberry	Celtis occidentalis
Mulberry	Red Mulberry	Morus rubra
Sassafras	Sassafras	Sassafras albidum
Thorn	Hawthorn	Crateagus sp.
Cherry	Black cherry	Prunus sp.
Plum	Plum	Prunus americana
Birch	River birch	Betula nigra
Whitewood	Tulip poplar	Liriodendron tulipifera
Iron wood	Iron wood	Carpinus caroliniana

of stones and stakes as markers all argue that surveyors were not marking the few trees in a grassland area but rather they were surveying in mostly forested lands.

Thus, if the survey record is taken at face value, the data suggest that during the first generation of colonial settlement, between 1730 and 1755, the forests on the floor of the northern Shenandoah Valley were 71% oak, 14% hickory, 6% pine, 3% walnut, and 6% other species. These species were distributed widely throughout the valley floor and interspersed with more open grassland areas, particularly along stream courses, thus exhibiting a distribution pattern that was almost the reverse of Samuel Kercheval's conclusion in 1833.

These general observations are similar to what is found in modern-day forests, which remain dominated by oaks in particular. Yet the historic picture is at odds with the anticipated frequencies of specific species. The limited references to hickories and chestnuts and the unusually high number of references to walnuts are particularly striking. Either these observations demonstrate a general surveyor bias or chestnuts and yellow poplars especially have been overestimated in reconstructions of the "potential natural vegetation" of the central Appalachians (Schantz and Zon, 1924; Kuchler, 1964). The decline in frequency of white oak, red oak, and hickory during the second half of the century and the increase in frequency of pine and chestnut oak may indicate a change in the composition of the forest as a consequence of intensified settlement. Alternatively, the data may simply reflect an increase in the occurrence of land grants on mountain slopes and in shale regions where chestnut oak and pine, respectively, would be more common. Similarly, the decline in the frequency of saplings mentioned in the surveys may indicate decreased timber "poaching" on ungranted or unoccupied land, which would be in the early stages of forest succession. The forests during the 1730s and 1740s, on the other hand, may have been in early stages of succession because they had been subjected to extensive burns or to the first wave of livestock grazing. If fire were suppressed or beaver or deer reduced in abundance over the course of the century, then saplings may have become less abundant than before European settlement. Finally, it is possible that surveyors developed a bias against saplings once they learned that saplings were often harder to mark and less likely to last than larger trees.

Patterns and Causes of Open Land

Another approach to the problem of open savannas in the Shenandoah Valley is to assess the possible explanations for the creation or maintenance of open land. If no reasonable mechanism can be proposed to account for extensive unforested areas, then a large-scale continuous prairie is unlikely. If, however, a plausible mechanism can be identified, it would support the idea of extensive grassy areas covering the Shenandoah early in the eighteenth century.

The most common native vegetation in the eastern United States is unquestionably forest. Only limited areas of the eastern United States inland from coastal wetlands do not support forest growth naturally. Low areas with high water tables along the Coastal Plain are covered by marsh grasses; areas with perched water tables on the Piedmont and in the Great Valley also support grassy marshes; and some mountain slopes are covered with boulders and rocks, having no vegetation at all. Fires, storms, floods, wild-animal grazing, and aboriginal activities could remove forest cover temporarily, but the natural process of forest succession would lead to the reestablishment of mature forests within a few decades. Without a recurring process to prevent forest regrowth, open lands would not persist for extended periods of time.

Only two explanations—climate and soils—could account for both the creation and maintenance of open grasslands. Although such grasslands occur naturally in parts of the western Middle West and on the Great Plains, there is no evidence to suggest that climate in the Shenandoah Valley during the late seventeenth and early eighteenth centuries resembled that of the Great Plains.

The soils of the Shenandoah Valley are well known. Extensive areas are underlain by limestones, which generate well-drained soils and little surface runoff. Although these conditions argue against soils as a causal mechanism for creating prairies, scattered springs along limestone-shale contact zones or in association with silts and clays commonly lead to the formation of small marshy areas. And, as noted previously, localized travertine-marl precipitates can cause or enhance floods, temporarily rendering floodplains inhospitable to trees. Again, however, none of these factors is likely to retard forest succession indefinitely.

If we can judge from current environmental patterns, cyclonic storms account for one-third of the annual precipitation of 36–40 inches received in the valley, and floods are common throughout the entire western Chesapeake drainage basin from January through April. Even extreme flooding, however, is infrequent and, if extensive areas were to be deforested by floods, they would be quickly recolonized. The death or fall of a few canopy trees notwithstanding, severe weather conditions would be insufficient to account for more than scattered, small, unforested locales.

Fire, as we have noted, can also be discounted as a force for the sustained appearance of grassy areas. Fires contribute to more open forest understories by killing seedlings and saplings, but lightning-induced fire would have to occur during a prolonged drought period when a sufficient quantity of flammable fuel would be present. There is no evidence of such a drought period during the eighteenth century. Even if aboriginal fire practices for land clearance or hunting took place in the Shenandoah Valley immediately prior to European occupation, such fires would ignite ground cover rather than tree crowns, and it would require an intensive program of fire management to create extensive grassy areas. Because of the absence of Native Americans at contact time, there was no human impediment to normal forest succession.

Grazing animals, on the other hand, could create and maintain open land if they were present in sufficient numbers. Four species of indigenous mammals had the capacity to affect the distribution and regrowth of forests in the Shenandoah Valley: buffalo (*Bison bison*), elk (*Cervis canadensis*), white-tailed deer (*Odocoileus virginianus*), and beaver (*Castor canadensis*).

The abundance of bison in western Virginia during the eighteenth century is difficult to assess for lack of evidence. Despite an occasional placename reference to "buffalo," there are only three documented references to bison in colonial Virginia and none in the Shenandoah Valley. The nearest reference is to the New River Valley in southwestern Virginia in the 1660s. Had bison been common in the region, it is likely that aboriginal populations would have taken up residence there.

Elk and deer, however, were relatively abundant in western Virginia. Deer in particular, with their substantial dependence on woody browse rather than on grass, could have affected the early stages of forest succession or the recolonization of meadows (Thomas and Toweill, 1982). Yet current evidence from the increasing deer herds of the western Chesapeake watershed suggests that they prefer herbaceous vegetation when grazing in meadows, on a scale which does not prevent the reestablishment of forest.

Beaver may have been common in the Shenandoah Valley during the early eighteenth century, although not as abundant apparently as on the northern Piedmont. Through food gathering and dam building, beaver can make a significant impact on the vegetation of riparian habitats (Barnes and Dibble, 1986; Naiman et al., 1988). Beaver activity can replace stream-side woodland with grassy meadows, marshes, shrub thickets, and bogs. Yet the valley's predominantly limestone structure, with its limited surface drainage, restricted opportunities for the formation of the type of streams preferred by beavers. Our initial analysis, therefore, suggests that there were no outstanding natural mechanisms present in the Shenandoah Valley in such force or continuity as to sustain extensive grassy areas indefinitely.

Unresolved Environmental Research Issues

Although we are confident that our evaluation of the documentary and survey records for the northern Shenandoah Valley has provided supportable conclusions about landscape and environment in the western Chesapeake watershed, some research issues remain unresolved, and our survey is to this extent a work in progress. One issue that needs further exploration concerns the modification of our data that would follow if we were to take account of the 200-odd subdivisions of the original grants which occurred between 1760 and 1800. It should also be possible to construct a composite cadastral map of Frederick County. When completed, a composite ground-cover map would provide the most precise and accurate geographic information we are ever likely to have about vegetation in colonial America during the pioneer contact period.

The resolution of two further issues depends upon constructing a land-cover map for the colonial period. Until we know the distribution of open, grassy areas in relation to stream courses, we will have neither an understanding of the frequency of grasslands in upland areas nor the means of differentiating among such terms as barren, bog, glade, marsh, and meadow. Colonists did not employ these terms as precisely as they tend to be understood today.

As Frederick County became more effectively occupied during the eighteenth century, the relationship between settlement site selection, demographic change, and land-use modification presents a further set of problems. We have initiated research on this relationship and have already begun to dismantle another set of "environmental truths." The current state of our knowledge concerning settlement sites and soil types indicates that site choices were not compartmentalized along ethnic lines but that most pioneer settlers preferred limestone soils and consciously avoided shale soils. Although this hypothesis seems to be supported in Frederick County, the issue of soil selection is more complex. By mapping 31 original surveyed tracts along the limestone-shale contact zone that traverses the valley floor from north to south, we have discovered that 11 were primarily on limestone land and only 3 were mainly on shale, but 17 were located along the limestone-shale interface. This pattern strongly indicates that early settlers were conscious of the ecological advantages provided by edge habitats. While the limestone areas were first exploited for their agricultural potential, the shale lands proved to be superior for mill sites because of their greater surface drainage and steeper stream gradients. It is no coincidence, therefore, that by the end of the colonial era a high correlation existed between the distribution of mills and the presence of shale land. These findings constitute the basis of another research project.

BIBLIOGRAPHY

Alvord, C. W., and L. Bidgood. 1912. *The First Explorations of the Trans-Allegheny Region by the Virginians, 1650–1674* (Cleveland: Arthur H. Clark).

Anburey, T. 1789. *Travels through the Interior Parts of America.* 2 vols. (Boston: Houghton Mifflin).

Barnes, W. J., and E. Dibble. 1986. The effects of beaver in riverbank forest succession. *Canadian Journal of Botany* 66: 40–44.

Bartram, J. Bartram Papers, v. 1, folder 54, Pennsylvania Historical Society, Philadelphia, Pa.

Bayard, F. M. 1798. *Travels of a Frenchman in Maryland and Virginia with a Description of Philadelphia and Baltimore in 1791,* 2nd ed., trans. B. C. McCary (Paris: Batilliot Freres).

Berkeley, E., and D. S. Berkeley, eds. 1992. *The Correspondence of John Bartram, 1734–1777* (Gainesville: University Press of Florida).

Briceland, A. N. 1987. *Westward from Virginia: The Exploration of the Virginia-Carolina Frontier, 1650–1710* (Charlottesville: University Press of Virginia).

Burnaby, A. 1798. *Travels through the Middle Settlements in North America in the Years 1759 and 1760,* 3rd ed. (London: T. Payne).

Catesby, M. 1771. *The Natural History of Carolina, Florida, and the Bahamas.* 2 vols. (London: Benjamin White).

Craig, A. J. 1968. *Vegetational History of the Shenandoah Valley, Virginia.* Geological Society of America, Special Paper 123.

Cresswell, N. 1924. *The Journal of Nicholas Cresswell, 1774–1777* (New York: Dial Press).

Darlington, W. M., ed. 1849. *Memorials of John Bartram and Humphry Marshall with Notices of Their Botanical Contemporaries* (Philadelphia: Lindsay & Blakiston).

Day, G. M. 1953. The Indian as an ecological factor in the northeastern forest. *Ecology* 34: 329–46.

Ewan, J., and N. Ewan, eds. 1970. *John Bannister and His Natural History of Virginia, 1678–1692* (Urbana: University of Illinois Press).

Fithian, P. V. 1934. *Journal, 1775–1776,* ed. R. G. Albion and L. Dodson (Princeton: Princeton University Press).

Fontaine, J. 1972. *The Journal of John Fontaine: An Irish Huguenot Son in Spain and Virginia, 1710–1719,* ed. E. P. Alexander (Williamsburg: Colonial Williamsburg Foundation).

Foote, W. H. 1850. *Sketches of Virginia, Historical and Biographical* (Philadelphia: Lippincott).

Gardner, W. M. 1986. *Lost Arrowheads and Broken Pottery: Traces of Indians in the Shenandoah Valley* (Strasburg, Va.: Thunderbird Museum).

———. 1986. The Palaeoindians of the Shenandoah Valley, Virginia. *Archeology* 38: 28–34.

Garren, K. H. 1943. Effects of fire on vegetation of the southeastern United States. *Botanical Review* 9: 617–54.

Gottmann, J. 1969. *Virginia in Our Century* (Charlottesville: University Press of Virginia).

Herman, J. S., and D. A. Hubbard, eds. 1990. *Travertine-Marl: Stream Deposits in Virginia.* Commonwealth of Virginia, Division of Mineral Resources, Publication 101 (Charlottesville).

Hinke, W. J., and C. E. Kemper, eds. 1904. Moravian diaries of travels through Virginia. *Virginia Magazine of History and Biography* 12: 134–53.

Hofstra, W. R. 1991. Land policy and settlement in the northern Shenandoah Valley. In *Appalachian Frontiers: Settlement, Society, and Development in the Preindustrial Era,* ed. R. D. Mitchell (Lexington: University Press of Kentucky), 105–26.

Holland, C. G. 1979. The ramifications of the fire hunt. *Quarterly Bulletin of the Archeological Society of Virginia* 33: 134–40.

Hughes, S. S. 1979. *Surveyors and Statesmen: Land Measuring in Colonial Virginia* (Richmond, Va.: Virginia Surveyors Foundation and Virginia Association of Surveyors).

Jones, H. 1956. *The Present State of Virginia* (Chapel Hill: University of North Carolina Press for the Virginia Historical Society).

Kemper, C. E. 1921. Documents relating to early projected Swiss colonies in the valley of Virginia, 1706–1709. *Virginia Magazine of History and Biography* 29: 1–17.

Kercheval, S. [1833] 1925. *A History of the Valley of Virginia* (Strasburg, Va.: Shenandoah Publishing House).

Kuchler, A. W. 1964. *Manual to Accompany the Map of Potential Natural Vegetation of the Coterminous United States.* American Geographical Society, Special Publication No. 36 (New York).

Ladurie, E. L. 1971. *Times of Feast, Times of Famine: A History of Climate since the Year 1000,* trans. B. Bray (New York: Doubleday).

Lamb, H. H. 1972. *Climate: Past, Present, and Future* (London: Methuen).

La Rochefoucauld-Liancourt, Duc François Alexandre Frédéric de. 1800. *Travels through the United States of North America . . .* 4 vols., 2nd ed., trans. H. Neuman (London: R. Phillips).

Lederer, J. 1672. *The Discoveries of John Lederer,* ed. W. P. Cumming (London: Samuel Herick).

Little, S. 1974. Effects of fire on temperate forests: Northeastern United States. In *Fire and Ecosystems,* ed. T. T. Kozlowski and C. E. Ahlgren (New York: John Wiley and Sons), 255–65.

Lutz, H. J. 1930. Original forest composition in northwestern Pennsylvania as indicated by early land survey notes. *Journal of Forestry* 28: 1098–1103.

Marquis de Chastellux, F. J. 1786. *Travels in North America in the Years 1780, 1781 and 1782.* 2 vols., trans. H. C. Rice Jr. (Paris: Prault).

Marye, W. B. 1955. The Great Maryland Barrens. *Maryland Historical Magazine* 50: 11–23, 120–42, 234–53.

Maxwell, H. 1910. The use and abuse of forests by the Virginia Indians. *William and Mary Quarterly,* 1st ser., 19: 73–103.

McAtee, W. L. 1938. Journal of Benjamin Smith Barton on a visit to Virginia, 1802. *Castanea* 3: 85–177.

Meade, W. 1857. *Old Churches, Ministers, and Families of Virginia.* 2 vols. (Philadelphia: Lippincott).

Michaux, A. 1865. *The North American Sylva: Or a description of the forest trees of the United States, Canada, and Nova Scotia.* 5 vols., trans. J. J. Smith (Philadelphia: Rice, Rutter, and Co.).

Mitchell, R. D. 1977. *Commercialism and Frontier: Perspectives on the Early Shenandoah Valley* (Charlottesville: University Press of Virginia).

Mitchell, R. D., and W. R. Hofstra. 1994. "Scattered for the Benefit of the Best Lands": The Evolving Landscapes of the Early Shenandoah Valley. Ms.

Naiman, R. J., et al. 1988. Alteration of North American streams by beaver. *BioScience* 38: 753–62.

Porter, F. W. 1975. From backcountry to county: The delayed settlement of western Maryland. *Maryland Historical Magazine* 70: 329–49.

Prunty, M. C. 1965. Some geographic views of the role of fire in settlement processes in the South. *Proceedings of the Tall Timbers Fire Ecology Conference* 4: 161–68.

Robison, W. C. 1960. Cultural plant geography of the middle Appalachians. Ph.D. diss., Boston University.

Ruffin, E. 1825. Inquiry into the causes of the formation of prairies, and the peculiar constitution of soils which favors or prevents the destruction of forests. *Farmer's Register,* 330–55.

Russell, E. W. B. 1983. Indian-set fires in the forests of the northeastern United States. *Ecology* 64: 78–88.

Schantz, H. L., and R. Zon. 1924. Natural vegetation. In U.S. Department of Agriculture, *Atlas of American Agriculture,* pt. 1, *The Physical Basis of Agriculture,* section E (Washington, D.C.: Government Printing Office).

Simmons, I. G. 1993. *Environmental History: A Concise Introduction* (Cambridge, Mass.: Blackwell).

Smyth, J. F. D. 1784. *A Tour in the United States of America.* 2 vols. (London: G. Robinson).

Spurr, S. H. 1951. George Washington, surveyor and ecological observer. *Ecology* 32: 544–49.

Stilgoe, J. R. 1982. *Common Landscape of America, 1580–1845* (New Haven: Yale University Press).

Thomas, J. W., and D. E. Toweill. 1982. *Elk of North America: Ecology and Management* (Harrisburg, Pa.: Stackpole Books).

Turner, B. L., et al. 1990. *The Earth as Transformed by Human Action: Global and Regional Changes in the Biosphere over the Past 300 Years* (New York: Cambridge University Press).

Vankat, J. L. 1979. *The Natural Vegetation of North America* (New York: John Wiley and Sons).

Wayland, J. W. 1907. *The German Element of the Shenandoah Valley of Virginia* (Charlottesville: University Press of Virginia).

Weld, I. 1807. *Travels through the States of North America.* 2 vols. (London: John Stockdale).

CHAPTER TEN

Human Influences on Aquatic Resources in the Chesapeake Bay Watershed

VICTOR S. KENNEDY AND KENT MOUNTFORD

Technological improvements made the Chesapeake Bay a major fishing center in the early 1880s. Subsequent exploitation, with increasing degradation of the environment, changed the biotic structure of the Bay. The history of two species—the anadromous shad that depended on the rivers and the oysters indigenous to the estuary—is examined, from the time they began to be exploited to the present, and proposals for protecting the ecosystem are considered.

The Chesapeake Bay, ten leagues broad, and four, five, six, and even seven fathoms deep, flows gently between its shores: it abounds in fish when the season of the year is favorable.

—ANDREW WHITE, S.J., 1634

European colonists sailing on Chesapeake Bay in the early seventeenth century sometimes ran aground on reeflike oyster beds, many of which broke the Bay's surface at low tide. If the impact tossed a colonist overboard during the summer, he or she may have been less likely than now to encounter the stinging sea nettle jellyfish, at least not in the numbers that plagued bathers in the late twentieth century and today. Subsequent oyster harvesting depleted the stock of oysters and broke up the reefs, eliminating them as navigational hazards. As we will see, the decline in oyster abundances and the possible increased abundances of sea nettles may be interrelated.

Changes in animal abundances are one example of the effects of increased human immigration and population growth on the watershed of Chesapeake Bay. Humans over the past few centuries have influenced the movement of materials into the Bay, changed methods of exploiting aquatic animal resources, and affected the distributions and abundances of aquatic plants and animals.

Movement of Material into the Bay

Grace Brush writes in 1986 that three categories of materials—sediments, nutrients, and toxins—have affected the Bay ecosystem over time. Initially, American Indians in the Bay's watershed altered forest communities by using fire to drive game, by clearing land for agriculture, and by removing understory when gathering firewood. This increased the runoff of sediment and nutrients to the Bay and its tributaries. The influx of European settlers and subsequent population growth in the Bay's watershed greatly amplified the movement of these substances into the aquatic ecosystem. Industrial development added toxic runoff to the mix.

Sediment

Land erosion and the subsequent deposition of sediment is common in estuaries, even in the absence of human activity. Additional sediment can be carried by currents into Chesapeake Bay from adjacent coastal waters. Thus, natural processes gradually alter the Bay over several millennia, producing a slowly changing mix of open water, marsh, and dry land.

Henry Miller (Chap. 6) and Jay Custer in 1986 reviewed anthropological data and each concluded that Indians had little effect on the Bay's ecology, except perhaps on a small scale near their encampments. Thus, although Indians cleared vegetation, their influence on land erosion may have been minimal. With the seventeenth-century increase in European settlement, land clearing accelerated, as demonstrated by Brush in 1986 (see also Chaps. 3, 8, and 11, in this volume). Sherri Cooper and Grace Brush tell us that soil erosion and deposition into tributaries in the central portion of the Bay doubled after about A.D. 1760, when intensive agriculture began. This increased deposition filled in navigation channels as well as colonial harbors such as Port Tobacco, Piscataway, and Joppa Town, which are now located inland.

Eroded sediment accumulates in reservoirs and behind dams. Grant Gross and his colleagues have shown that tropical storms like Agnes in 1972 and Eloise in 1975 scour thousands of metric tons of sediment from behind dams and reservoirs and deliver these extremely large sediment loads to the Bay. These sediments can completely bury bottom-dwelling organisms, including oyster beds that have been scraped almost level with the Bay bottom by oyster harvesters. The sediments carry large quantities of nutrients, which can spur unwanted aquatic plant growth and lower concentrations of dissolved oxygen to lethal levels.

Suspended sediments hinder penetration of light, which is essential for photosynthesis, especially by submersed grasses. Loss of light leads to decreased productivity of these aquatic plants and thus of the animals that depend on them. Data compiled by Richard Batiuk and his colleagues indicate that increased deposition of sediment and increased nutrients led to a decline in Bay grasses from an estimated 247,000 acres at the time of European settlement to about 38,000 acres in the 1980s.

Nutrients

Movement of nutrients to the Bay was affected not only by more rapid runoff from cleared land but also by use of domestic animal wastes in farming. As Lorena Walsh reports (Chap. 11), colonial animal husbandry meant increases in manure-producing draft animals, cattle, pigs, and poultry, though this manure probably had little effect on the Bay because farms were relatively widely dispersed over the large acreage of the watershed. However, as agriculture intensified, topsoil was eroded and soil fertility declined. Court Stevenson and his colleagues reviewed historical records and reported in 1999 that this led to use of natural fertilizers such as guano, imported from South America beginning around 1820, then to use of manufactured fertilizers. As human populations increased and became concentrated in urban and suburban communities, lawn and garden fertilizers as well as sewage effluents added thousands of metric tons of nitrogen and phosphorus, which caused eutrophication (overfertilization) of the system. In recent decades, manure from livestock feedlots and confined poultry-raising systems has contributed to the Bay's nutrient loads. All these nutrients have created substantial water-quality and habitat problems.

Walter Boynton and his colleagues used data from studies of mature forested lands to estimate nutrient inputs to the Bay watershed over time. They judged that inputs of nitrogen have increased about 4 to 8 times and of phosphorous, about 10 to 30 times, since the precolonial period. Although modest increases of nutrients might at first have increased plant and animal productivity without detrimental effects on water clarity or oxygen concentrations, eventually the larger quantities of nutrients would stimulate blooms of algae (the microscopic plants, such as floating phytoplankton, upon which Bay animals depend for food and oxygen). Such blooms make the water cloudier, hindering light transmission. Roger Newell calculated that the large numbers of oysters present in the Bay a century ago, before harvesting reduced their numbers, could filter much of the Bay's suspended material from the water over oyster beds in a few days, thereby helping to keep water clarity high. With limited inputs of nutrients and with large populations of oysters, the precolonial Chesapeake and its tributaries were probably much clearer than now. For example, Thomas Hariot extolled Virginia's crystal rivers when he viewed them from the perspective of English streams he knew.

Oxygen concentrations in water near the Bay bottom may have been higher in the past. Microorganisms that bloom in response to increased nutrients are eventually eaten or die, and their remains sink slowly into the deeper stratified layer of the Bay. Donald Pritchard and Jerry Schubel (Chap. 4) describe how lighter, fresher water lying on top of heavier, saltier water results in stratification of the water column. This stratification causes estuaries to have low concentrations of dissolved oxygen in bottom waters in spring and summer in the presence of increased nutrients. This is because the deeper waters, where large quantities of dead microorganisms are decaying, are cut off from the atmosphere, and gas exchange that would deliver new oxygen to bottom water is hindered. In recent decades, monitoring in summer by the USEPA Chesapeake Bay Program shows that decaying microorganisms cause serious or total loss of dissolved oxygen in roughly half of mainstem and tributary waters deeper than about 22 feet.

Cooper and Brush postulate that oxygen in deeper Bay water was less regularly depleted in the precolonial period than in the colonial period. There is evidence of even greater depletion after 1940. Water with low oxygen concentrations is avoided by fish and blue crabs and is lethal to bottom-dwelling organisms, which serve as food for many fish species. Thus, if oxygen concentrations were higher than now, there may have been substantially more deeper-water habitat and more available food on the Bay bottom for numerous estuarine species in the precolonial and colonial periods.

Toxins

European immigrants introduced toxic materials to the Bay as industrial activities such as mining (see Chap. 1), petroleum refining, and manufacturing proliferated. In the twentieth century, chemically synthesized fertilizers as well as manufactured pesticides and herbicides began to be used on farms and later on lawns and gardens. Disposal of commonplace as well as exotic chemicals into the Bay increased after World War II until the deleterious effects of such actions were recognized and laws passed to regulate such disposal. Unfortunately, physical circulation patterns of the estuary, as described by Pritchard and Schubel in this volume and by William Odum, ensure that much toxic material will remain within the estuary. Toxic compounds often attach to the surface of suspended particles, which eventually settle to the bottom of the estuary. Once in the sediment, this toxic material may be chemically and biologically recycled and may continue to leach back into the water for decades. But removal activities such as dredging can also release the toxins back into the water, so efforts to remove toxic sediments may be more harmful than leaving them to be entombed beneath settling sediment.

We conclude that increased sedimentation within the Bay and its tributaries was probably the first widespread effect of human entry into the watershed, with increased nutrient runoff occurring next, followed by production of toxins. Significantly, these effects do not replace one another but coincide and are cumulative. Reversal of such cumulative effects has begun and will require sustained, continuing effort to be successful.

Changes in Fishing Technology

Besides changing the estuary by increasing the amount and kinds of materials running off the land, humans affected the Bay by harvesting—and sometimes overharvesting—its biological resources. The appearance of new or improved capture technologies periodically intensifies exploitation of Bay organisms.

James Wharton examined colonial writings to develop a history of fishing in Virginia. He reports that William Strachey in the early 1600s described the Indians' use of canoes; nets made from tree bark, deer sinews, or handspun grass; fishing poles with a line and a bone hook; arrows attached to a line and shot into the water to impale and retrieve fish;

and ingenious multicompartmented fish traps made of reeds. Robert Beverly wrote about Indian children using pointed sticks as harpoons, the use of reed structures as shoreline weirs or cross-channel dams, the employment of a woven reed pot to snare fish in rapids, the capture of sturgeons by means of a noose placed over the fish's tail, and night fishing from a log canoe using spears and a grass dip net and with the help of a fire on a hearth in the middle of the canoe (see Fig. 10.1).

Wharton discovered that, by contrast with Indian harvesting gear and techniques, the fishing gear of the early settlers was often inadequate. The settlers also were unskilled and some were lazy (one group let their nets rot), so they periodically faced starvation. In the supply of material imported from England in 1620, only one trammel net (a type of gill net; see below) was included among the thousands of items purchased for the new colony (records of the Virginia Company of London, in Hatch 1957). There was, however, a substantial quantity of preserved fish of several kinds, evidence that the colonists were not yet able to exploit sufficiently the Bay's resources. Many species of fish migrate from the ocean into the Bay in spring and summer to spawn and feed, and fish like shad and river herrings could be harvested easily from shallow streams while they were spawning. However, in the fall and winter many species move out of the Bay or into deeper water in the Bay. Colonists without the appropriate equipment would have faced winter periods when fish were unavailable.

Miller (Chap. 6) examined records of the late seventeenth and early eighteenth centuries and found that the main fishing implements in homes were hooks and lines, with only about 5% of the homes equipped with fish gigs or nets. Eventually, as Walsh shows (Chap. 11), over one-third of the smaller estates in York County, Virginia, and St. Mary's County, Maryland, had fishing or oystering equipment after the Revolutionary War.

Near-shore exploitation of oysters was practiced by Indians and colonists where possible. Eventually, new or improved gear allowed harvesting of deeper-dwelling populations. For example, Miller reported that oyster shell in colonial trash dumps in St. Mary's City changed in shape over time. Shells from seventeenth-century sites were those of rounded, inshore "cove" oysters, which were probably harvested from shallow water. An early eighteenth-century dump contained elongated shells of "channel oysters," which must have been raked or tonged from deeper sites. Coincidentally, oyster tongs appeared in household inventories in Maryland in the early 1700s and a visiting Swiss writer reported their use in 1701 (Wharton, 1957: 37–38). Eventually, oyster dredges be-

Fig. 10.1. Early depiction of Indian fishing methods by Theodore De Bry (Hariot, 1588). Note fishing weirs, spears, a rake (to collect oysters?), dip net, fire in canoe, and various vertebrates and invertebrates. Courtesy of the John Work Garrett Library of the Johns Hopkins University.

came available in the 1800s to harvest the extensive beds in the Bay and deeper tributaries (Figs. 10.2 and 10.3).

Increased exploitation began to deplete the oysters. Even limited harvest efforts of a small number of settlers could have measurable local effects on the resource. Bretton Kent's work on colonial oyster middens in the St. Mary's City region of Maryland found that oysters averaged 3 inches in length around 1645, when the human population numbered about 100. This average decreased to about 1.2 inches around 1695, when the local population had doubled, then increased again to about 3 inches by 1710 after the human population in the region declined to the 1645 level.

After 1800, use of the Bay's biological resources increased. In 1884, George Goode reported that until the 1870s, the fisheries of the Chesapeake Bay were of limited commercial importance. Fishing occurred only during the spring and fall using gill-nets (long curtains of netting that drift with currents or are attached to stakes) and haul-seines (Fig. 10.4). Harvests improved in Chesapeake Bay with the introduction of the pound net, as described by Edward Earll. This net (Fig. 10.5) is somewhat like the Indians' compartmented fish traps except that it is made of netting suspended on poles. It was first used in the Bay in 1858, but the first one that was deployed was badly constructed and failed. A second effort to use pound nets in the early 1870s was also a failure, but the design used in a third attempt in 1875 was so successful in Mobjack Bay, Virginia, that it was torn down by rival fishermen. However, some fishermen who had observed the pattern of the pound stakes used that knowledge to build 12 new nets in Mobjack Bay in 1876, and the practice spread rapidly to help make Chesapeake Bay a major center for U.S. fisheries in the early 1880s.

Not only did the pound net increase fishery harvests, but it captured species that were almost unknown to regional fishermen. One such species was Spanish mackerel, a summer visitor in the Bay which was generally unfamiliar to local fishermen before 1875. The pound fishery for Spanish mackerel in the Bay rapidly became the largest in the United States in the late 1800s.

For many Bay fisheries, fishing gear did not undergo much further change, as shown by the continuing use of only a few gear types in the striped bass fishery over many decades, described by Edward Raney. Even though a limited number of fish-capture devices were developed, there were efforts to make gear more efficient. For example, Charles Stevenson reported in 1899 that drift gill nets began to replace the less

Fig. 10.2. "Opening of the Oyster Season." From *Harper's Weekly* 26, no. 1343 (10 September 1882): 284, depicting aspects of the New York oyster industry. Note that the upper illustrations are actually examples of the use of scissorlike tongs, not of dredging. Courtesy of the Calvert Marine Museum, Solomons, Md.

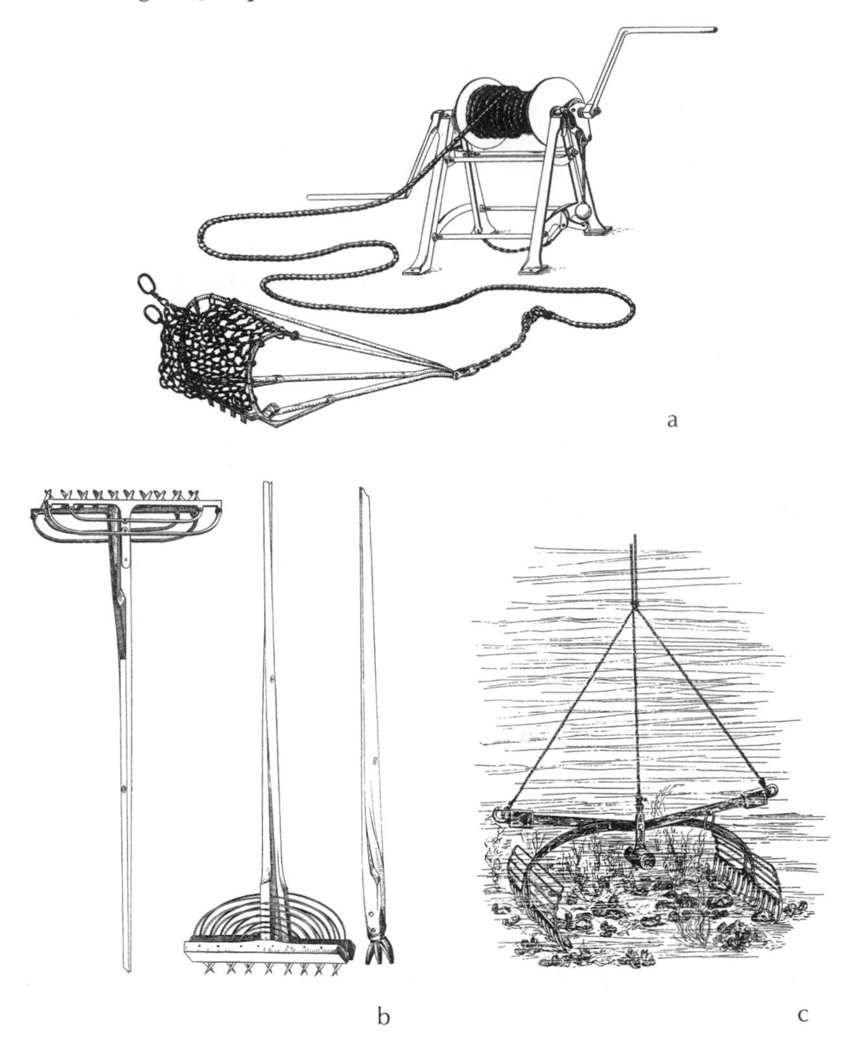

Fig. 10.3. Chesapeake Bay oyster dredge (*a*) and tongs (*b*); from Goode, 1887, pl. 237, 238. Deepwater (patent) tongs (*c*), from Smith, 1891, pl. XLIV.

effective (and more expensive) stake gill nets after about 1870. We have already noted the effect of relatively efficient pound nets on Bay fisheries. In the early 1950s, the invention of the hydraulic dredge allowed the Bay's submerged soft-clam beds to be exploited. Perhaps the most technologically advanced is the modern Atlantic menhaden fishery, with airplanes to spot menhaden schools, fast boats to deploy a purse seine around a tar-

Fig. 10.4. A haul seine and a minimal catch (date unknown). Note the number of laborers required to haul the seine on shore. Courtesy of the Calvert Marine Museum, Solomons, Md.

geted school, and power winches to haul the seine and its catch onto the mother ship. These tools have enabled the fishery to become among the largest in the Bay—even though their use is banned in Maryland, where pound nets are the approved gear.

The banning of purse seines in Maryland's menhaden fishery is one example of legislated inefficiencies in fishing gear to prevent rapid depletion of fishery stocks. Another example is that of Maryland's oyster industry, which is governed by rules that mandate use of sail on traditional dredging boats for most of the weekly fishing period; that designate oyster bars as tonging, patent tonging, or dredging bars; and that restrict highly efficient scuba-diving by oyster harvesters.

Changes in Species' Distributions and Abundances

Human activities in the Bay watershed have had consequences for the distributions and abundances of organisms. As Timothy Silver shows

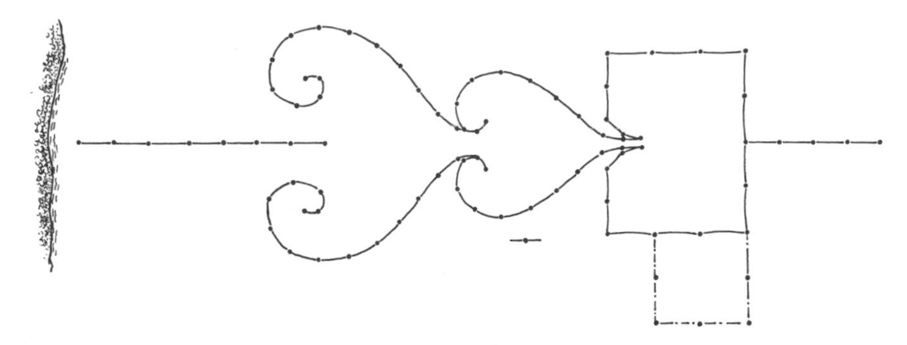

Fig. 10.5. (*top*) Diagram of a pound net used in the Spanish mackerel fishery in Chesapeake Bay (redrawn from Goode, 1887, pl. 148). (*bottom*) Brailing (retrieving) fish with dip nets from within the confines of a small pound net. The net hangs from the poles in the background, and interior portions are lifted to confine the trapped fish in a shallow pool to allow for brailing. Courtesy of the Calvert Marine Museum, Solomons, Md.

(Chap. 8), the removal of vegetation affects hydrology. Runoff from storms on denuded land is hastened, which leads to increased flooding. Rapid runoff means that rain has less time to percolate (sink) into the soil, so the water table is lowered. This decreases water flow in streams and from springs. Robert Biggs has estimated that when the Susquehanna River was draining a fully forested basin, which would have slowed runoff and enhanced percolation, it had peak river flows 25–30% lower than they are today; base flows in dry weather would have been higher than they are now because of a higher water table. Thus, seasonal variations in river flow were probably smaller than today, when there is more developed and paved land in the watershed to enhance runoff and hinder percolation.

Changes in river flow affect the saltiness (salinity) of the estuary, so seasonal variations in salinity may also have been less extreme during the precolonial period. Biggs concluded that year-round salinity values in the upper Bay and upper reaches of tributaries before deforestation were higher than today. This implies that salt-tolerant plants and animals extended further up the estuary than they do now. For example, Indian shell middens on the Potomac River occur upstream beyond where oysters can survive today, and ancient and extensive oyster shell deposits occur beneath the mud in the upper Bay north of the present range of oysters in the Bay. Miller in 1986 reported on studies of numerous seventeenth-century archaeological sites in which bones of fish that require high-salinity water (black drum, red drum, sheepshead) were found in areas that are now relatively low in salinity.

Under normal conditions and even when unaffected by human actions, populations of organisms vary in abundance from year to year. Food availability, climate, water-circulation patterns, predation, and so on, influence this annual variation. Commercial and recreational fishing add to the complexity and the variability. When a fishery is first established, the tendency is to overfish the resource, sometimes to commercial extinction.

Technological innovations like those described above have increased the exploitation of Bay fisheries in the past two centuries. Before that time, fishing was limited and generally artisanal in scope. Custer reports that intensive use of Chesapeake Bay living resources by humans does not seem to have begun until about 5,000 years ago. Exploitation of coastal resources increased thereafter, coincident with the slowing of the rate of sea level rise. Indians' movement from interior settlement sites to coastal locations led to use of oysters, the periwinkle snail, soft clams, and ribbed mussels as food, the remains of which, along with those of turtles and fish, appear in shell middens of the time. After about 1000 B.C., the proportion of oysters in the middens increased and the diversity of other shellfish species decreased, perhaps indicating a specialization in food use. The presence of roasting basins at sites dated after A.D. 1000 may indicate large-scale drying of shellfish, especially oysters. Some sites in the watershed seem to have focused on anadromous fish. Harrison Wright noted the great abundance of net-sinkers near Wyoming, Pennsylvania, along the north branch of the Susquehanna River and the use of American shad vertebrae as an impressing tool to mark pottery. He concluded that Indians fished American shad in great quantities at this location.

As noted earlier, there is little or no evidence that Indian activity had other than limited local effects on the aquatic resources of Chesapeake Bay. Habitations were often seasonal and decentralized, especially in winter. When foraging for oysters depleted local near-shore supplies, for example, a family group could easily move to a less-depleted site. Certainly, Europeans who arrived in the Chesapeake Bay region were impressed by the bounty of natural resources, although one needs to be cautious because the accounts of early colonial writers (well described by Wharton) are not those of trained biologists. Some reports also may have been deliberately exaggerated to attract immigrants; Wharton notes the lament of a settler in 1623 who "had not so much as the scent" of "all the deer, fish, and fowl [that] is so talked of in England . . ." (1957: 27). Nevertheless, the inventories of Thomas Hariot in 1588, Robert Beverly in 1705, and William Byrd in 1737 (see Beatty and Mulloy, 1940) provide a broad outline of the flora and fauna encountered by settlers (although some descriptions are unrecognizable to the modern reader).

The arrival and population growth of European settlers stimulated major changes in the Bay's watershed. Wharton writes that in 1612 Governor Dale of Virginia made the first attempt to keep statistics on fish catches. Although this effort was short-lived and no quantitative data exist on abundances of aquatic organisms in the colonial period, anecdotal evidence suggests that some species became significantly depleted. For example, many early accounts of fish in the Bay region (summarized by Wharton) refer to the presence of great quantities of sturgeon. (There are two sturgeon species, shortnose and Atlantic, but the colonists made no distinction between them in their writings.) These were eaten by the settlers and were exported to Europe as a source of meat, caviar, and isinglass. Numbers of the two species of sturgeons in the Bay and elsewhere are now so low that the shortnose sturgeon is listed as endangered while the Atlantic sturgeon is a candidate for listing under the Endangered Species Act.

During the middle and late 1800s, exploitation of the Bay's aquatic resources accelerated as human populations grew, distant markets expanded, and technology improved. Some catch records began to be kept at that time and reveal the extent of harvesting pressure. Spencer Baird reported that 1 million barrels of salt fish had been produced from over 25 million American shad and 750 million river herring captured in the Potomac River in 1833; catches had declined to less than 1% of these numbers by 1878. Similarly, striped bass catches in the Potomac dwindled from 316,000 pounds in 1866 to 50,000 pounds in 1878. Baird blamed the harvest declines on dams and other river obstructions, as well as aquatic pollutants (sawdust, coal tar from gas production, toxic chemicals) and the rapid growth in the number and size of fishing nets.

Eugene Cronin described increased exploitation of other fish and shellfish since the late nineteenth century (Fig. 10.6). Oyster harvests peaked in the late 1800s and fell to very low levels thereafter. Catches of striped bass were low and constant until after World War II, when they tripled in quantity before declining rapidly in the late 1970s until a fishing moratorium was imposed in the 1980s; this moratorium allowed the species to recover so that it again supports a fishery. Blue crab harvests increased throughout the first half of the twentieth century before fluctuating thereafter. There is great concern that the fishery may be in trouble because of harvest levels that since 1994 have been lower than the high yields of the previous decade. A similar concern has been expressed about the menhaden stocks in the Bay. Use of the escalator dredge to harvest soft clams produced an initial large increase in harvest followed by a general decline, as is typical for relatively unmanaged new fisheries; harvests of soft clams in the 1990s were about 10% of levels in the 1960s when the fishery was new.

By the late nineteenth century, Bay fisheries for shad, river herrings, and oysters were among the most important in the United States. They have subsequently declined precipitously. Indeed, in 1980, Maryland placed a moratorium on commercial and recreational fishing for American shad and there were calls in the early 1990s for a moratorium on oystering throughout the Bay. Let's take a closer look at commercial fisheries for these resources.

Shad and River Herrings

Four commercial species of anadromous fish (American shad, the less common hickory shad, and the river herrings, alewife and blueback) served in the past as important food resources for Indians, European settlers, and their descendants. Their fishery was described by George Goode (1884) and Hugh Smith (1899), and especially by Charles Steven-

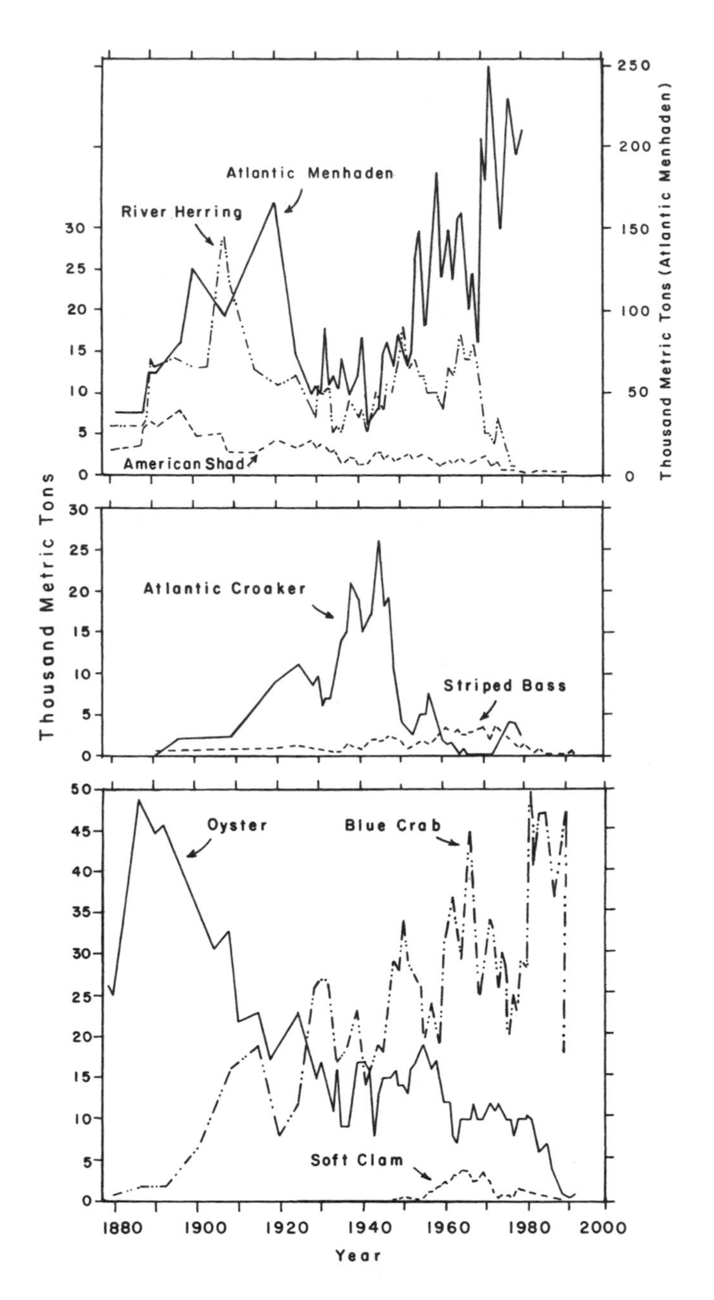

Fig. 10.6. Commercial landings of fish and shellfish in Chesapeake Bay (predominantly) and the Atlantic waters of Maryland and Virginia. (*top*) Filter-feeding fish: American shad, river herring (alewife and blueback combined), Atlantic menhaden. (*middle*) Carnivorous fish: Atlantic croaker and striped bass. (*bottom*) Shellfish: Oyster, soft clam, and blue crabs. Modified and updated after Cronin (1986).

son (1899), whose account we depend on for much of what follows. These fish make annual spring spawning migrations from the ocean into Bay tributaries. The alewife usually arrives in the Bay in March, moving far upriver to spawn in smaller and smaller branches of streams (hence it is also known as branch herring). The two species of shad arrive around the same time as the alewife or shortly after. A few weeks later the blueback herring arrives, moving to spawn in the streams just above the tidewater, although usually not as far upstream as the branch herring. The blueback used to appear in such enormous numbers that it overwhelmed the market, and thus it was also called glut herring.

In the late eighteenth century and early nineteenth century, anadromous fish were caught by seine along their migration routes, yielding thousands of fish annually to riverside settlements. American shad were able to swim for hundreds of miles up rivers such as the Susquehanna, reaching Binghamton, New York, some 300 miles from the river mouth. Such migrations were subsequently hampered by dams built to improve navigation and to provide hydropower for sawmills and gristmills and for electrical generation. By 1899, American shad could go no farther upriver in the Susquehanna than Clark's Ferry, Pennsylvania, about 80 miles from the river mouth.

Depletion of anadromous fish populations was due not only to barriers across rivers. There were additional problems posed by dams and the resultant concentration of people in their vicinity. Below dams, sawdust from sawmills and chemicals from tanneries, paper mills, and factories polluted the water or covered spawning grounds. Land clearing led to flooding and erosion of soil onto spawning grounds. It also meant that streamflow decreased significantly after freshets ended. Thus, streams were often too shallow after spring runoff to be navigated by anadromous fish, except perhaps by branch herrings. Agriculture-related pollutants ran into streams, and inappropriate or damaging methods of capture of young fish resulted in high mortalities and loss of these fish as future spawners.

Another significant influence on the fate of American shad was the change from local river-course fisheries centered on riverside settlements to fisheries concentrated farther downstream and in the estuary itself. With newly profitable distant markets to exploit, shipping facilities became concentrated downstream. The downstream fisheries intercepted fish before they reached their spawning grounds and the result was overfishing. Yields of American shad fell in every river until 1880, when artificial culture became extensive and temporarily reversed the downward trend. By 1896 the East Coast yield of American shad had risen to 2.5 times the 1880 harvest, although that increase was due to not only artificial culture but also greater use of more efficient gear.

In 1896, the shad fishery in the Chesapeake Bay watershed yielded about 40–50% of the total shad landings in the United States, making the Bay fishery the most extensive and valuable on the East Coast. More than 8,000 people were involved in Maryland and Virginia, with an additional 1,000 employed as shore-based laborers and transporters. Of the gear used, there were about 4,000 boats, more than 13,000 stake nets, about 4,000 drift nets and 2,100 pound nets, and 132 haul seines. Total length of the netting represented by these numbers was about 850 miles, or more than four times the length of the Bay. Gear use varied by location within the Bay and its tributaries. The most extensive pound-net fishery on the Atlantic coast in 1896 extended in the Bay's main stem from the James River northward to the Potomac River, whereas the most valuable drift-net fishery on the Atlantic south of Delaware Bay extended 25 miles from the mouth of the Susquehanna south in the Bay's main stem to Pooles Island in the upper Bay (Fig. 10.7). Within the Bay's tributaries, stake and drift nets and haul seines were employed extensively, with pound nets used in a few locations.

These kinds of fishing gear were also used to capture river herrings. Smith reported that about 900 people and about 500 boats were employed in the river herring fishery in Maryland and Virginia in 1896. These numbers do not include people and boats involved in the shad fishery, in which river herrings were a significant by-catch. Within the U.S. fisheries that focused solely on river herring, Maryland harvested 13% and Virginia harvested 9% of the total catch in 1896. The nation's largest pound-net harvest (25 million fish) was made in Virginia, whereas the largest haul-seine catch (21 million fish) occurred in Maryland. The Potomac River was the leading source of river herring in the nation, followed by the Susquehanna River.

Around the beginning of the twentieth century, harvests of shad and river herrings declined, in spite of strong efforts at culturing and stocking. Culture efforts could be of only limited use because they did nothing to change the major reasons for the declines in the species, namely, the presence of river barriers, deteriorated spawning habitat, and overfishing. Significant efforts to restore the fishery in the 1990s included use of hatcheries to provide fry for stocking in restored habitat, the transplanting of adults ready to spawn, and removal of some dams and construction of fish-passage facilities around other dams.

Fig. 10.7. Laying out a drift gill net for shad fishing at night on the Susquehanna River (Goode, 1887, pl. 163).

The Eastern Oyster

Perhaps the best-documented fishery in Chesapeake Bay is that of the eastern oyster. Ernest Ingersoll and Charles Stevenson provided detailed histories of the early fishery and Victor Kennedy and Linda Breisch updated that history to the late 1980s. The fishery provides an important example of the heavy exploitation of the Bay's natural resources over time, notwithstanding scientists' continued warnings. Initially, the oyster fishery declined primarily because it was managed to satisfy political demands rather than as a renewable resource. However, over the past 40 years the fishery's great decline has been from oyster disease.

Oyster abundances in the colonial period impressed explorers and writers. Wharton (1957: 37) quotes the Swiss writer Louis Michel, who wrote in 1701 that "the abundance of oysters is incredible. There are whole banks of them so that the ships must avoid them. A sloop, which was to land us at Kingscreek, struck an oyster bed, where we had to wait about two hours for the tide. They surpass those in England by far in size, indeed they are four times as large. I often cut them in two, before I could put them in my mouth." Charles Stevenson noted that one of the complaints filed in 1680 by a group of settlers in Kent Island, Maryland, was that their provisions had run out so they were forced to eat oysters collected from the Bay shore. Unfortunately, anecdotes such as these are all that are available concerning exploitation of oysters by early colonists.

Oyster beds became overfished in the more populous New England region by the early 1800s, so Connecticut schooners that used heavy dredges to harvest oysters began to exploit Virginia beds around 1808. Passage of legislation in Virginia in 1811 banning such dredging compelled the schooners to move up-Bay to Maryland waters. Subsequently, Maryland's first oyster-related laws in 1820 also prohibited such dredging and established residency requirements for firms transporting oysters out-of-state. As the shallower beds became depleted in the Bay, dredging was again permitted and a type of deepwater (patent) tongs was invented (described by Smith in 1891).

Oysters were not used just for food. The shells were burned to provide agricultural fertilizer (lime) and were used in building roads; they also were ground up to provide chicken grit for farmers. Thus, oysters were increasingly in demand as population growth continued in the Bay watershed. Quantitative data on harvests began to be collected in Maryland around 1839, when about 700,000 bushels of oysters were estimated to have been landed. Subsequently, large reefs were discovered in Tangier Sound, allowing the fishery to expand rapidly. The building and improvement of national turnpikes and westward expansion of the Baltimore and Ohio Railroad helped increase demand, which was met by increased exploitation. A. J. Nichol wrote that the need to prepare the catch for regional consumption as well as for export from Maryland led to the establishment in Baltimore alone of about 80 processing firms (raw and steam packers and steam canners) by 1868. Dexter Haven and his colleagues described similarly intense exploitation of Virginia's oyster resource at the turn of the century.

Interest in estimating oyster harvests was high even more than a century ago. William Brooks reported that about 5 million bushels of oysters were gathered in Maryland and 2 million in Virginia in 1865. He calculated that annual production in the Bay as a whole had reached 17 million bushels by 1875, continuing to rise over the next few years before starting its long decline. Brooks also calculated that the average annual harvest from the Bay between 1834 and 1890 was 7 million bushels, or 392 million bushels for that 56-year period.

Brooks wrote that warnings about the effects of overfishing were constantly ignored, even while people bemoaned downturns in harvests. He spent 1882 and 1883 resurveying 59 oyster beds in Tangier Sound, where a survey had been performed three years earlier by the U.S. Coast Survey. He found that 75% of the beds had fewer oysters per square yard than in 1878–79 and that the ratio of living oysters to empty shells had declined substantially. Brooks concluded that the Tangier Sound beds were deteriorating and that management practices should encourage conservation of the resource based on scientific principles.

Kennedy and Breisch found that Brooks was generally ignored by the managers and politicians of his day. Oyster harvests in Maryland peaked with a catch of about 15 million bushels in the 1884–85 fishing season; there was a later, smaller peak in Virginia. Thereafter, there was a general decline to a fairly constant harvest level of two to four million bushels from about the late 1920s to the 1950s. Since then, the presence in the Bay of two virulent diseases of oysters has depleted harvests in some years to less than 100,000 bushels annually in Maryland and Virginia together.

The potential consequences of this extensive harvest over time are large. In sedimentary ecosystems such as estuaries, shelled organisms like oysters are "ecological engineers" that provide substrate for use by organisms that cannot settle on or survive in sediment and that require hard material to live on. Thus, oyster reefs serve as attachment and living space for newly settling oysters, as well as for barnacles, mussels, sea squirts, and assorted smaller invertebrates. The numerous crevices created among the oyster shells provide shelter for other invertebrates and fish. Loss of oyster reefs means loss of biological abundance and diversity associated with such structures.

Depletion of oyster populations also may have affected food webs in the Bay. Oysters feed by removing suspended material such as phytoplankton from water passing over the reef. Newell proposed that the decline in oyster abundances from the levels of the 1800s to the minimal abundances available today has resulted in a corresponding decline in the removal of phytoplankton from the water by oysters. This caused a decrease in the amount of nutrient material packaged in oyster feces and deposited on the estuarine bottom and a probable change in the amount and the size spectrum of phytoplankton, especially in the summer. This may have led to a shift from Bay food webs dominated by bottom-dwelling organisms in clear-water conditions to food webs dominated by microscopic plants and bacteria that live suspended in the now-turbid water. For example, Cooper and Brush found evidence in cores of sediments taken from the Bay bottom that bottom-dwelling diatoms (microscopic plants) of the precolonial period were replaced by free-floating diatoms, components of the phytoplankton. Such a shift may have resulted in an increase in the abundance of zooplankton (microscopic animals) that feed on the phytoplankton, and a subsequent increase in the larger predators that feed on the zooplankton. These larger predators include sea nettles, leading Newell to propose that their numbers are higher now than they might have been when oysters were abundant. Thus, our colonist who was unlucky enough to fall into the Bay in the summer would at least have been likely to encounter fewer sea nettles.

It may be significant that the accounts of the natural history of Bay waters by Byrd, Beverly, and Hariot which we have examined make no mention of floating jellyfish, even though Swiss writer Michel reported being able to see many kinds of fish through the clear water of the Bay. We found one seventeenth-century report of a stinging jellyfish in the compendium by P. Force, who relates an observation in 1688 of a sea nettle off the Bay mouth, but this may be the Portuguese man-of-war *Physalia physalis,* a tropical species often found in the Gulf Stream, not the Bay sea nettle *Chrysaora quinquecirrha.*

We found only one reference to sea nettles in the literature of the eighteenth century, from the *Maryland Gazette* of 1750: "ANNAPOLIS July 25. Some few days since, James Mitchell, a laboring man of this place, going out of a flat into Rappahannock River in Virginia, got intangled in a great number of Sea Nettles and was drowned." Anaphylactic shock as a response to marine jellyfish stings is uncommon but not unheard of today (cited in *Maryland Historical Magazine* 17, no. 4 [1922]: 374–75).

During the Civil War, Confederate prisoner John King wrote from Point Lookout, Maryland, in 1864: "(In) the middle of July . . . 'Sea Nettles' appeared . . . (growing) from the size of a penny to that of a breakfast plate. . . . It was amusing to watch the bathers swim under one of these, the sting resembling that of a nettle. . . . After a windy night the beach at the edge of the water was slimy where the queer things had been left by the receding tide."

The two Barrie brothers wrote of summer sailing trips around the Bay at the turn of the twentieth century and reported on numerous swimming events, including an hour's swim in July 1901 at Solomons, Maryland. They mention being stung by sea nettles on only one occasion and report that they were unable to swim on 26 July (1906?) at Oxford, Maryland, because the water was white with jellyfish. On such summer sailing trips in recent decades, one could not have ignored the presence of sea nettles, except perhaps in the summer of 1972 after Tropical Storm Agnes turned much of the Bay into a freshwater sea for a few months.

Raymond Cowles (1930: 331) refers to unpublished observations by Lewis Radcliff, who undertook research cruises in Chesapeake Bay from October 1915 through September 1916 (about three decades after the peak oyster harvests of the 1880s), and who found sea nettles from the Bay mouth to Sandy Point in October 1915. Sea nettles were infrequent in December, not recorded in January through June, "very abundant" in July, and still abundant in September 1916. Significantly, Radcliff reported that a Virginia fisherman known to him would take up his nets in July, anticipating a "run" of sea nettles and "burning" from use of his nets (the sea nettles would adhere to the net material and retain their ability to sting even when dead).

Oyster biologist Reginald Truitt, writing for the Conservation Department (1926: 41), reported that sea nettles were very troublesome in 1925, becoming so abundant after July 8 that "not in recent years, nor in the memory of the old sailors, had there been such vast numbers of them. Normally nettles are sufficiently common to be feared by bathers and are seen rather frequently, but during the past season bathing in the greater part of the Chesapeake Bay was not attempted because of them."

Paula Johnson (1988: 96) refers to the present-day use by Patuxent River crabbers of "substantial clothing—long sleeves, rubber boots, gloves, and sunglasses—to protect themselves from the stinging tentacles of sea nettles that drift into the [crab] pots and become entangled in the [trot] lines" during the hottest months of summer. Again, there appears to be no reference to such protection for crabbers or fishermen in the nineteenth century before the century-end decline in oyster abundance. We believe that there would have been more references to sea nettles before the end of the nineteenth century if they were as abundant as they are now.

Introduced or Exotic Species

Plants

A number of non-native plant species have been introduced and become established in the Bay, as reported by Court Stevenson and Lori Staver. Among these have been the Eurasian water milfoil and the water chestnut, both of which broadly occupied Chesapeake tributaries, and, most recently, hydrilla, which is found mainly in the freshwater tidal

Potomac River. These plants provide habitat for fish and invertebrates, but they also hinder boating and fishing where they have filled the waterways.

Animals

Interest in fish culture in North America increased in the mid-1800s as habitats deteriorated and fish stocks declined. J. T. Bowen reported that because carp culture had been practiced successfully in Asia and Europe for centuries, carp became a food fish of interest to Spencer Baird, the first U.S. commissioner of fish and fisheries. W. L. Fairbanks and W. S. Hamill found that specimens were imported by a German fish culturist in May 1877. The fish were held temporarily in ponds at Maryland's state hatchery on Druid Hill in Baltimore and later in ponds constructed on the grounds of the Washington Monument in Washington, D.C. The program of providing cultured fish was a huge success, with small carp distributed all over the country. Complaints about the introduction of the species began in 1883, because carp cause turbidity in the creeks they inhabit, can displace native species, and are not acceptable to every palate. Consequently, plantings declined, ceasing in 1896. Nevertheless, carp remain common in Chesapeake Bay tributaries.

Two species of bivalves appeared in the brackish and freshwater regions of the Bay and its tributaries in the last fifty years, presumably as a result of human action. Sewell Hopkins and Jay Andrews note that the Atlantic rangia clam has been collected in soft sediments in brackish and fresh waters in the Bay since the early 1960s. This species inhabits a part of the estuary that has limited species diversity, and it does not appear to have a negative influence on the ecosystem. Paul Dresler and Robert Cory report that the Asiatic clam entered fresh waters in the Bay in the 1970s. Unlike the Atlantic rangia clam, the Asiatic clam is an industrial pest, often clogging water-intake systems in power plants and other water-use facilities. Finally, the European zebra mussel is expected eventually to invade the Bay's freshwater tributaries because it has been reported in rivers surrounding the Bay's watershed. It too is expected to clog water-intake systems and to cost municipalities and industrial facilities millions of dollars annually in preventive measures.

Pollution and Mitigation

Pollution, particularly by nutrients from land-based sources, has played a major role in the declining condition of Chesapeake Bay. According to Timothy Hennessey, recognition of damage to the Bay ecosystem came in the middle 1970s and brought about restoration efforts by the Chesapeake Bay Program. A multijurisdictional partnership involving Maryland, Virginia, Pennsylvania, and the District of Columbia has crafted several agreements among federal and state governmental programs which have committed to significant reductions in incoming nutrients wherever management controls can work. One goal was that controllable discharge of nutrients to the Bay was to be reduced 40% from 1985 levels by the year 2000. That goal has not been met, but it has been reaffirmed, with no timeframe established.

Other goals directed at improving habitat and living resources have been set. Chief among these is a recognition that reductions in nutrients and suspended sediments in the water are necessary before conditions will be favorable for the return of submersed aquatic vegetation. Vegetated areas exceeded 60,000 acres in the 1990s. Richard Batiuk and his colleagues declared that a target of 114,000 acres of restored vegetation was a reasonable short-term objective, with an ultimate objective of some 600,000 acres being ecologically attainable. In 1992, a goal for dissolved oxygen in bottom waters was set to foster habitat improvement for fish and bottom-dwelling species, as reported by Steve Jordan and his colleagues.

Fisheries managers have also begun to look realistically at habitat restoration and harvest restrictions. A relatively short moratorium on striped bass fishing gave that species a measurable boost and permitted resumption of recreational and commercial harvesting. The fishing moratorium and habitat-restoration efforts on behalf of American shad have produced spawning fish in increasing numbers most every year, although those numbers are very much smaller than the stream-clogging runs of the nineteenth century.

For oysters, the restoration of the former three-dimensional structure of the ancient oyster reefs has been proposed. The vertical structure of an unfished oyster reef enhances turbulence over the oysters, increasing the likelihood of successful feeding, reducing exposure to low oxygen, and preventing accumulation of smothering sediments. New reefs have been built of oyster shell in the Piankatank and Great Wicomico Rivers in Virginia (http://www.vims.edu/fish/oyreef/rest.html). They emerge from the water's surface at ebb tide, even as Michel had seen in 1701.

Although habitat restoration and nutrient reduction will not return Chesapeake Bay to the conditions John Smith found in 1607–8, if these and other mitigation and conservation efforts can be achieved in the face of predicted increases in population and development in the watershed, they may allow the Bay once again to "abound in fish when the season of the year is favorable."

ACKNOWLEDGMENTS

We thank S. Gaber for his assistance with archival research in the context of this project, D. Orner for harvest statistics, and L. E. Cronin, P. Foer, M. Leffler, T. C. Malone, J. A. Mihursky, A. Schwartz, and P. Vogt for their comments on earlier drafts. Contribution Number 3230 from the University of Maryland Center for Environmental Science.

BIBLIOGRAPHY

Baird, S. F. 1880. Human agencies as affecting the fish supply, and the relation of fish culture to the American fisheries. 6. Influence of civilized man on the abundance of animal life. United States Commission of Fish and Fisheries, Part VI. *Report of the Commissioner for 1878:* XLV–LVII.

Barrie, R., and G. Barrie Jr. 1909. *Cruises Mainly in the Bay of the Chesapeake* (Philadelphia: Franklin Press).

Batiuk, R. A., R. J. Orth, K. A. Moore, W. C. Dennison, J. C. Stevenson, L. W. Staver, V. Carter, N. B. Rybicki, R. E. Hickman, S. Kollar, S. Bieber, and P. Heasly. 1992. *Chesapeake Bay Submerged Aquatic Vegetation Habitat Requirements and Restoration Targets: A Technical Synthesis.* USEPA Chesapeake Bay Program CBP/TRS 83/92 (Annapolis, Md.).

Beatty, R. C., and W. J. Mulloy. 1940. *William Byrd's Natural History of Virginia or the Newly Discovered Eden* (Richmond, Va.: Dietz Press).

Beverly, R. [1705] 1947. *The History and Present State of Virginia* (reprint, Chapel Hill: University of North Carolina Press).

Biggs, R. B. 1981. Freshwater inflow to estuaries, short- and long-term perspectives. In *Proceedings of the National Symposium on Freshwater Inflow to Estuaries,* v. 2, *Coastal Ecosystems Project,* ed. R. D. Cross and D. L. Williams. United States Fish and Wildlife Service FWS/OBS-81-04 (Washington, D.C.), 305–21.

Bowen, J. T. 1970. A history of fish culture as related to the development of fishery programs. In *A Century of Fisheries in North America,* ed. N. G. Benson. American Fisheries Society, Special Publication 7 (Washington, D.C.), 71–93.

Boynton, W. R., J. H. Garber, R. Summers, and W. M. Kemp. 1995. Inputs, transformations, and transport of nitrogen and phosphorus in Chesapeake Bay and selected tributaries. *Estuaries* 18(1B): 285–314.

Brooks, W. K. [1891] 1996. *The Oyster: A Popular Summary of a Scientific Study.* Reprint, with an introduction by K. T. Paynter Jr. (Baltimore: Johns Hopkins University Press).

Brush, G. S. 1986. Geology and paleoecology of Chesapeake Bay: A long-term monitoring tool for management. *Journal of the Washington Academy of Sciences* 76: 146–60.

Conservation Department. 1926. *Third Annual Report of the Conservation Department of the State of Maryland* (Baltimore).

Cooper, S. R., and G. S. Brush. 1993. A 2,500-year history of anoxia and eutrophication in Chesapeake Bay. *Estuaries* 16: 617–26.

Cowles, R. P. 1930. A biological study of the offshore waters of Chesapeake Bay. *Bulletin of the U.S. Bureau of Fisheries* 46: 277–381.

Cronin, L. E. 1986. Chesapeake fisheries and resource stress in the nineteenth century. *Journal of the Washington Academy of Sciences* 76: 188–98.

Custer, J. F. 1986. Prehistoric use of the Chesapeake estuary: A diachronic perspective. *Journal of the Washington Academy of Sciences* 76: 161–72.

Dresler, P. V., and R. L. Cory. 1980. The Asiatic clam, *Corbicula fluminea* (Müller), in the tidal Potomac River, Maryland. *Estuaries* 3: 150–51.

Earll, R. E. 1883. The Spanish mackerel, *Cybium maculatum* (Mitch.) Ag.; Its natural history and artificial propagation, with an account of the origin and development of the fishery. United States Commission of Fish and Fisheries, Part VIII. *Report of the Commissioner for 1880:* 395–424.

Fairbanks, W. L., and W. S. Hamill. 1932. *The Fisheries of Maryland* (Baltimore: Maryland Development Bureau of the Baltimore Association of Commerce).

Force, P. 1846. *Tracts and Other Papers Relating Principally to the Origin, Settlement, and Progress of the Colonies in North America* . . . , v. 4 (Washington, D.C.: published privately by W. Q. Force).

Goode, G. B. 1884. *The Fisheries and Fishery Industries of the United States,* section 1, *Natural History of Useful Aquatic Animals* (Washington, D.C.: U.S. Commission of Fish and Fisheries).

———. 1887. *The Fisheries and Fishery Industries of the United States,* section 5, *History and Methods of the Fisheries: Plates* (Washington, D.C.: U.S. Commission of Fish and Fisheries).

Gross, M. G., M. Karweit, W. B. Cronin, and J. R. Schubel. 1978. Suspended sediment discharge of the Susquehanna River to northern Chesapeake Bay, 1966 to 1976. *Estuaries* 1: 106–10.

Hariot, T. [1588] 1971. *A Briefe and True Report of the New Found Land of Virginia* (facsimile ed., New York: Da Capo Press of Plenum).

Hatch, C. E. 1957. *The First Seventeen Years: Virginia, 1607–1624* (Charlottesville: University Press of Virginia).

Haven, D. S., W. J. Hargis Jr., and P. C. Kendall. 1978. *The Oyster Industry of Virginia: Its Status, Problems and Promise.* Virginia Institute of Marine Science Special Report 168 (Gloucester Point, Va.).

Hennessey, T. M. 1994. Governance and adaptive management for estuarine ecosystems: The case of Chesapeake Bay. *Coastal Management* 22: 119–45.

Hopkins, S. H., and J. D. Andrews. 1970. *Rangia cuneata* on the East Coast: Thousand-mile range extension, or resurgence? *Science* 167: 868–69.

Ingersoll, E. 1881. *The History and Present Condition of the Fishing Industries. The Oyster Industry.* U.S. Census Bureau, Tenth Census (Washington, D.C.: Department of the Interior).

Johnson, P. J., ed. 1988. *Working the Water: The Commercial Fisheries of Maryland's Patuxent River* (Charlottesville: University Press of Virginia).

Jordan, S., C. Stenger, M. Olson, R. Batiuk, and K. Mountford. 1993. *Chesapeake Bay Dissolved Oxygen Goal for Restoration of Living Resource Habitats.* USEPA Chesapeake Bay Program TRS 88/93 (Annapolis, Md.).

Kennedy, V. S. 1989. The Chesapeake Bay oyster industry: Traditional management practices. In *Marine Invertebrate Fisheries: Their Assessment and Management,* ed. J. F. Caddy (New York: John Wiley), 455–77.

Kennedy, V. S., and L. L. Breisch. 1983. Sixteen decades of political management of the oyster fishery in Maryland's Chesapeake Bay. *Journal of Environmental Management* 16: 153–71.

Kent, B. 1988. *Making Dead Oysters Talk* (Baltimore: Maryland Historical Trust).

King, J. R. 1917. *My Experience in the Confederate Army and in Northern Prisons. United Daughters of the Confederacy, Clarksburg, W.V.* (available at http://www.innova.net/~vsix/elmiradoc19.htm).

Miller, H. M. 1986. Transforming a "splendid and delightsome land": Colonists and ecological change in the Chesapeake, 1607–1820. *Journal of the Washington Academy of Sciences* 76: 173–87.

Newell, R. I. E. 1988. Ecological changes in Chesapeake Bay: Are they the result of overharvesting the American oyster, *Crassostrea virginica?* In *Understanding the Estuary: Advances in Chesapeake Bay Research.* Chesapeake Research Consortium Publication 129 (Baltimore), 536–46.

Nichol, A. J. 1937. *The Oyster-packing Industry in Baltimore. Its History and Current Problems.* Bulletin, Chesapeake Biological Laboratory, University of Maryland, Solomons, Md.

Odum, W. 1970. Insidious alteration of the estuarine environment. *Transactions of the American Fisheries Society* 99: 836–47.

Raney, E. C. 1952. The life history of the striped bass, *Roccus saxatilis* (Walbaum). *Bulletin of the Bingham Oceanographic Collection* 14: 5–97.

Smith, H. M. 1891. Notes on an improved form of oyster tongs. *Bulletin of the U.S. Fish Commission for 1889,* 9: 161–65.

———. 1899. Notes on the extent and condition of the alewife fisheries of the United States in 1896. U.S. Commission of Fish and Fisheries, Part XXIV. *Report of the Commissioner for 1898:* 33–43.

Stevenson, C. H. 1894. The oyster industry of Maryland. *Bulletin of the U.S. Fish Commission for 1892,* XII: 205–97.

———. 1899. The shad fisheries of the Atlantic Coast of the United States. U.S. Commission of Fish and Fisheries, Part XXIV. *Report of the Commissioner for 1898:* 101–269.

Stevenson, J. C., J. I. Marusic, B. Ozretic, A. Marson, G. Cecconii, and M. S. Kearnery. 1999. Shallow water and shoreline ecosystems of the Chesapeake Bay compared to the northern Adriatic Sea: Transformation of habitat at the land-sea margin. In *Ecosystems at the Land-Sea Margin: Drainage Basin to Coastal Sea,* ed. T. C. Malone, A. Malej, L. W. Harding Jr., N. Smodlaka, and R. E. Turner (Washington, D.C.: American Geophysical Union), 29–79.

Stevenson, J. C., and L. Staver. 1993. *Submersed Aquatic Vegetation of Chesapeake Bay: A Literature Review and Synthesis.* Report to U.S. Fish and Wildlife Service (Annapolis, Md.).

Wharton, J. 1957. *The Bounty of the Chesapeake. Fishing in Colonial Virginia* (Charlottesville: University Press of Virginia).

White, Father Andrew. 1634. A relation of the colony of the Lord Baron of Baltimore, in Maryland, near Virginia. In P. Force (1846), 4(12): 1–87.

Wright, H. 1884. On the early shad fisheries of the North Branch of the Susquehanna River. United States Commission of Fish and Fisheries, Part IX. *Report of the Commissioner for 1881:* 619–42.

CHAPTER ELEVEN

Land Use, Settlement Patterns, and the Impact of European Agriculture, 1620–1820

LORENA S. WALSH

The environment of tidewater Maryland encouraged new settlement patterns. On prime tobacco lands, wealthy seventeenth-century settlers used bound laborers and hand-cultivation techniques that, while they protected the soil, created a different social structure than found elsewhere in the watershed. Changes in markets and productivity and new agricultural techniques and crops altered the appearance of the landscape and resulted in soil depletion and erosion in the watershed.

Settlement Patterns

In the tidewater Chesapeake, Old World settlers learned to use abundant natural resources in distinctive ways. Local ecologies strongly affected settlement patterns and agricultural systems and were in turn reshaped by the newcomers who claimed dominion of the land. This chapter deals with two main topics: the outlines and social context of European American and African American settlement patterns, and the ways in which dominant crops and agricultural techniques shaped and changed the landscape on individual plantations.

In the seventeenth century, a combination of abundant land, scarce labor, an estuarine environment, and the requirements of tobacco culture produced a pattern of dispersed farmsteads scattered loosely along the banks of rivers and streams. This system is shown in Augustine Herrman's well-known map of 1670, where plantations occur between one-quarter mile and one-and-a-half miles apart (Herrman, 1670; Earle, 1975:

19). The pattern Herrman pictured 200 years ago was corroborated in the late twentieth century by the location of seventeenth-century archaeological sites (Smolek, 1984).

Anglo-American farmers selected homesites and farm sites with three things in mind. First in importance were fertile, relatively level, and well-drained soils. Native Americans farmed the very best agricultural soils found in the region, and Anglo-American tobacco planters learned to identify (and sometimes to covet) these most promising sites by closely observing the places where Indians located their villages. New planters drew also on English agricultural traditions as well as on direct observation of the lay of the land and of the forest cover various soils supported (Smolek, 1984; for a local example, see Walsh, 1988). Prime to good soils, especially in southern Maryland, follow the waterways. Regional soil maps demonstrate this general pattern (U.S. Department of Agriculture, 1972). Even Coastal Plain soils can vary markedly in agricultural potential and productivity within a short distance, as more detailed areal soil surveys show (U.S. Department of Agriculture, 1973, 1974, 1978). Archaeologists have demonstrated that early planters not only chose to settle in the better areas but selected the best soils in those areas. More than 90% of known seventeenth-century Maryland sites are located on or near prime, or at least good, tobacco soils (Smolek, 1984).

A second consideration was the water transport needed for the bulky staple crop. Along the Bay, most shoreline elevations are low, and settlers could make landings wherever ravines intersected rivers and creeks. In Maryland and Virginia seventeenth-century homesites were located at a median distance of about 600 feet from the modern shoreline of navigable water. Only about 10% of early sites are found more than 5,000 feet inland (Smolek, 1984). Because their farms lay within a coastal zone just over a quarter of a mile wide, most planters could wrestle their hogsheads to a landing using only the labor available in their households. This was wise, since public and private roads were few and badly maintained, carts and draft animals even scarcer. Ship captains ordered sailors to roll hogsheads from distant plantations to landings, but the owners of such inland tobacco often suffered considerable losses in transit. Close water access was important, but rather than determining farm location in and of itself, it was often a fortuitous by-product of the riverside location of good soils. (For discussions of soil and water access, see Walsh, 1977: ch. 7; Beverley, 1705: 57; Carr, 1974; Menard, 1975a: ch. 2; Middleton, 1953: pt. 1; and Earle, 1975: ch. 2.)

In general, inland areas lacked access to water transport and had few soils suitable for the crucial staple. Some inland soils would yield decent crops of maize, but only inferior tobacco. So long as new settlers could move to undeveloped waterfront tracts farther up the rivers, they continued to take up more distant prime tobacco lands rather than choose inferior interior parcels closer to hand. Even in early settled counties like St. Mary's, Maryland, uplands remained largely unoccupied, sometimes for 50 to 75 years after initial settlement by Europeans (Fig. 11.1). Latecomers, who arrived in a county after the best lands had been taken up, or who settled in undeveloped places with less agricultural potential, were also discriminating. Seventeenth-century plantations in areas that generally offered only poor soils were sited on or near localized spots of good cropland, as were plantations established later on uplands (Smolek, 1984; Walsh, 1988: 230).

A third consideration in selecting a homesite was availability of good drinking water. In the Chesapeake, thousands of ravines dissect the shoreline of the Bay, and most have springs. The location of springheads may have shaped the specific location of seventeenth-century house sites in Maryland, where, unlike in Virginia, settlers did not dig wells. Dwellings were usually built on knolls adjacent to and between the areas of ravines. Householders did not have to go far to draw fresh water, and the knolls, which are usually well drained, were the best places to build easily erected, earth-fast dwellings, whose frames lay directly on the ground or were erected in postholes (Smolek, 1984).

Tobacco culture also dictated dispersed settlement. Planters sought to acquire a great deal more land than they could develop in the short run. Acutely short of labor and capital, they chose to make the most of more abundant land. They rotated fields in long fallow, after three to four years of cropping in tobacco and perhaps a year or two more in corn, then allowed fields to regain fertility by resting them for about 20 years. This necessitated a high ratio of land to labor. Men who had or expected to have children to provide for chose choice waterfront tracts with enough land to afford a living, over many years, for at least one heir. Those who engendered more offspring counted on buying additional undeveloped lands to endow all family members, or at least all sons (Carr, Menard, and Walsh, 1991; for a summary of inheritance practices, see Carr, 1989).

The Anglo-American settled landscape reflected these considerations. Seventeenth-century settlers carefully measured the bounds of waterfront land that was essential for raising and marketing their crops and was where their dwellings stood. Rivers and creeks defined the principal outlines of individual tracts, but interior lines were often vague. Land-

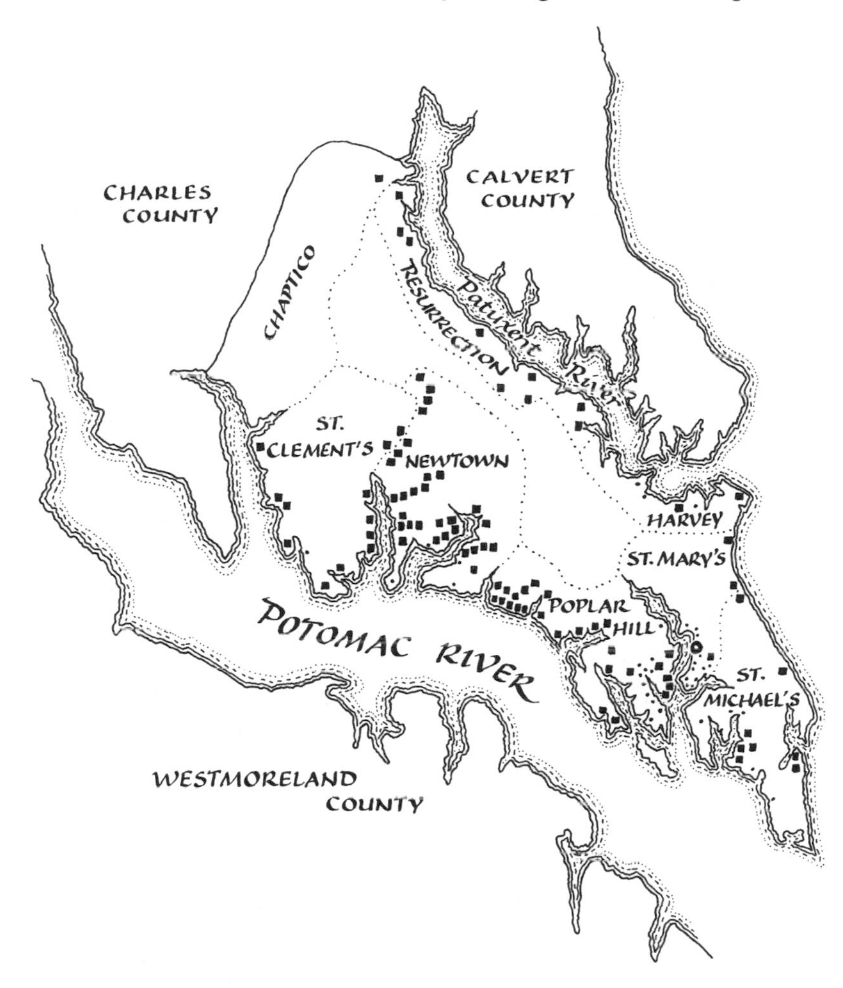

Fig. 11.1. St. Mary's County, Maryland, communities c. 1700. Poplar Hill includes St. George's Hundred; St. Michael's includes St. Inigoe's Hundred. (St. Mary's County rent rolls and tract maps; drawing, Gayle Henion)

Key: .... approximate community boundaries
- • tracts taken up by 1642
- ■ tracts taken up by 1660
- o site of initial settlement

owners were less concerned about back boundaries, because at first they used upland portions of their farms only as timber reserve and unfenced livestock range. Early surveyors did not trouble themselves to measure exact distances through a maze of snake-infested swamps, bramble thickets, and ravines. Landowners "most Commonly sold and others bought land . . . without Survey by Instruments bounding by guess from heads of Creeks to heads of Creeks or other parts as they agreed and Sometimes by paths sometimes mentioning Courses & Distances and Sometimes not" (Williamson, 1722).

By the end of the seventeenth century in older tidewater Maryland counties, all waterfront land with any agricultural potential had been taken up. As more householders cleared more new fields, some now farther inland, boundary conflicts proliferated. Landowners commissioned more accurate resurveys of early grants, and county courts convened juries of inquiry to adjudicate conflicting bounds. The Maryland legislature decided in 1699 that it was entirely reasonable that a landowner retain title to all the land included in an early grant if, upon resurvey, the tract could be made to conform to its stated geographic bounds by increasing the measured length of any or all of its lines by as much as 50% (*Archives of Maryland*, 1883–1972: v. 22, 481–94).

Social Landscapes

This dispersed settlement pattern produced not only a distinctive geographic landscape but also a differentiated social landscape that reflected the varying returns to agriculture that local natural resources afforded or denied. When an area was first settled, title to the best of the new lands was acquired by some men of ordinary means as well as by others who had recently immigrated with some capital or who had already developed estates in older areas. Early arrival conferred distinct advantage for everyone, but men with wealth, even those living elsewhere, had the advantage of both information and connections. Most such men had either visited the new area or knew others who had. And they had little difficulty in persuading surveyors—usually fellow gentlemen or men who had expectations of achieving that status—to look out for their interests (Walsh, 1988: 226–27).

On Maryland's western shore, the various hundreds (civil administrative subdivisions of a county) took on distinctive local characteristics early in their development by Anglo-Americans. In colonial Maryland (unlike Virginia), parishes (alternative ecclesiastical subdivisions that had taxing and some administrative responsibilities) did not become functional units until 1692. In St. Mary's and Charles Counties, hundred and parish boundaries differed, while in Anne Arundel County the bounds of the two units coincided much more closely. Data for St. Mary's and Charles are presented here by hundred, because available studies use that unit of analysis. The Anne Arundel data are presented by parish for the same reason. Note, however, that information from rent rolls, collected for hundreds rather than parishes, has been retabulated, where necessary, to reflect parish boundaries. Hundreds—and, to a greater extent in Anne Arundel, parishes—were administrative subdivisions that reflected not only physical boundaries but also community networks, and individual and group patterns of interaction (Walsh, 1977: ch. 5; 1988; cf. Rutman, 1973, 1980; Beeman, 1977; and Hammett, 1977: ch. 10). The sorts of men who initially took up the land in these subdivisions, the size of original grants, and the varying agricultural potential of those lands established limits that continued to affect the makeup of local populations for many generations.

Everywhere, wealthy planters acquired a goodly share of the finest tobacco lands. To develop their farms they needed bound labor—either European servants who had to work for three to seven years to pay off their passage from the Old World or enslaved Africans who served for life. So long as servants were the primary source of bound labor, wealthy men had a relative but not preponderate advantage over common landowners. For example, until the mid-1680s in St. Mary's, Charles, and Anne Arundel Counties—three of the major tobacco-producing counties on Maryland's western shore—there was a relatively broad distribution of bound labor among households of inventoried decedents (Table 11.1).

The shift from servant to slave labor altered the social landscape, just as it lessened the chances for ordinary planters to advance in wealth and status. After tobacco prices slumped in the 1680s, rich men had a strong advantage over smaller planters, for they were able to grow enough tobacco with which to purchase more workers and to further develop their holdings. Slaves cost more than servants, and at first only the rich could buy them (Menard, 1977). Until the last quarter of the eighteenth century, most rich men lived in prime tobacco areas, and so most slaves lived there as well. Two circumstances contributed to a growing slave population in good tobacco areas. First, wealthy planters simply bought more slaves than did poorer cultivators. Second, given about a two-to-one ratio of men over women among imported slaves, those who lived on larger

Table 11.1 Hundreds, soil quality, tract size, labor ownership, and wealth distributions in Charles, St. Mary's, and Anne Arundel counties, 1658–1775

Hundred or parish	Years	Soil type	% in tracts of 1–200 acres	% Servant owner	% Slave-owner	% TEV £1–94	% TEV £226+	% Tenant
Charles County								
Wicomico	1658–87	Good	51	45	9	59	14	33
	1688–1709			39	39	67	19	39
	1710–20			16	16	74	7	46
Pickawixen	1658–87	Good	79	48	25	60	10	25
	1688–1709			43	23	73	9	26
	1710–20			26	37	63	23	17
Port Tobacco	1658–87	Good/poor	69	34	9	88	6	33
	1688–1709			38	23	38	19	43
	1710–20			23	27	68	12	37
Nanjemoy/ Riverside	1658–87	Good/poor	66	51	2	73	7	15
	1688–1709			24	4	76	6	43
	1710–20			20	1	68	7	40
Chingomuxen	1658–87	Poor	80	35	0	100	0	100
	1688–1709			19	0	84	0	31
	1710–20			11	0	79	7	32

St. Mary's County								
Resurrection	1658–87	Excellent	75	67	30	72	16	19
	1688–1709			32	23	48	15	26
Harvey	1658–87	Excellent	57	48	12	52	32	10
	1688–1709			45	35	45	25	40
St. Clement's	1658–87	Good	57	32	14	51	7	27
	1688–1709			30	13	50	14	0
Newtown	1658–87	Good/poor	80	33	10	81	4	24
	1688–1709			19	13	78	7	23
Chaptico	1658–87	Good/poor	68	57	14	71	0	0
	1688–1709			12	6	89	5	38
Poplar Hill/ St. George's	1658–87	Good/poor	77	26	6	80	3	29
	1688–1709			22	15	88	7	18
St. Mary's	1658–87	Good/poor	80	30	14	68	30	30
	1688–1709			22	11	82	11	18
St. Inigoe's/ St. Michael's	1658–87	Good/poor	45	45	14	76	9	25
	1688–1709			20	11	80	7	35
Beaverdam Manor	1688–1709	Poor	74	25	0	100	0	100

(continued)

Table 11.1 *(Continued)*

Hundred or parish	Years	Soil type	% in tracts of 1–200 acres	% Servant owner	% Slave-owner	% TEV £1–94	% TEV £226+	% Tenant
Anne Arundel County								
St. James	1658–87	Excellent	65	53	15	66	18	11
	1688–1709			39	34	58	20	30
	1710–75			14	69	35	43	30
All Hallows	1658–87	Good	48	55	22	51	24	36
	1688–1709			35	35	56	23	31
	1710–75			16	59	45	33	24
Middleneck	1658–87	Good/poor	65	49	5	73	8	32
	1688–1709			38	40	51	34	40
	1710–75			26	62	37	39	22
Westminster	1658–87	Good/poor	82	54	2	67	9	11
	1688–1709			34	13	64	14	34
	1710–75			25	49	54	30	29

Source: Charles, St. Mary's, and Anne Arundel Counties inventories, tract maps, and soil surveys, as well as the 1705 rent roll.
Note: The analysis is based on 2,224 inventories. Inventories seldom state the hundred or parish of residence, so decedents were placed in administrative units by locating their land on a county tract map or, if they did not own land, based on the residence of landowners with whom they associated. Percentage of tenants are those among inventoried decedents only. Information on landownership is missing for some of the decedents. The real incidence of tenancy was greater than these figures suggest. TEV = total value of inventoried personal property; values are in £ sterling constant value. Data for Charles and St. Mary's Counties are presented by hundred; the Anne Arundel data are by parish.

plantations had a better chance of finding a mate and eventually having children than did enslaved men and women who lived in places where other blacks were few and scattered. In areas where slaves were from the outset concentrated in high numbers on large plantations, natural increase began earlier than among more geographically dispersed forced migrants. Their native-born offspring, in turn, increased in numbers even more rapidly (Menard, 1975b; Lee, 1986; Kulikoff, 1986; pt. 3; Walsh, 1983).

By 1700 we can distinguish three kinds of areas within the three counties. Figs. 11.1, 11.2, and 11.3 show the location of the various hundreds (or parishes), and Tables 11.1 and 11.2 demonstrate the relationship among tract size, soil quality, wealth distributions, levels of tenancy, and proportions of households owning bound labor. First were areas with prime tobacco lands generally held in large parcels. In hundreds where the land had been initially granted in large tracts, either to rich planters or, in a few cases, to the Roman Catholic clergy, landowners chose to develop their estates with a combination of bound laborers and white tenants. (Maryland's proprietors did not permit the Roman Catholic Church or the Jesuit Order to hold land outright. However, individual priests, or for a time in the seventeenth century Catholic laymen, held title in their own name to lands in fact belonging to the church.) None was sufficiently wealthy to buy enough slaves to develop all his holdings. But all were in a good position to attract white tenants, because nonlandowners stood a better chance of getting ahead if they rented the most productive tenancies rather than more marginal land.

Descendants of the first landowners continued these arrangements, and there was little subdivision of holdings. Large planters sought to retain their "special good land" in the family; should sale of some land become necessary, they disposed of tracts of lesser value. Because the best land tended to be sold in large parcels and because the price of good land continued to rise, only those who already had substantial capital could afford to buy the few choice tracts that came on the market. Once such an area had been parceled out among early claimants, there were few opportunities for newcomers to acquire freehold farms (Walsh, 1977: ch. 7). During the eighteenth century, the prime tobacco areas first laid out in large tracts were peopled with a combination of rich planters and their slaves and a high proportion of relatively poor tenant farmers who usually did not own bound labor. Wealth among planter households was inequitably distributed, with tenants at the low end, rich landholders and slaveholders at the top, and few households in the middle. Resurrection

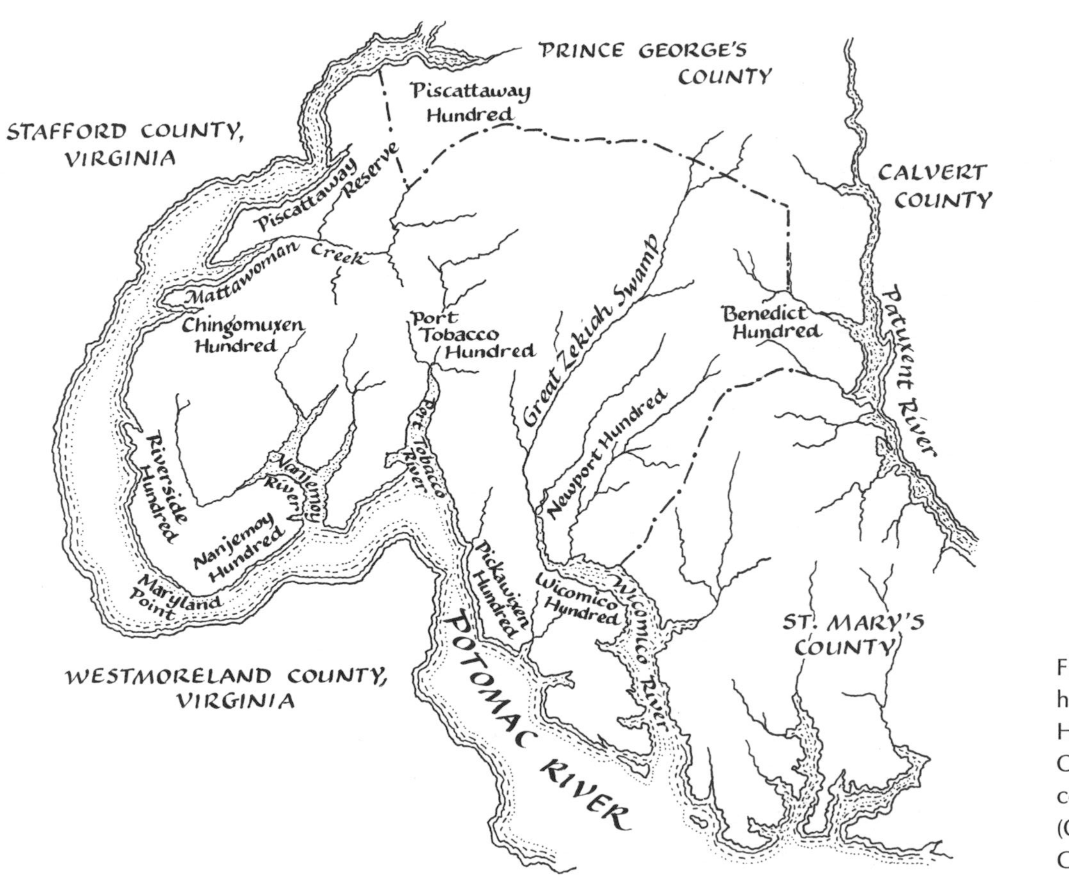

Fig. 11.2. Charles County, Maryland, hundreds as of 1696. Piscattaway Hundred, originally part of Charles County, became part of the new county of Prince George's in 1696. (Charles County rent rolls; drawing, Gayle Henion)

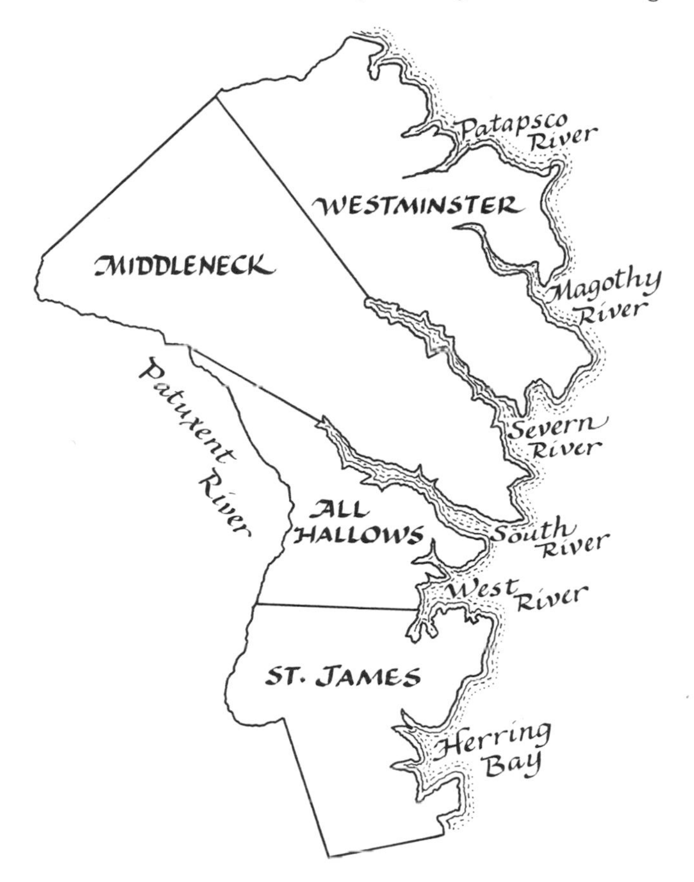

Fig. 11.3. Anne Arundel County, Maryland, parishes c. 1700. (Anne Arundel County rent rolls and tract maps; drawing, Gayle Henion)

Hundred in St. Mary's County, Wicomico Hundred in Charles, and St. James Parish in Anne Arundel are examples.

The second type of area in the counties was prime tobacco land that was first taken up in more numerous grants of smaller acreage. Early settlers of varying initial wealth developed fertile farms of middling size and were able to pass on the improvements to their heirs. Again, farms were seldom subdivided. Middling planters first worked their best tobacco soils and reserved any surplus of good land for their children. Mean household wealth tended to be higher than the overall county average, with slaves relatively numerous (there were fewer very large holdings than in areas of the first sort, but about half of the householders owned

Table 11.2 Tract size, wealth, tenancy, and labor force distribution, St. Mary's and Anne Arundel Counties, 1703–1776

Hundred or parish	Mean tract size in 1705[a]	% Slaveowner	% with 10 or more slaves	% TEV £1–94	% TEV £226+	% Known tenant
			St. Mary's County			
St. Mary's	189	37	7	65	18	19
Newtown	203	40	8	62	16	18
Poplar Hill/ St. George's	232	32	9	70	14	21
Resurrection	273	58	17	45	34	23
St. Clement's	360	46	9	53	27	21
Harvey	405	51	18	59	21	23
Chaptico	467	49	12	55	22	36
St. Inigoe's/ St. Michael's	843	38	2	67	13	32
Beaverdam Manor	159[b]	36	1	74	7	100
			Anne Arundel County			
Westminster	150	49	13	54	30	29
Middleneck	181	62	22	37	39	22
All Hallows	277	59	20	45	33	24
St. James	291	69	28	35	43	30

Source: St. Mary's and Anne Arundel Counties inventories, as well as the 1705 rent roll.

Note: The analysis is based on 2,910 inventories. Inventories seldom state the hundred or parish of residence, so decedents were placed in administrative units by locating their land on a county tract map or, if they did not own land, based on the residence of landowners with whom they associated. Percentage of tenants are those among inventoried decedents only. Information on landownership is missing for some of the decedents. The real incidence of tenancy was greater than these figures suggest. TEV = total value of inventoried personal property; values are in £ sterling constant value. Data for St. Mary's County are presented by hundred; the Anne Arundel data are by parish.

[a]Includes all tracts patented in a given hundred or parish by 1705.

[b]As of 1763.

slaves in the 1700s) and tenant farmers fewer than in places where tracts were larger. Most families had insufficient land to rent much of it to tenants, instead developing their farms with servants and slaves. Examples include St. Clement's Hundred in St. Mary's County, Pickawixen Hundred in Charles, and Middleneck Parish in Anne Arundel.

Third were subdivisions in which there was a smaller proportion of good tobacco land. In these areas, poorer planters often predominated among the original settlers. They and their heirs were less likely to advance in wealth than were men who started out with more capital and who farmed better land. In hundreds such as Poplar Hill and St. George's in St. Mary's County; Nanjemoy, Riverside, and Chingomuxen in Charles; and Westminster Parish in Anne Arundel, mean and median household wealth were lower than the county average, farms smaller in size, land more equitably distributed among resident householders, and bound laborers much less numerous. The white population grew fastest in these more poorly endowed subdivisions, because marginal lands remained unclaimed the longest, or at least undeveloped and available for sale at modest prices. Conversely, over time the black population expanded most rapidly in prime tobacco areas where conditions for family formation among the enslaved were most favorable and opportunities for whites to establish new households most limited.

In the first quarter of the eighteenth century, subdivisions of a fourth sort appeared in southern Maryland, as the growing white population outstripped the supply of good tobacco land. The new subdivisions were located in areas of relatively poor soils, usually in the interior, that were farmed largely by tenants for the benefit of absentee landlords. Beaverdam, a proprietary manor in St. Mary's County, is one such area. There, average wealth per household was the lowest in the county, the range of wealth among households minimal, and few of the tenants owned any bound labor.

Eventually the interior parts of tidewater counties were surveyed and patented, either by resident landowners or by absentee landlords, and the grid of occupation was completed. For example, by 1763 almost all lands in the interior of St. Mary's County had been patented, although not necessarily improved. As farms multiplied in the uplands, landowners became more concerned with establishing clearer interior bounds. The bounds of older farms were more exactly measured; "vacancies," small parcels of unclaimed land within or adjacent to early surveys, were identified and patented by neighboring landowners; and interior tracts were claimed and improved where the agricultural potential of the land warranted. The only exception was swamps. Although useful for hunting, trapping, and livestock range, such areas did not yield enough revenue to offset property taxes. Landholders who owned adjacent tracts declined to patent wetlands, instead making informal agreements with their neighbors for mutual, tax-free use. Many swamps were not taken up until the early nineteenth century, when rising timber prices induced some landowners to lay claim to them.

Does this model for the evolution of the social landscape, originally developed for southern Maryland, also apply in Virginia? I tested it for York County, on the peninsula between the James and York Rivers, an area occupied by Europeans very early in the seventeenth century. The county specialized in the more valuable sweet-scented strain of tobacco through the late 1780s, after which farmers turned to a more mixed agriculture, with special emphasis on corn. The answer is mostly yes. Size of landholdings had less predictive value there, for early white settlers took up land in more lockstep fashion and in holdings of more uniform size. Years of intermittent war between the Europeans and the larger local Native American population induced the Europeans to locate their settlements more compactly for reasons of security.

Charles Parish in southern York County (the present Poquoson area), as seen in Fig. 11.4, has more swamps than fertile farmland, and in the seventeenth and eighteenth centuries a high proportion of poor planters lived there. Consequently, as Table 11.3 shows, in this parish there were fewer servants, and later fewer slaves, than in the rest of the county. The land was so bad, however, that average farm size was larger than in the better-endowed parts of the county. Yorkhampton Parish, in the middle of the county, had more good land and more wealthy planters than the county average. Most of these planters owned both servants and slaves early in the 1600s and often had large concentrations of slaves in the 1700s. Bruton, a later-settled, moderately fertile upriver parish, was home to a higher proportion of planters of middling wealth, most of whom did not own servants and who were slower to buy slaves than their Yorkhampton neighbors.

Colonial Chesapeake Agricultural Systems

Seventeenth-century planters in both colonies exploited land, which was abundant and cheap, and they sought to make the most of labor, which was scarce and dear. In the process of learning how to survive in an often hostile environment (see Chap. 7) and in finding a staple which

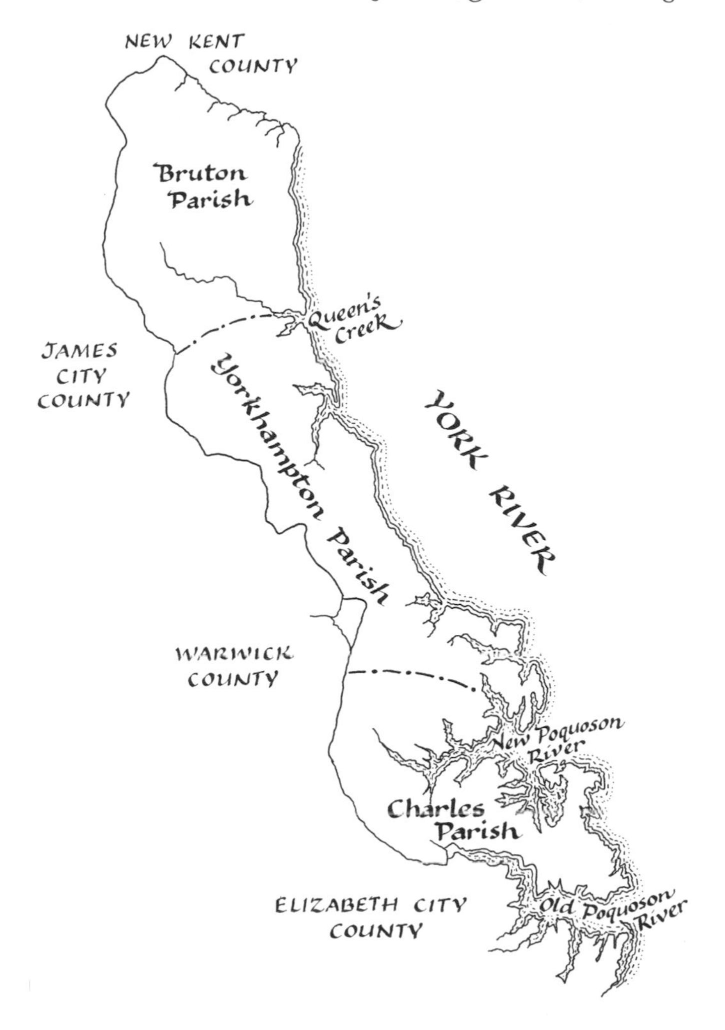

Fig. 11.4. York County, Virginia, parishes c. 1700. (York County rent rolls and tract maps; drawing, Gayle Henion)

they could exchange for imports, the colonists developed a new system of husbandry. Abandoning most European agricultural practices, planters adopted girdling and slash-and-burn clearing, long fallows, and hoe culture from Native Americans, as well as embracing maize and tobacco. At first they concentrated on maximum production of tobacco from fresh lands. The Europeans' main innovations were the introduction of domestic livestock and the use of metal tools. The annual work cycle was al-

Table 11.3 Parish, soil quality, tract size, labor ownership, wealth, and land distributions in York County, Virginia, 1636–1776

Parish	Years	Soil type	Mean tract size in 1704[a]	% Servant owner	% Slave-owner	% with 10 or more slaves	% TEV £1–94	% TEV £226+	% Known tenant	% Known landowner
Charles	1636–87	Poor	339	17	6	0	72	6	6	33
	1688–1709			12	21	3	85	9	9	50
	1710–54			2	64	21	48	34	18	66
	1755–76			0	56	17	51	32	35	54
Yorkhampton	1636–87	Good	287	62	54	0	23	31	0	85
	1688–1709			20	49	9	51	20	17	71
	1710–54			3	70	14	37	33	13	73
	1755–76			0	87	30	22	46	20	72
Bruton	1665–87	Good	288	35	23	0	58	15	8	50
	1688–1709			16	42	10	61	19	10	60
	1710–54			10	60	11	49	27	22	61
	1755–76			0	71	29	29	54	14	79

Source: York County probate inventories and 1704 rent roll.

Note: The analysis is based on 669 inventories. Inventories do not always state the parish of residence, so decedents were placed in parishes by locating their land on a county tract map or, if they did not own land, based on the residence of landowners with whom they associated. Percentage of tenants are those among inventoried decedents only. Information on landownership is missing for some of the decedents. The real incidence of tenancy was greater than these figures suggest.

TEV = total value of inventoried personal property; values are in £ sterling constant value.

[a]Includes all tracts patented in a given parish by 1704.

most wholly shaped by the seasonal demands of tobacco. Production of food crops—primarily maize—was usually limited to the requirements of self-sufficiency, and almost all essential manufactures were imported. Until the last quarter of the century, small farmers, who often owned a few indentured servants, were the major producers. Then, as the supply of European servants dwindled, richer planters turned to slaves as the primary source of bound labor. Large plantations became more common and wealth more concentrated (see Chap. 8; Walsh, 1989, 1993; Carr, Menard, and Walsh, 1991; Miller, 1986; Earle, 1975: ch. 2, 1988; Kirby, 1991).

The invasion of European farmers, and later of African workers, did affect the Chesapeake environment, although not in the gross ways the traditional historiography has usually stressed (e.g., Craven, 1926; for more recent evaluations, see Silver, 1990, and Kirby, 1991). An important change was the widespread introduction of domestic livestock and the related diminution of populations of large wild animals. Archaeological studies of faunal remains that settlers deposited in trash pits demonstrate that the settlers relied heavily on deer and fish for meat until extensive hunting reduced deer herds and an increase of domestic livestock provided more reliable sources of protein (Miller, 1984, 1988; Silver, 1990: ch. 6). In the Chesapeake as well as New England, trade between Native Americans and European colonists in meat, furs, and "vermin" scalps also contributed to further reductions of indigenous animal populations.

Hogs and cattle were let loose to forage at will in the woods and swamps, as were horses by the 1680s—planters fenced crops, not animals. Little or no winter forage was provided, and given the animals' high rates of natural increase, planters could more readily afford losses to predators, inadequate winter food, and spring mirings than they could spare labor for penning and forage production. Cattle and hogs surely affected both wooded areas and wetlands. How they changed local ecologies through selective destruction of particular kinds of plants, soil compaction, and the introduction of European weeds needs further study. Silver (Chap. 8) has noted the effects of overgrazing on the forests of the Coastal Plain and eastern Piedmont by the end of the eighteenth century and these animals' contributions to local soil erosion. The appearance of sheep beginning about 1690 prompted further change, both from their grazing habits and from bounties that encouraged the destruction of wolves whose depredations at first made sheep rearing impossible. These changes were repeated over time as Europeans and their descendants moved farther inland (for New England, see Cronon, 1983).

The effects of habitat change on smaller animals and marine life is another open question. The activities of farmers and domestic livestock probably crafted conditions that encouraged some species and discouraged others. For example, near universal use of worm fences—constructed of split rails, from 9 to 14 feet long, laid lengthwise between vertical stakes crossed at the top to form a fork—provided extensive cover for a variety of animals (Stilgoe, 1982: 188–92). Campaigns to reduce the number of squirrels and crows were probably less successful than those against wolves. As for marine life, local human populations were perhaps not so large that fishing and crabbing had much impact until the third quarter of the eighteenth century when some commercial fishing began. (This was on a much smaller scale than that of the later nineteenth century [see Chap. 10] and was pursued with traditional technologies.) We do know that in places where they were most concentrated, seventeenth-century settlers could deplete local oyster beds (Miller, 1986; Kent, 1988).

Similarly, European herbaceous plants—introduced deliberately for food, medicine, or ornament, or inadvertently in shipments of animals or seeds—found new niches in the landscape, competing with and sometimes displacing smaller native plants. Yentsch and Reveal (Chap. 12) have catalogued a number of these, including dandelion, tumbleweed, and purslane. Similarly, Brush (Chap. 3) has used increases in ragweed pollen, a plant that colonizes freshly disturbed soil and spreads as forests are cut down, for dating sediments deposited since European settlement.

Until about the middle of the eighteenth century, Chesapeake agricultural techniques changed little. The spread of European and African American plantations throughout the Tidewater slowly changed both the regional and local landscape as more virgin forest was cut down, but there was minimal impact on environmental quality. Although tobacco remained the predominant crop, hoe culture and long fallows prevailed. Hill and hoe culture caused very limited soil erosion, because the patchwork of man-made hills and tree stumps and roots left in the fields retarded both runoff and the leaching of vital plant nutrients. Only a small proportion of available farmland was cultivated at any one time, and rests of about 20 years that followed 6 to 8 years of cultivation preserved much of the long-term fertility of the soil. (By the mid-1700s, three to four years in tobacco were often followed by three to four years in maize.) As Sharrer explains (Chap. 14), long field rotations also served to keep in check pathogenic organisms, which may have contributed as much to declining yields as did the depletion of essential minerals. Much of the land remained either in wood or in regenerating old fields (Walsh, 1989, 1993;

Carr, Menard, and Walsh, 1991; Miller, 1986; Earle, 1975: ch. 2, 1988). Planters limited the number of laborers employed on particular tracts to suit the amount of arable land that could be cultivated in 20-year field rotations. Those who prospered enough to purchase more workers bought additional lands for them to farm. Development continued to be extensive rather than intensive (see also Chap. 13).

Changes on individual farms followed a predictable pattern. The first fields were on the best soils the tracts afforded, and after these were temporarily exhausted with crops of tobacco, then corn, new fields were cleared. Planters expected that some or all of the farmstead components would follow shifting fields. Tobacco barns, fences, and often dwellings were flimsy, impermanent affairs. When they abandoned old fields, planters dismantled worm fences and tobacco houses and reassembled them at new clearings. So long as they did not use plows, or used plows only to put in an acre or so of wheat, there was no reason to lay off fields in geometric shapes. It was much more efficient to hill and fence only the best land, excluding stretches of infertile or poorly drained soils. Planters were somewhat more reluctant to abandon a good house site, but they had limited attachment to actual homes. Cheaply and speedily constructed earth-fast dwellings had a short life span in this humid, termite-infested region. Families often abandoned dwellings too decayed to be worth repairing, or recycled the usable materials, or else built entirely new houses of similar small size and impermanent construction in the same general area rather than take the time and spend extra money to build more lasting structures (Carson et al., 1981). The resulting landscape appeared unsightly and slovenly to European observers, who were accustomed to a labor-rich, land-poor society. Slash-and-burn clearings, abandoned old fields, irregularly shaped and irregularly fenced patches of standing corn and tobacco, free-ranging livestock, slapdash dwellings and barns, and ill-defined upland farm boundaries made sense only to those who had learned to function in a labor-poor, land-rich region.

Four kinds of evidence support the contention that seventeenth-century agricultural practices were not particularly destructive of local soils. First, while most seventeenth- and early eighteenth-century travelers' accounts dwell on the unkempt appearance of almost all farms, they do not mention signs of soil erosion (Miller, 1986; Earle, 1975, 1988). Second, the size of the tobacco crops a single worker could produce in the tidewater Chesapeake increased through 1690. Had planters in older counties been permanently exhausting their lands, increasing yields are unlikely (Walsh, 1989). Third, where records of crop yields per laborer are available for the same plantation over two generations or more of owners, there is little evidence of decline on well-managed plantations. Tobacco yields fell after the best virgin lands had been cropped, but they could be maintained at respectable levels for several generations, either with long fallows or increased manuring. (Such plantations are described in Walsh, forthcoming.) Fourth, most planters were careful not to use more labor or keep more livestock than their farms could support with long rotations. Landlords, once tenements were developed, limited the amount of additional land tenants might clear, forbade subleasing, and often restricted the number of hands that tenants could use to work the land (Carr, Menard, and Walsh, 1991; Bliss, 1950; Walsh, 1985).

After the middle of the eighteenth century on Maryland's western shore, changes in crop mix and agricultural techniques accounted for most alterations in the landscape. As planters responded in midcentury to growing international markets for maize and English grains, they added wheat and surplus maize to their crop mix, made more use of plows to raise the grains, and applied manure to augment yields or to continue fields longer in production. From the 1730s, maize crops in southern Maryland reached 10 or more barrels per worker, a level of deliberate market production where about half of the crop was marketable surplus. Wheat output remained low, reflecting the low yields per acre most planters could effect with poor plows, weak draft animals, and acidic soils. As more wheat was grown, rusts, smuts, and Hessian flies became increasingly common and increasingly destructive (Chap. 14). The labor demands of corn and tobacco limited the amount of land planters could cultivate in wheat. However, even very small crops of wheat raised market income because, unlike maize, about 90% of the net wheat crop was sold (Walsh, 1989). Rising wealth among landowners encouraged construction of more substantial and permanent dwellings (Carson et al., 1981), and richer landowners became less likely to move their dwellings to follow shifting fields. Large landowners, however, could and did move slave housing to fit field rotations.

Planter efforts to counter diminishing returns to tobacco culture dominate the story of the eighteenth-century tidewater Chesapeake. These included import-replacement strategies, the addition of maize and wheat as major revenue crops, efforts to maintain more livestock to produce manure that would allow sustained yields, a gradual shift from hoe to plow culture, more division of labor by gender, and fuller utilization of the entire labor force year-round. By adding grains as major cash crops, Chesapeake slaveowners were able to maintain gross revenues per laborer in constant values at mid-seventeenth century levels. Evidence from individual farms suggests that annual gross revenues per worker were very roughly £15 sterling constant value from the 1640s to the early 1680s, probably fell to about £10 from 1680 to 1740 and rose again to around £15 in the third quarter of the eighteenth century (Carr and Walsh, 1988; Walsh, 1989).

Nonslaveowners had less flexibility. They often did not have enough labor or draft animals to put in more than an acre or two of wheat. Most could, however, grow some extra maize and supplement family income by small sales of livestock, dairy products, produce, and cloth, as well as from occasional wage labor or craftwork. By midcentury, land values rose as population increased. Poor folk had a harder time acquiring land, and permanent as well as temporary tenancy increased. Yet, there is evidence that even landless but well-situated tenants shared to some degree in growing regional prosperity in the third quarter of the century (Kulikoff, 1986: ch. 4; Walsh, 1985).

Larger slaveowners took the lead in making greater use of plows. So long as laborers used hoes alone to prepare land for planting, there were constraints to expansion, as larger crops could only be grown if hills had been prepared to grow them in. Plowing released labor time in spring which permitted planters to raise twice as much corn, and plows also aided in weeding during the growing season. The extra corn could be fed to the animals that pulled the plows and that when penned produced manure that was then used to fertilize corn and tobacco hills. Land for wheat could be prepared in off-seasons, and the grain seeded and plowed or harrowed in after the tobacco was harvested and before the corn needed gathering. Planters required extra labor only for the short harvest, and wheat growers had no difficulty hiring free whites or the slaves of planters who did not grow much grain (Carr and Menard, 1989).

The Changing Landscape after the Revolution

During and after the American Revolution, markets tilted decidedly in favor of grains, and many tidewater planters abandoned tobacco culture for wheat. This shift in crop mix significantly changed the appearance of the landscape, and had a decidedly adverse effect on the local environment. (For an example of the contrasting pre- and postwar agricultural systems and their consequences, see Chap. 13.) Tobacco planters turned wheat and maize farmers proceeded to plow ever more extensively with little regard for land contours and to clear and cultivate ever more marginal lands. Yields of wheat per acre were low, so more land had to be cultivated for this cash crop than with tobacco. Additional acres had to be devoted to pasture and meadows to feed increasing numbers of draft animals. Fields were more completely cleared and often laid out in geometric shapes to facilitate plowing. The amount of land in agricultural production in southern Maryland may have increased from about 2% of the total in 1720 to nearly 40% in the early 1800s. Fallows were shortened or abandoned altogether. More complete plowing of fields and removal of tree stumps and roots accelerated erosion, carrying off plant nutrients and, before long, a great deal of soil. Timber shortages became common, brought about by increased clearing, by the need to fence more fields, and by increasing sales of firewood to town markets (Walsh, 1992; Miller, 1986; Froomer, 1978; Papenfuse, 1972; for timber sales see Chap. 8 and Kirby, 1991).

The result was widespread soil depletion and massive soil erosion that clearly dates to the period of expanded grain culture. Many streams navigable by ships in the 1770s became silted up by the early 1800s. Enough silt and chemical nutrients were washing into the rivers to change the predominant species of fish in some Chesapeake waters. Some bottom-oriented species, highly regarded as food fish in the seventeenth and eighteenth centuries, were seriously depleted by the mid-nineteenth century. By 1820 changing land-use patterns were significantly altering the estuarine ecology and aquatic resources of the Bay. Brush's sediment cores show a marked increase in siltation toward the end of the eighteenth century which followed on the heels of expanded grain culture (see also Brush, 1986). Although Kennedy and Mountford (Chap. 10) could locate no quantitative data on the abundance of aquatic organisms before the late 1800s, they do find anecdotal evidence for significant depletion of some fish species, such as sturgeon, in the eighteenth century. This is corroborated by marked changes in the varieties found in period archaeological deposits (Miller, 1986).

Between 1790 and 1818, some planters made substantial profits from these changes, especially large land- and slaveowners who realized economies of scale in grain culture. Between 1790 and 1807 gross revenues per laborer among slaveowners rose to about £25 constant value, and from 1810 to 1818 averaged over £35. Doubtless, at the time they believed that increased profits justified extractive agricultural methods. Nonslaveowners or planters with only a hand or two realized lesser returns per worker. Many smallholders opted to migrate west, and enough moved out that the white population ceased to grow. After international grain markets slumped in 1819–20 and cheap western tobacco captured much of the European market, southern Maryland farmers entered a long period of hard times, and the region became something of a backwater (Walsh, 1989; the best source for early nineteenth-century southern Maryland agriculture is Marks, 1979).

Changes in crops and farm techniques had profound implications for the lives of blacks in the early nineteenth century. Wheat farmers needed fewer hoe hands than tobacco planters, and some slaveowners began to reduce or at least to stabilize their labor forces. Some slaves were freed, but many more were forced to migrate west or south to cultivate tobacco where its culture was expanding—Kentucky, Tennessee, and, for a time, Georgia—or else were sent to work in cities or in the new cotton south. Work routines also changed dramatically, shifting from seasons of intense effort alternating with seasons of greater ease, to full year-round employment. Tasks, especially for men, became more varied and specialized. Slave women, on the other hand, who continued to toil in the fields at the most unskilled tasks, gained little and perhaps lost more from the changing organization of labor (Carr and Walsh, 1988; Walsh, 1993).

Finally, and somewhat paradoxically, it was during the post-Revolutionary years, when new agricultural practices were causing a measurable deterioration of estuarine resources, that tidewater Chesapeake residents also began to exploit these resources more intensively. During the eighteenth century, most farmers relied primarily on domestic livestock for their meat supply, and they fished, crabbed, or gathered oysters only occasionally, and largely for their own tables. Apparently most did so from the shore, as few, other than large planters, owned boats or seines (Miller, 1988). By 1800, over a third of small planter and minor slaveowner estates in St. Mary's County, Maryland, and York County, Virginia, included small boats; probably the majority of farmers who actually lived on the water had a canoe or bateau. More than a third also had either fishing equipment (lines, hooks, or gigs, and occasionally a seine or gill net, sometimes owned in partnership with another household) or oystering gear (tongs—often called "paws"—and rakes) or both (Walsh, 1991).

Newly expanding urban areas began to offer a tempting market for marine resources and afforded small farmers a welcome supplement to limited agricultural revenues. Water-related activities also offered free blacks a means of subsistence, as most could not afford the tools and livestock necessary for cultivating a full-sized farm. Those who managed to rent a few acres of land near the water had the best chance for growing a little tobacco, raising garden crops to feed the family over part of the year, and making a few additional dollars from sales of fish, crabs, and oysters.

Before the Revolution, the influence of tobacco was predominant. Afterward, a period of marked change in crop mix and agricultural techniques dramatically altered the local landscape and significantly depleted natural resources. Still, local circumstances that had arisen in the earlier seventeenth century continued for many generations to affect what sort of people lived in particular neighborhoods, the typical farm size, and the distribution of resources among resident householders in the various county subdivisions. The relationship among changes in land use, land ownership, exploitation of marine resources, and the composition of local populations in the early nineteenth century needs further research. In places where local populations ceased to grow again until the early twentieth century, the hand of history continued to weigh heavily.

BIBLIOGRAPHY

Manuscript sources for the Maryland portions of this study include rent rolls; patents; plats; boundary commission proceedings in county court records and provincial court papers; patented and unpatented certificates of survey; tract maps for St. Mary's County for 1642, 1705, and 1763; tract maps for parts of Anne Arundel County; biographical files for residents of the three counties compiled through stripping of relevant record series; and computer files of probate inventories, all housed in the Maryland State Archives, Annapolis. The processed materials are part of a larger study of social and economic developments in the colonial Chesapeake which has been funded by grants to Historic St. Mary's City from the National Science Foundation (GS32272); by the National Endowment for the Humanities to Historic St. Mary's City (RO6228-72-468; RO10585-74-267); jointly by the Endowment to Historic Annapolis, Inc., and Historic St. Mary's City (RS0067-79-0738; RS20199-81-1955); and by a research grant to Lorena S. Walsh from the Economic History Association. Probate inventory data is available for Charles County from 1658 to 1720 and for St. Mary's and Anne Arundel Counties throughout the colonial period.

Similar biographical files and tract maps for York County, Virginia, were assembled by the Historical Research Department at the Colonial Williamsburg Foundation supported by the National Endowment for the Humanities (RO20869-85). Photostat and microfilms of the manuscript sources are available at the State Library of Virginia, Richmond.

Archives of Maryland. 1883–1972. Ed. W. H. Browne et al. 72 vols. (Baltimore: Maryland Historical Society).

Beeman, R. R. 1977. The new social history and the search for "community" in Colonial America. *American Quarterly* 29: 422–43.

Beverley, R. [1705] 1947. *The History and Present State of Virginia,* ed. L. B. Wright (Chapel Hill: University of North Carolina Press).

Bliss, W. F. 1950. The rise of tenancy in Virginia. *Virginia Magazine of History and Biography* 58: 427–41.

Brush, G. S. 1986. Geology and paleoecology of Chesapeake Bay: A long-term monitoring tool for management. *Journal of the Washington Academy of Sciences* 76: 146–60.

Carr, L. G. 1974. "The metropolis of Maryland": A comment on town development along the Tobacco Coast. *Maryland Historical Magazine* 69: 124–45.

———. 1988. Diversification in the colonial Chesapeake: Somerset County, Maryland, in comparative perspective. In *Colonial Chesapeake Society,* ed. L. G. Carr, P. D. Morgan, and J. B. Russo (Chapel Hill: University of North Carolina Press), 342–88.

———. 1989. Inheritance in the colonial Chesapeake. In *Women in the Age of the American Revolution,* ed. R. Hoffman and P. J. Albert (Charlottesville: University Press of Virginia), 155–208.

Carr, L. G., and R. R. Menard. 1989. Land, labor, and economies of scale in early Maryland: Some limits to growth in the Chesapeake system of husbandry. *Journal of Economic History* 49: 407–18.

Carr, L. G., R. R. Menard, and L. S. Walsh. 1991. *Robert Cole's World: Agriculture and Society in Early Maryland* (Chapel Hill: University of North Carolina Press).

Carr, L. G., and L. S. Walsh. 1988. Economic diversification and labor organization in the Chesapeake, 1650–1820. In *Work and Labor in Early America,* ed. S. Innes (Chapel Hill: University of North Carolina Press), 144–88.

Carson, C., N. F. Barka, W. M. Kelso, G. W. Stone, and D. Upton. 1981. Impermanent architecture in the southern American colonies. *Winterthur Portfolio* 16: 135–96.

Craven, A. O. [1926] 1965. *Soil Exhaustion as a Factor in the Agricultural History of Virginia and Maryland, 1606–1860* (reprint, Gloucester, Mass.: Peter Smith).

Cronon, W. 1983. *Changes in the Land: Indians, Colonists, and the Ecology of New England* (New York: Hill and Wang).

Earle, C. V. 1975. *The Evolution of a Tidewater Settlement System: All Hallows' Parish, Maryland, 1650–1783.* University of Chicago Department of Geography Research Papers, no. 170.

———. 1988. The myth of the southern soil miner: Macro-history, agricultural innovation, and environmental change. In *The Ends of the Earth: Perspectives on Environmental History,* ed. D. Worster (New York: Cambridge University Press), 175–210.

Froomer, N. L. 1978. Geomorphic change in some Western Shore estuaries during historic times. Ph.D. diss., Johns Hopkins University.

Hammett, R. C. 1977. *History of St. Mary's County, Maryland* (Ridge, Md.: by the author).

Herrman, A. [1670] 1982. Virginia and Maryland, 1670 [1673]. In *The Hammond-Harwood House Atlas of Historical Maps of Maryland, 1608–1908,* ed. E. C. Papenfuse and J. M. Coale III (Baltimore: Johns Hopkins University Press), 14–15.

Kent, B. W. 1988. *Making Dead Oysters Talk: Techniques for Analyzing Oysters from Archaeological Sites* (Annapolis: Maryland Historical Trust, Historic St. Mary's City, and Jefferson Patterson Park and Museum).

Kirby, J. T. 1991. Virginia's environmental history: A prospectus. *Virginia Magazine of History and Biography* 99: 449–88.

Kulikoff, A. 1986. *Tobacco and Slaves: The Development of Southern Cultures in the Chesapeake, 1680–1800* (Chapel Hill: University of North Carolina Press).

Lee, J. B. 1986. The problem of slave community in the eighteenth-century Chesapeake. *William and Mary Quarterly* 3d ser., 43: 333–61.

Marks, B. E. 1979. Economics and society in a staple plantation system: St. Mary's County, Maryland, 1790–1840. Ph.D. diss., University of Maryland.

Menard, R. R. 1975a. Economy and society in early colonial Maryland. Ph.D. diss., University of Iowa.

———. 1975b. The Maryland slave population, 1658 to 1730: A demographic profile of blacks in four counties. *William and Mary Quarterly* 3d ser., 32: 29–54.

———. 1977. From servants to slaves: The transformation of the Chesapeake labor system. *Southern Studies* 16: 355–90.

Middleton, A. P. 1953. *Tobacco Coast: A Maritime History of Chesapeake Bay in the Colonial Era* (Newport News, Va.: The Mariner's Museum).

Miller, H. M. 1984. Colonization and subsistence change on the seventeenth-century Chesapeake frontier. Ph.D. diss., Michigan State University.

———. 1986. Transforming a "splendid and delightsome land": Colonists and ecological change in the Chesapeake, 1607–1820. *Journal of the Washington Academy of Sciences* 76: 173–87.

———. 1988. An archaeological perspective on the evolution of diet in the colonial Chesapeake, 1620–1745. In *Colonial Chesapeake Society,* ed. L. G. Carr, P. D. Morgan, and J. B. Russo (Chapel Hill: University of North Carolina Press), 176–99.

Papenfuse, E. C., Jr. 1972. Planter behavior and economic opportunity in a staple economy. *Agricultural History* 46: 297–311.

Rutman, D. B. 1973. The social web: A prospectus for the study of the early American community. In *Insights and Parallels: Problems and Issues of American Social History,* ed. W. L. O'Neill (Minneapolis: Burgess Publishing), 57–89.

———. 1980. Community study. *Historical Methods* 13: 29–41.

Scott, J. 1807. *A Geographical Description of the States of Maryland and Delaware* (Philadelphia: Kimber, Conrad).

Silver, T. 1990. *A New Face on the Countryside: Indians, Colonists, and Slaves in South Atlantic Forests, 1500–1800* (Cambridge: Cambridge University Press).

Smolek, M. A. 1984. "Soyle light, Well-Watered and on the River": Settlement patterning of Maryland's frontier plantations. Paper presented at the Third Hall of Records Conference on Maryland History, St. Mary's City, Maryland.

Stilgoe, J. R. 1982. *Common Landscape of America, 1500 to 1845* (New Haven: Yale University Press).

U.S. Department of Agriculture, Soil Conservation Service. 1972. General soil map, Southern Maryland Resource Conservation and Development Project, Charles, Calvert, and St. Mary's Counties (Hyattsville, Md.: U.S. Department of Agriculture).

U.S. Department of Agriculture, Soil Conservation Service, in cooperation with Maryland Agricultural Experiment Station. 1973. *Soil Survey of Anne Arundel County, Maryland* (Washington, D.C.: Government Printing Office).

———. 1974. *Soil Survey of Charles County, Maryland* (Washington, D.C.: Government Printing Office).

———. 1978. *Soil Survey of St. Mary's County, Maryland* (Washington, D.C.: Government Printing Office).

Walsh, L. S. 1977. Charles County, Maryland, 1658–1705: A study of Chesapeake social and political structure. Ph.D. diss., Michigan State University.

———. 1983. Anne Arundel County population. In "Annapolis and Anne Arundel County, Maryland: A Study of Urban Development in a Tobacco Economy, 1649–1776." Final report to the National Endowment for the Humanities, RS 20199–81–1955.

———. 1985. Land, landlord and leaseholder: Estate management and tenant fortunes in southern Maryland, 1642–1820. *Agricultural History* 59: 373–96.

———. 1988. Community networks in the early Chesapeake. In *Colonial Chesapeake Society,* ed. L. G. Carr, P. D. Morgan, and J. B. Russo (Chapel Hill: University of North Carolina Press), 200–241.

———. 1989. Plantation management, 1620–1820. *Journal of Economic History* 49: 393–406.

———. 1991. Report on selected agricultural and food-related items in York County, Virginia, and St. Mary's County, Maryland, inventories, 1783–1820. Colonial Williamsburg Research Department, Williamsburg, Va.

———. 1992. Chesapeake planters and the international market, 1770–1820. In *Lois Green Carr, The Chesapeake and Beyond—A Celebration* (Crownsville, Md.: Maryland Historical and Cultural Publications).

———. 1993. Slave life, slave society, and tobacco production in the tidewater Chesapeake. In *Cultivation and Culture: Labor and the Shaping of Slave Life in the Americas,* ed. I. Berlin and P. D. Morgan (Charlottesville: University Press of Virginia), 170–99.

———. Forthcoming. *"To Labour for Profit": Plantation Management in the Chesapeake, 1620–1820* (Charlottesville: University Press of Virginia).

Williamson, S. 1722. Deposition in *Mason* v. *Cheseldyne,* St. Mary's County, Provincial Court Papers, Maryland State Archives, Annapolis, Maryland.

CHAPTER TWELVE

Chesapeake Gardens and Botanical Frontiers

ANNE E. YENTSCH AND JAMES L. REVEAL

The watershed was a frontier where culture contact between different groups introduced plants into the New World, irrevocably altering its flora and spreading plant scions into British gardens and forests. Archaeological studies of early settlements show the layout and content of their gardens. Records and correspondence describe the exchange of species and provide an inventory of early garden plants. Changing garden plants and structures from the 1700s to the 1900s demonstrate evolving needs and attitudes of Chesapeake gardeners.

From 1600 to 1700, the Chesapeake was a border zone, a frontier between the Old and New Worlds. The effects of settlement reached beyond human society to touch on other dimensions of the natural world, revealing the interrelationship between humans and the broader ecosystem. Exploitation of the natural environment for profit initially concentrated on native resources, such as timber (cedar, white oak, bald cypress) and furs (deer and beaver skins), and eventually on staple crops such as corn and tobacco. Gardens created by colonial families, whether for profit or pleasure, encapsulate many of the reciprocal links between people and the natural world of the 1600s and 1700s. Thus, our focus is not corn or tobacco but the other cultivated plants—their variety and origins, and how over time plants whose origins spanned the globe came to be grown in Chesapeake gardens. Although some attention is given to garden design, placement, and aesthetics, the primary emphasis is the introduction of plant species to the region and the export of Chesapeake species to England. One aspect of the reciprocity was species shifting and more fluid border frontiers, which plants crossed quickly, if not always with ease.

As the first English ships sailed into Chesapeake Bay, their passengers and crew saw a bewildering array of natural objects. Observing the range of flora and fauna, explorers wrote of its landscape as evidence of a land of plenty. One hundred years earlier, French explorers, scrutinizing the tree-lined shoreline, gave it a French name connoting its amazing forests. Land birds flew overhead, broad-leaved trees and shrubs bordered the shoreline, deer browsed, yet people sometimes also saw bears or gray wolves stalking the land and heard their cries and yowls. Ashore, the newcomers saw a wealth of wildflowers. Some, like the orange-scarlet trumpet creeper, climbed high into trees, winding among leaves and branches; elsewhere, virgin's-bower, or clematis, twined. Maidenhair fern grew less conspicuously, nestled low, hidden in the shade; in the proper season *Mertensia virginia* burst forth in startling blue blooms. The spring flora of May 1607 visibly proclaimed its richness to the English. George Percy wrote excitedly of the land as "all flowing over with fair flowers of many sundry colors and kinds" (Quin, 1967: 17). Few doubted that agriculture would be successful there.

But the British colonists were not the first people to realize the Chesapeake's agricultural potential. Centuries earlier, Native Americans entered the region, settled, burned the forest, hunted its birds and animals, fished, harvested shellfish, and gathered wild plants from the woods, inland swamps, and coastal marshes. Gradually they began to cultivate corn, beans, squash, pumpkins, and tobacco. Most of these domesticated plants originally were native to other parts of the New World and were brought into the Chesapeake by prehistoric families. Many plants originated in Mexico, carried northward over generations as people migrated or exchanged plant knowledge with neighboring tribes. A botanical diaspora began; some plants survived their wanderings, some thrived, and some did neither. As plants were passed along or people moved about, the local environments shifted: tropical to subtropical to temperate; mountain to plain to coastal shore. These domesticated plants were tested by the heat, the cold, the wetness or lack thereof at each point along the way, resulting in a variety of unique but still genetically diverse cultivars. Those that reached the Chesapeake represented another phase in the gradual spread of indigenous New World plants; hence, species shifting began long before European colonization.

In 1607 most Chesapeake cultivars were unfamiliar to English explorers and colonists, yet the settlers wasted no time planting them in their gardens. Likewise, Caribbean peoples and natives of both North and South America also quickly incorporated European fruits and vegetables into their garden plots. The global dispersal of plants intensified. Columbus carried garlic aboard his ships in 1492 and 1493; it soon spread throughout much of the New World, augmenting the ancient and widely established use of chili peppers in food recipes. Orange seeds made themselves at home in the West Indies, becoming fruit-bearing trees. Southeastern Indians exchanged peaches introduced into Florida during the 1500s. Thereafter, "volunteer" peach trees marched north in a sporadic line dictated by the movements of native traders, passing through lands that later became Georgia and the Carolinas. No one knows when peaches reached Virginia, but Hugh Jones (1724: 315) wrote of both peaches and nectarines (clingstones) as "spontaneous" because Chesapeake Indians "had, and ever had had greater variety, and finer sorts of them than the English."

This glimpse of the Chesapeake's botanical history as intensive European contact began indicates some of the unique characteristics of the region. People usually view frontiers as defined by human migrations that cross into new lands to establish communities in strange territories. An aspect of frontier life is fewer barriers to cultural change—people are more open to experimentation. Frequently, there is cultural contact with new societies and a dynamic interchange of ideas, beliefs, and social practices. One result is that people, plants, and animals (whether insects, birds, fish, or mammals) encounter strange or exotic species and unfamiliar situations. What happens is not fully predictable; what happened is not always readily observable when read from distant points in time.

Yet change inevitably ensues, as it did in the Chesapeake when European colonists introduced new flora and fauna. They began a more intensive phase of species shifting than that of the prehistoric period. From the onset, although the Euro-American culture was conservative in many ways, its horticultural domain was experimental. European colonists learned how to farm by intermingling Old and New World cultivars. At the same time Chesapeake plants began to appear in Old World gardens. It was an era when men's fascination with and dependence upon the natural world led them to measure their worldly success, in part, by their mastery of nature. The planters, merchants, and mariners involved in colonization came from European cultures in which plants, their cultivation, and formal placement on the land represented a potent symbolic mechanism through which families "conversed" about their power, wealth, and social status. There was an intense, lively, and dynamic interest in botanical curiosities because of the role plants played in this cultural milieu. Yet people also took great care to separate themselves from the wilderness—which represented an uncontrolled nature, or *wildness.* They exploited it and were not, in modern terms, environmentally sensitive.

It mattered little what went into gardens; it was more important that what came out was useful. In New England, escaping sorrel left kitchen gardens to plague farmers in their fields. Plants in the Chesapeake also moved out across the land. The aggressive spread of European plants across North America had begun. Many of the escapees—yarrow, lovage, comfrey, alfalfa, sweet clover, hound's-tongue, common thistle, forget-me-not, and the common white daisy—became so commonplace that later generations never questioned but that these were native plants and never thought they might have had non-American origins. Eventually, the culturally shaped landscape together with its globally derived contents became a shared inheritance. Whether plants were Asian, African, European, or American in origin mattered little in folk memory. Because recorded scientific observations of agriculture developed more slowly than the experimental flow of horticultural practice, modern texts sometimes posit later dates for the introduction of plants and animals into colonial ecosystems than archaeological or herbarium data reveal.

Thus, in researching the evolving Chesapeake garden, one encounters the accidental spread of plants, the exploitation of plants for profit, and the cultivation of plants for pleasure, food, and medicine. Formal gardens—in which there was deliberate manipulation of the landscape to express where one's family stood in the social, political, or economic hierarchy—began to grow side-by-side with the gardens of ordinary folk, whether black or white. The latter type of gardens were plots wherein the land was altered for practical purposes to feed people or provide fruits and vegetables for local markets. The inflow of plants continued, and the diversity of botanical specimens adapting to the Chesapeake environment expanded to include flowering bulbs obtained on the African continent and roses from the Far East. The influence of this botanical diaspora can be seen today in almost every nook and cranny of the region.

The Start of Euro-American Gardens

The garden history of the Chesapeake is a record of transplanted useful fruits, vegetables, and herbs; exotic specimens imported and grown to demonstrate extraordinary horticultural skill; and weeds that came along for the ride. The early years of British occupation in the Chesapeake were characterized by the importation of many plants, especially domesticated European species. It was a period when families cleared land and became acquainted with the region's weather and the vagaries of the soil; survival was paramount. The horticultural experimentation that took place was often unrecorded. Frontier life was difficult and attention was focused on subsistence activities.

Upon arriving in 1607, Jamestown settlers planted familiar crops and exotic ones: European strains of wheat, Chinese apricots (brought to England in the 1400s), West Asian figs, Mediterranean olives, and Brazilian pineapples. The transfer of plants was under way. At its start the flow was east to west, carrying along Old World species. To picture the earliest settlers' gardens one need only reflect on descriptions of postmedieval English gardens and consider, as one example, the types of roses traditionally grown—damask roses, cabbage roses, musk roses, and other varieties, like the holy rose, brought back during the Crusades. Many gardens consisted of raised planting beds set among a range of protective fencing, from ragtag wooden slats to tight, stout paling, depending on the energy of the landowners.

The first successful garden in Jamestown may have belonged to Sir Thomas Gates, who created a town garden near the governor's mansion and filled it with traditional European fruits—apple and pear trees—as well as other plants whose identities went unrecorded. He benefited from the mistakes of the first gardens, and his success prompted others, in 1620–21, to try exotic fruits and vegetables again: lemons, almonds, figs, olives, ginger, sugarcane, pomegranates, plantains (bananas), and prickly pears. Within five or six years, Joan Pierce had gathered 100 bushels of ripe figs from her three- to four-acre garden. First and foremost, however, gardening in the Chesapeake was a practical activity that families, large and small, poor and rich, undertook to provide familiar vegetables for themselves. As part of the working landscape, gardens held few decorative elements. Most families initially grew European vegetables, fruits, pot herbs, and medicinal herbs as well as physic flowers used in healing.

At the onset garden cultivation was slightly irregular, not fully organized. In part this was because men outnumbered women, significant because English tradition dictated a gendered division of garden labor. Thus men knew how to raise cereal crops but not the herbs and kitchen vegetables that their female relatives grew. Further, the Chesapeake's main money crop—tobacco—was labor intensive. It demanded the labor of all household members just when women's root, herb, and salad gardens needed extra nurture, weeding, and watering. A conflict was inevitable, and there is much in the history of Chesapeake foodways that suggests families resolved it in favor of tobacco production.

Any plant that would grow and could be used was adopted, first by experimentally inclined planters and later by more conservative ones. Among the food plants that thrived best were those originally grown by Native Americans, because they had had hundreds of years to adapt to the region. The first generations of Chesapeake families were unfamiliar with sunflowers, lima beans, scarlet runner beans, and bush and common string beans; with peppers, corn, and summer and winter squash; and with fruits like strawberries, cranberries, and blueberries. Soon, however, European and Native American plant materials were thoroughly intermingled. The colonists set aside the Indian names, giving plants new English ones, and began to weave them into English recipes.

Science still provides sparse data about the early contents of Chesapeake gardens. The recovery of archaeological evidence of old plant remains, their identification, and the establishment of chronologies is still in its infancy. Plants of the past can be identified through pollen analysis, but the art of looking at pollen samples enclosed in small spaces and deposited over very short spans of time is a recent advancement. Historical archaeology reveals more readily the grand garden outlines and terracing because raised planting beds leave only fragmentary soil stains in the ground, whereas the root holes left by tree-lined allées are easy to read. Landscape historians note that the bulk of written documentation describes elite gardens.

Among the common folk, garden lore passed down by word of mouth. Today, gardeners participate in informal exchange, passing plants from neighbor to neighbor, taking cuttings with and without permission, and scrambling over fences to look closely at a new plant. Mechanisms similar to those that maintain heirloom plants in the present surely existed in the early Chesapeake. In fact, they may have been one of the major means whereby misconceptions about horticulture in the New World were set aside.

Misconceptions about New World Horticulture

The first gardens prospered initially but were short-lived. Too often, the English grew plants that needed other environmental conditions to flourish. They introduced the plants suddenly, too quickly for long-term survival; many species were exotics whose needs and habits the English knew little about. The colonists failed to recognize that many domesticated plants had spread across Europe so gradually that their origins were not simply ancient Greece or Rome (both countries served as symbolic ancestors for the English and hence were considered to be appropriate plant donors). They were unaware that many had grown in Egyptian and Near Eastern gardens and that others had been nursed by skilled gardeners, even royal gardeners, until the plants had adapted to different climates. No one knew the many years it took for New World cultivars to migrate northward.

Other problems arose because English colonists believed vigorous plant growth was determined by heat, which they related to latitudinal location. Horticulturists assumed that environments across the Earth at similar latitudes would support similar food crops. Although few planters expected that environmental conditions of the Chesapeake would exactly parallel those of England, many thought the Chesapeake growing season would be similar to that found in warm, temperate regions of Spain given that both areas lay at the same latitude. Fruits such as figs, oranges, lemons, limes, peaches, and pomegranates, grapes for wine, olives, and almonds grew well in Spain. Consequently, English settlers looked forward to mild weather and the successful propagation of a wider range of plants than they could grow at home. This expectation can be seen in John Lawson's 1709 account: "When we consider the Latitude and convenient Situation of Carolina, had we no farther Confirmation thereof, our Reason would inform us, that such a Place lay fairly to be a delicious Country, being placed in that Girdle of the World which affords Wind, Oil, Fruit, Grain, and Silk, with other rich Commodities, besides a sweet Air, moderate Climate, and fertile Soil" (79). Hugh Jones, writing in 1724, provided a biblical explanation for this fecundity: "The Country is in a very happy Situation," Jones claimed, "between the extremes of Heat and Cold . . . . Certainly it must be a happy Climate, since it is very near the same Latitude *with the Land of Promise*" (296; emphasis added).

Climatic differences between the two hemispheres made it impossible to take horticultural knowledge based on British experience and apply it without change. Chesapeake weather was both hotter and colder. Colonists and plants alike found themselves living in a region of climatic extremes. People grew to expect violent spring and summer thunderstorms, unpredictable frosts, a bone-chilling late winter wind, humid August weather, weeks of drought, hard-driving rains, and little snow. Spring came sooner; summer lasted longer. As a result, some English plants thrived (garlic and chives; English peas of various sorts; and roots such as beets, leeks, shallots, parsnips, turnips, carrots, onions, and radishes). Some did not, and it is a certainty that most exotic or "tender" species did not do well when first planted in New World gardens and that many were placed in gardens whose locations were unsuitable for good growth.

Garden Placement in the 1600s

Early colonists placed country gardens wherever they were best suited in terms of soil conditions and the positions of outbuildings. Some families chose an exposure that would catch the early spring sun; others followed the lay of the land. This was easier done in rural areas than in the towns, where many had sparse knowledge of their plants' needs for heat and light. Still, archaeologists have unearthed outlines of fenced-in garden plots at almost all the small towns established in the early Chesapeake. Their findings are complimented by legislative records. Virginia legislators mandated garden fences to keep out animals such as swine, deer, moles, rabbits, raccoons, and woodchucks. Their 1624 law required all freemen to "fence in a Quarter of an Acre of Ground, before Whitsuntide next ensuing, for planting vines, herbs, roots, and the like, under the Penalty of ten Pounds of Tobacco a man" (Stith, 1768: 321). And in 1633, Cecilius Calvert, the first Lord Baltimore, wrote out instructions concerning St. Mary's City, Maryland, requiring the back lot of each house to have land for a garden (22).

Garden Plants in the 1600s

Additional information on horticulture is provided by a few explorers and travelers. Most, however, like Captain John Smith, did not provide species-specific botanical information, using instead broad descriptive categories or referring to cultivars simply as "garden stuffe." This pattern is exemplified in Smith's 1629 consideration of the Jamestown woods that had been turned into pastures or gardens wherein grew "all manners of herbs and roots we have in England in abundance and as good grass as can be" (Arby, 1884: 887). Additional narratives in this genre include John Josselyn's 1672 book, *New-England Rarities Discovered,* where he ex-

---

Opposite: Fig. 12.1. Plat drawn in 1697 showing location of a peach orchard in Calvert County, Maryland (Maryland State Archives, Charles County Court Records v. 1, MDHR 8132).

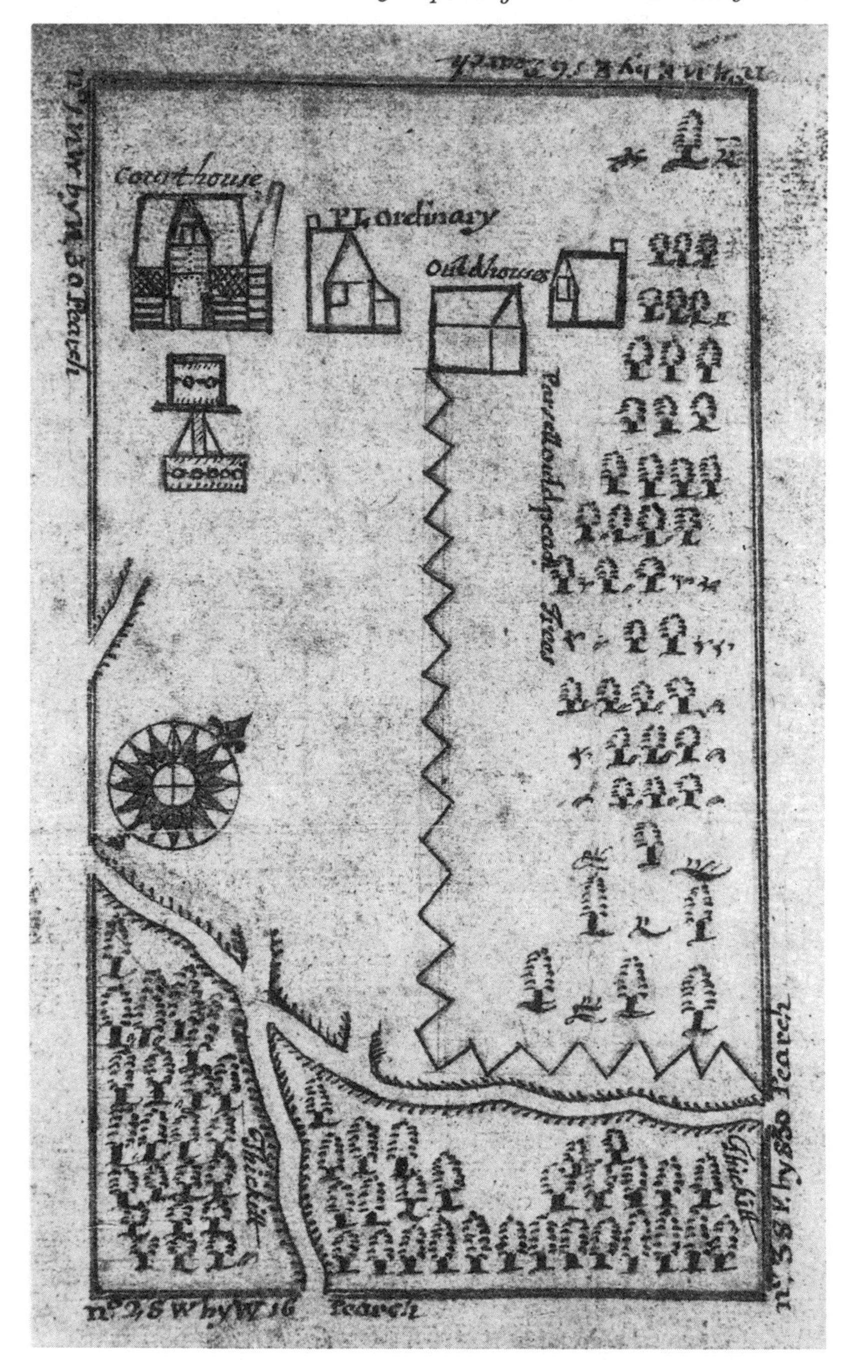
Court house
P L ordinary
Old houses

plained that asparagus prospered mightily, requiring neither hotbeds nor extensive fertilizing. Asparagus also flourished in the colonial Chesapeake, where archaeologists have recovered remains of their planting beds. Even today, asparagus connoisseurs look forward each spring to the fresh stalks grown on the Eastern Shore. Yet many familiar European plants did not do well in North America. Although retaining its sweet taste, the European broad bean degenerated into a small, or pygmy, bean in the Carolinas. Anise seeds did not grow to maturity; bloodwort was a sorry plant, and fennel had to be taken up and kept in a warm cellar each winter in Connecticut. Josselyn found that neither rosemary nor southernwood grew and that rue would hardly grow. Lavender was also problematic, while gilly flowers converted to Yorkshire fennel. Some of these events vexed Chesapeake gardeners. It made life problematic because many were tried and true European herbs whose medicinal uses families knew well.

Fortunately most fruits prospered. In a 1697 letter, Thomas Lawrence wrote that Maryland hogs ate more and sweeter peaches than English duchesses could obtain (Jordan, 1978). Virginia's Hugh Jones boasted in 1724 of giant peaches over a foot in circumference. Other people complained because peach trees grew like weeds. In 1688 the Rev. John Clayton informed English botanists that Chesapeake planters could "sleep" while apples, pears, peaches, and apricots grew. And, he continued, "after their setting or engrafting there needs no more labour but your prayers, that they may prosper, and now and then an eye to prevent their casualties, wounds or diseases" (1688: 49). Land plats showed orchards; probate court records listed estates whose walks were lined with cherry, apple, or peach trees. Among the Chesapeake's colonial period archaeological assemblages, archaeobotanists identify remains of traditional European fruits. Naomi F. Miller of the University of Pennsylvania found remnants of apricots, cherries, grapes, peaches, and plums at the Calvert site in Annapolis, Maryland. Advertisements by seedmen, however, reveal more about the varieties (early, late, large, small, etc.) that were grown than is seen in the archaeological record.

Weedy Plants in the 1600s

Numerous letters, diaries, and shipping manifests identify many of the bulbs and seeds shipped to the New World. Men also shipped small plants and trees in wooden containers. But weeds—plants that grow out of place which no good gardener sows or gathers—also crossed the ocean

Fig. 12.2. Thistles were both native to North America and introduced from Europe, moving out quickly to grow in unexpected places (author's collection).

to thrive in the new ecosystem. Dandelion and tumbleweed were among the earliest European weeds to appear. Native Americans called the coarse plantain "Englishman's Foot" because it grew on low-lying land wherever the settlers or their cattle trod. Weed seeds arrived mixed among food seeds or stuck in the cracks of animal hooves. They also made the passage inside the digestive tracts of cows, and purslane soon appeared wherever cattle dung was spread. Horses, sheep, goats, swine, poultry, and even people were unknowing carriers. Finally, many plants now considered weeds—yarrow, tansy, feverfew, and chicory, for example—escaped from kitchen gardens to grow in meadows, on cleared land, or alongside dirt roads.

How Chesapeake Plants Left the Region

Plants also escaped from the Chesapeake with human assistance. The shipment of plants eventually went west *and* east. By the 1640s, there are incidental records of Virginians sending Atanasco lilies that had grown in swampy land along rivers to plant collectors in London. A more scientific exchange of Chesapeake plants began in 1637 when John Tradescant the Younger arrived in the Tidewater. The reasons he traveled the world to collect plants illustrates the fascination horticulture held in seventeenth-century society and provides a context for the botanical diaspora that continues to this day.

Many British families did not want to move abroad, yet they still wanted to grow New World plants. Others were simply curious about what grew elsewhere. The discourse surrounding plants was central among various sectors of British society. This focus was among the factors that motivated John Tradescant the Elder to open "Tradescant's Ark" near London in 1629, one of England's oldest natural history museums. Gardeners sent exotic plants from around the world for its planting beds. The first groups contained specimens from the St. Lawrence River, where French plant collectors were active. Since southern, temperate plants and tropical plants were sparsely represented in the museum's collections, John Tradescant the Younger left England in the 1630s to gather living specimens and seeds of the Chesapeake plants to improve the museum. He returned to the Chesapeake region in 1642 and 1654 for more species.

Although no precise list of the Tradescants' garden survives, their trees included specimens from Virginia such as the bald cypress, Atlantic white cedar, red mulberry, black locust, hackberry, tulip tree, bladdernut, and red maple. They planted bergamot, fox grape, and wild grape. John Tradescant also brought back trumpet honeysuckle, thimbleberry, pokeweed, silkgrass, virgin's-bower, Virginia rose, and Virginia creeper, which is now weedy in much of Europe. The Tradescants promoted the use of Chesapeake herbs, including the false Solomon's seal, columbine, bloodroot, wild lupine, cardinal flower, balmony, goldenrod, and coneflower (i.e., Virginia leopard's bane). They introduced Tradescant's aster, now common in gardens across the world, and the evening primrose, which today has numerous cultivated forms.

The Ark's public could also see poison ivy, spiderwort (now known as the insectivorous pitcher plant), Jerusalem artichoke or "Virginia potato," and medicinal snakeroot—whose availability English apothecaries were happy to advertise. Along the James River, John Tradescant the Younger also obtained a maidenhead fern, the noble liverwort of Virginia (actually, a flowering plant), and a fragrant Carolina jasmine that John Parkinson, a noted English botanist, soon grew. In essence, the contents of the Ark encapsulated the vitality of the plant exchange between the Old World and the Chesapeake.

New World "exotics" helped fuel the growing desire of the English and European elite to have Old and New World plants in their formal pleasure gardens. This desire prompted more adventurous plant collectors to comb the Far East for plants as well. James Cunningham arrived in London in 1700 with a wealth of wondrous oddities from China which grew readily in the English climate. But not all English wanted foreign plants, and in 1712, Joseph Addison, a well-known plant critic writing in the London *Spectator,* declared he was "more pleased to survey my row of coleworts and cabbages than to see the tender plants of foreign countries kept alive by artificial heat." This did not stop Mark Catesby, who returned to Charleston, South Carolina, in the 1720s to find herbaceous plants for perennial beds and flowering borders. While working on his *Natural History of Carolina, Florida, and the Bahama Islands,* he asked John Clayton of Williamsburg for help. Clayton responded by collecting plants on the tidewater peninsula, finding samples that were either unknown or had been grown in English gardens decades ago but were later lost from cultivation. Clayton spent the next 30 years collecting and bringing into cultivation hundreds of plant species. Many of these soon graced the homes of his Chesapeake friends. His seeds were even sent to Holland, where they were widely distributed throughout Western Europe and South Africa, and, in some cases, marketed back to the American colonies, where they were sold as new and exotic.

Changes in Chesapeake Gardens after 1650

Plants from elsewhere in the New World and from India, Africa, and the Far East joined the expanding botanical population of the Chesapeake. Men imported them for experimental reasons, to see whether they could be turned into moneymaking crops, or carried them as food stores on sea voyages. Many herbs and vegetables from East Africa, India, China, Japan, the Malay Peninsula, and East Indies nevertheless remained in their native countries. This happened because plant collectors rarely spoke native languages fluently and dealt mainly with men, who either were not close to the soil or lacked knowledge of these more humble plants because that was retained by women. Some African food plants were successful in Chesapeake gardens, but because their seeds could not survive the winter cold, they never became naturalized. Others had very short growing seasons. Yams were grown seasonally but required special treatment. If their roots were stored inside, in root cellars near a hearth, they could survive the winter to provide stock for spring planting. Families learned that such plants required more labor. Eventually, many farmers gave up trying to cultivate them. Other plants—rice, sesame, sugarcane, ginger, indigo, and sea-island cotton—required warmer weather than the region provided, although they flourished in the Carolinas and Georgia. These last crops left little lasting impression on Chesapeake gardens and are not usually discussed in accounts of regional agriculture; but they were here once, and farmers tried to cultivate them throughout the 1600s and early 1700s, using every ounce of wisdom they possessed.

As time passed, English settlers learned what would grow in the region and what would not; they became better able to identify natural commodities and more skilled in reporting what they observed. As crops became more reliable, upwardly mobile planters accrued wealth. This social group turned their attention to those aspects of their country estate that would make their wealth and social status visible. By the mid-1600s a few families had built brick mansions, bought luxuries as well as necessities, and created formal gardens near their homes. They grew certain plants simply because these were pleasingly beautiful and sweetly scented. Their favorite ornamental plants continued to be ones that reminded them of England or that became popular in England. So, it is likely they grew the old-fashioned, deeply perfumed (i.e., medieval) roses, lilies, lovage, mallow, marigolds, cowslips (primroses), violets, peonies, and daffodils, iris, larkspurs, and foxglove. As years progressed, wealthy families ordered plant varieties from England that included tulips, hyacinths, crocuses, and anemones. Michaelmas daisies must have been grown too.

The Chesapeake elite established the first regional aristocratic gardens from c. 1650 to 1680. Their formal design and English planting themes influenced Chesapeake gardens until after the American Revolution. Initially, these formal gardens retained medieval design elements. Owners separated the house and garden, and most placed the garden beside the main house. Diane McGuire (1992) writes of them as gardens and houses or gardens near houses in contrast to houses in gardens. Enclosed, individual, social spaces, the early formal gardens were, like their medieval antecedents, sanctuaries from the wilder, natural forests that surrounded them. Occasionally an archaeological site such as Bacon's Castle in Surry County, Virginia, produces evidence of a feature which speaks to this in an evocative way—one which only the rich possessed. Formal garden features, constructed of brick and known as exedras, provided a cool, refreshing spot to view the garden and contemplate its beauty.

Chesapeake planters supported the overseas shipment of curious and rare Chesapeake plants because it also enhanced their social prestige, at home and abroad. In the process, the importance of rare or decorative plants changed. Men with deep interests in horticulture recognized that they possessed an unusual commodity that others sought; some became better organized in their collection of native plants for European gardens and herbariums. A few individuals, including some of England's first botanists, also were instrumental in the transport of Chesapeake plants to England and Europe. The importation of botanical specimens to England brought significant alterations to the gardens of English aristocrats. The Royal Society in London in 1662 began to play a critical role in the plant diaspora; the efforts of its members helped produce the first systematic descriptions of Chesapeake flora and supported the collection of seeds, plants, and illustrations thereof.

Eighteenth-Century Chesapeake Gardens

Through the early 1700s, most ordinary families were content to grow a few simple plants for decoration. Among the herbs they cultivated, some are known today more for their flowers than for their medicinal or culinary properties. To the south John Lawson found gardens holding little except two or three types of roses, some spring violets and carnations plus prince's-feather and Tres Colors. Yet families had begun to in-

Fig. 12.3. Seventeenth-century garden at Bacon's Castle in Surry County, Virginia (courtesy of Nicholas Luccketti and the Society for the Preservation of Virginia Antiquities).

Fig. 12.4. Umbrella trees grew in Chesapeake towns such as Annapolis, where they were used for medicinal purposes. Today they are considered weeds and can be seen along the sides of country roads throughout the South. The seedpod depicted here was recovered from a water-logged deposit in an Annapolis well, where it was tossed during a period of yard cleaning and house renovation (illustration by Julie Hunter-Abbazia, from Yentsch, 1994).

corporate more native plants into their gardens than ever before. Flowering specimens of Native American plants now adorned gardens, and ordinary planters shaded arbors with honeysuckle, morning glories, and gourds. Runner beans and other vines such as Virginia creeper were planted for shade and decoration in the spring. Lawson (1709) described bushel beans planted around arbors or at support bases that they then climbed to cover the arbor roofs, creating a marvelous summer and early fall shade. The beans continued growing until the first frost drew close, when each died.

Gardeners learned by trial and error as demonstrated in letters between John Custis of Williamsburg and Peter Collinson of London in the 1730s and 1740s. Men carried specimen trees from the uplands and mountains to the coast to plant. The unicorn plant made its appearance in Annapolis; gentlemen planted umbrella trees (American magnolia) in Williamsburg. Decorative flowers, bushes, and trees—white dogwood, pink dogwood, redbud, wild plum, wisteria, lady slipper, passion flower, Indian iris, trumpet vine, sarsaparilla vine, Indian honeysuckle, and Carolina jasmine—were dug up in forests or wherever they grew and replanted where they could be admired. The people who did this work never thought to keep a record of their activities, although travelers did describe the exceptional plants. Gardeners were concerned to make plants grow and the process proceeded intuitively. Those with special interests in healing paid particular attention to the medicinal herbs known to Native Americans.

In Virginia, John Custis, a skillful gardener, successfully grew a wider range of plants than those Lawson recorded. He grew many plants obtained by exchanging seeds, cuttings, and bulbs. Varieties developed spontaneously that could better survive the growing conditions in the region (e.g., the noisette rose, Mary Washington). Even as the colonies were drawing apart politically from England, their gardens began to exhibit more Old World plants, while the practice of mingling vegetables and ornamentals—lily of the valley, bunch beans, balsam, marigolds, and radishes—gave its own visual flair to the planted landscape.

Men who had contacts in shipping circles or owned vessels obtained plant materials directly from England. Mariners were not noted for their skills in keeping plants alive, so the process was frustrating. A fortunate William Byrd I noted that his iris and crocus bulbs had made a successful voyage in 1683 and bloomed that year. More often seeds sprouted beforehand, roots desiccated, and the plant died. Custis was furious when a dog in steerage made a meal of his plants, tearing to bits his carnations and auriculas. Even with the best of intentions, plants barely escaped with their lives because sea captains, knowing little of individual plant requirements, were as apt to overwater, thus rotting the roots, as to underwater, thus starving the plants of moisture.

As the rough-and-ready life of the frontier moved westward and more

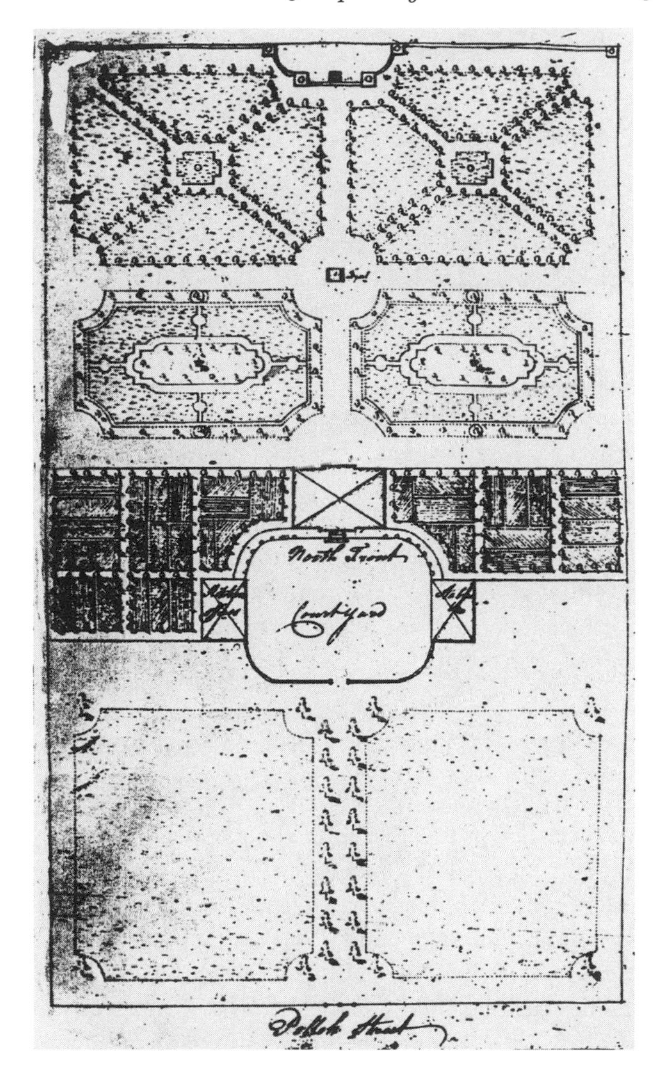

Fig. 12.5. The garden at Tyron Palace in New Bern, North Carolina, is similar to many of the formal gardens planted throughout the eighteenth-century by the Chesapeake elite (courtesy of Tyron Palace).

large-scale planters produced successful crops year after year, the land became tamer and more attention was given to gardens that were luxuries, planted for pleasure and not for profit. A division emerged (subsuming both form and content) that paralleled the society's ranking system, with native-born elite actively establishing formal gardens and filling them with fashionable trees, vines, shrubs, herbs, and flowers. The gardens were planted with exotic, "tender" plants, outlined by formal, symmetrical garden walks, and initially contained limited ornaments such as urns and statues to draw the eye and enhance the view. Not all were identical. William Byrd I in 1684 was among the first generation of Virginians to grow tulips and anemones. Although several generations of Byrds at Westover knew them well, these plants were rarely seen elsewhere. The family's exclusiveness extended into the plant world.

Relying on reciprocity among gardeners, Englishman Peter Collinson sent Virginian John Custis varieties of plants that would be uncommon in a New World garden: tulips, pinks, altheas, lilies, hyacinths, narcissuses, polyanthuses, fall crocuses, foxgloves, cyclamens, yellow asphodels, chrysanthemums, Persian lilacs, various roses, Italian tuberoses, and crown imperial lilies. Custis returned the favors in kind. Within the next 50 years, seedsmen like Robert Prince of Flushing, Long Island, began to list hundreds of rose varieties in their catalogs and to carry other rare specimens, although earlier in the century Englishman Batty Langley had listed less than two dozen roses among the plants recommended for British gardens.

Post-Revolutionary-Era Gardens

Most gardens remained conservative. They kept the terrace falls, ramps, and formal steps together with the parallel walkways which accentuated the strong axial flow and formal lines as well as the geometrically shaped parterres of older English gardens. At many, the terracing still survives. But beginning around 1785, the planter gentry incorporated newer garden styles within the old lines. The end of the Revolutionary War marks the start of an era of garden change. By the early nineteenth century men were designing different types of gardens and altering old carriage drives to provide parklike settings through which visitors entered. These settings were a sharp change from the earlier, open lawns, fields, and bowling greens in which mansions sat. The old, commanding vistas that kept mansions on display were now interrupted by tree plantings that shielded the houses and provided tantalizing glimpses of the terrain. The changes can be seen in regional maps and are depicted in a series of chair paintings by William Guy at the Maryland Historical Society. They have left a lasting imprint on any piece of land where an old house still stands, especially those built during the early 1800s.

The regional plant repertoire remained limited, especially in the gar-

Fig. 12.6. *Mount Vernon* (George Ropes, 1806) shows changes to the garden landscape made after the Revolution (courtesy of the National Gallery of Art; gift of Edgar William and Bernice Chrysler Garbisch).

dens of ordinary families. Such families consistently gardened in traditional ways, with modest experimentation and modest plants. Women peddled fresh produce such as greens and pot herbs in markets and door-to-door. The occasional street seller of today who markets her garden wares on the city streets of Annapolis, Baltimore, or Washington is following a tradition that began around 1630 and grew to include women of both African and European descent. Some garden plots were extremely small (10′ x 12′), but others were more sizable. By the late 1700s a few craftsmen, like William Ferris, achieved local fame as master gardeners. Sarudy (1989) found that Ferris cultivated hyacinths, tuberoses, and polyanthus, growing them in movable pots and giving them the same special care as his other precious plants: mignonette, asters, impatiens, and the newly introduced chrysanthemum. Marvelous gardens were no longer the prerogative of the rich.

The nurture of a luxuriant garden remained time-consuming. In Maryland, Stephen Bordley spent three early morning hours every day with his head gardener—a slave—on his Wye Island plantation to keep their multi-acre garden in shape. Charles Carroll, who maintained a lavish garden at Doughoregan Manor outside Baltimore, had upward of 200 slaves, including gardeners Jack and Harry. This labor source was not available to the average person, such as Baltimore court crier William Bigger, who planted a minimalist garden—two beds and one tree-lined path. Merchant James McCannon, whose wife owned their single slave, kept a simple garden plot. Only wealthy families could maintain citrus trees or plants that required much moisture. Indentured servants or slaves provided the labor, and when little rain fell in May, June, or July, planters did not do the watering themselves, as documented in a letter by John Custis: "I kept 3 strong Nigros continually filling large tubs of water and put [the plants] in the sun and watered plentifully every night, made shades and arbors all over the garden almost." He recorded that most of his plants perished. Given the care they received, one can only imagine the damage to gardens of poor and middling families throughout the region.

The aristocratic gardens of the post-Revolutionary Chesapeake flourished only because slaves watered and nursed the plants, hoed the ground, pruned shrubs, built terraces, weeded, and mucked out stalls for fertilizer. Slaves shifted the potted orange trees in and out of glasshouses and in and out of the shade. In the spring and fall, they capped smaller, tender plants with bell-shaped glass jars to trap moisture and keep the frost away. At summer's peak, slaves swung scythes to cut the grass on bowling greens when it reached knee height.

The garden work of African and African American slaves was not restricted to aristocratic gardens. In towns and cities, widows and merchants needed slaves' assistance. William Ferris worked side by side with an old Negro woman in his Annapolis flower garden. Doubtless there were many others who had a similar arrangement, as the garden history of the Chesapeake shows strong African influences. Yet this is a heritage that is largely unrecorded and unexplored. One place to begin is with what we know about the gardens that slaves kept for themselves.

African Contributions to Chesapeake Gardens

Behind outbuildings, by the slave quarters, and wherever small planting beds could be made on the slave-master's grounds, black men and women grew their own vegetables, sometimes working by the light of the moon when their bodies were weary, to be sure their families were fed.

Fig. 12.7. Eighteenth-century garden with raised beds and bell jars that gave plants warmth and shelter from the cold. Insets show bell jars protecting a strawberry plant and spring lettuce and the profile of a glass fragment from the base of a jar. Although the glass is shown as translucent, it was actually semi-opaque (illustration by Julie Hunter-Abbazia, from Yentsch, 1994).

In these garden spaces, they planted crops with familiar African names, including those we now know as okra, dasheen, black-eyed peas (cowpeas), pigeon peas, white and yellow yams, red rice, millet, sorghum, and watermelon. It is a surety, however, that these are not the only African

Fig. 12.8. African calabash vines growing around a slave quarter (illustration by Julie Hunter-Abbazia, from Yentsch, 1994).

plants that became naturalized in New World settings; the list reflects more what we know about what these folks grew than the reality. As improvisational in their approach to gardening as they were to other aspects of daily life, Africans and African descendents grew plants that had been transported from the Americas to Africa and integrated into the African cuisine a hundred years earlier (peanuts, peppers, tomatoes, sweet potatoes, lima beans) as well as plants of European origin (cucumbers, cabbages, English peas, turnips, kale, and collard and mustard greens), plus some, like corn and grancy graybeard, native to the land. Perhaps some African plants came to grow as weeds in the Chesapeake in a manner akin to that of the African guinea fowl, which roosted near small slave quarters at night, eventually adapted to the wild, and is seen today in rural Maryland and Virginia.

Planting was not limited to edible vegetables, and horticultural experimentation was not uncommon. Slaves and free blacks raised ornamental plants (such as hollyhocks), medicinal plants, and some with supposed magical properties (e.g., the gaudy red and yellow torch lily known as red-hot poker). Bitter gourds that could be turned into cooking, eating, or drinking vessels climbed over walls. Men and women with special expertise traded and sold their produce. Word of individuals with special planting skills spread from one quarter to another, from plantations on the Eastern Shore to those that stood across the Bay. Although the information exchange was informal and followed networks established by slaves, it eventually extended further. Some planters began to appreciate the gardening skills of their slaves and to ask why they did things in certain ways, as an exchange between planter Landon Carter and his slave Jack Lubber illustrates.

In June 1771, Carter reminisced in his diary, "I walkt out this even to see how my very old and honest Slave Jack Lubber did to support life in his Extreme age; and I found him prudently working amongst his melon vines, both to divert the hours and indeed to keep nature stirring that indigestion might not hurry him off with great pain." Carter then had taken "notice of his Pea Vines a good store and askt him why he had not got them hilled." "Master," Jack Lubber replied, "they have not got age enough and it will hurt too young things to coat them too closely with earth." This answer, Carter felt, exemplified "the Prudence of Experience" (Sobel, 1987: 42).

Plantation owners had always been good at diversifying their activities to gain every ounce of profit possible. It was not long before insightful planters were growing African-based crops to sell in city markets. The journal of James Moss, on file at the Maryland State Archives, indicates that by end of the century slaves on his plantation had raised thousands of watermelons. They were planted annually in April and May, harvested in summer, and shipped, possibly on a schooner with the

African name *Mingo,* to Baltimore, where the watermelons were sold in markets and door-to-door for the edification of black and white alike.

Plant lists from this same era contain a number of species that originated in different parts of Africa, including its southernmost tip at the Cape of Good Hope, and then entered Chesapeake gardens. South Africa was the source for Upton Scott's many types of bulbs: crimson, scarlet and pink gladioli along with a fragrant night-blooming variety, amaryllis, scarlet lilies, a rose-red watsonia, pale lilac babiana, and the butterfly iris. Gerber daisies and pelargoniums (geraniums) began to decorate gardens. Some African plants actually reached London via the Chesapeake, including the Angola pea, about which British plantsmen were told that if it were planted in hills, harvested, and dried they would reap a cocoa-like beverage rich as chocolate. Coca-Cola, in fact, has an African derivation, although the beverage was developed in Georgia late in the nineteenth century, well before it reached European markets.

Thus the plant diaspora grew, expanded, and created ever-widening circles; the routes crossed one another's path and sometimes doubled back. The diaspora was multicultural in character and as inclusive as the region's immigrant origins would suggest.

Nineteenth-Century Influences

The passion for planting seen among many of the gentry was shared by husbands and wives. By the Revolution it had also begun to spill beyond class lines and descend through the society. Gardening was a popular activity because it was both pleasurable and provided food, spice, and seasonal variety to family dining. By the 1790s, seedsmen and nurserymen, traveling and stationary, lived and plied their trade in Annapolis, Norfolk, Baltimore, Richmond, and Williamsburg. The varieties of available plants and seeds increased threefold. Ornamental plants in gardens grew ever more varied. The shape of the elite landscape also changed because the aesthetics of situating homes and creating vistas became one in which privacy was privileged. No longer did wealthy families want their mansion homes on parade, visible to any and all that passed.

Meanwhile, more plants were added to the region's flora. Plant materials that came in with the China trade included chrysanthemums, peonies, magnolias, forsythia, wisteria, gardenias, and new species of pinks and roses. Southern European immigrants brought Mediterranean cuisine, and nurserymen catered to their tastes as well, adding artichokes, eggplants, oyster plants, and cantaloupe from seeds obtained in Tripoli.

As the California Gold Rush began and exploration of the Far West intensified, immigrants were introduced to the West Coast whose origins spanned the Pacific Rim. They, like the immigrants to the eastern seaboard, brought along new plants. As these plants were grown, observed, eaten, or simply enjoyed by men and women who had moved west to California, word of their desirability spread eastward and Chesapeake families attempted to grow plants sent back from California.

Gardens altered greatly throughout the 1800s and gave the land new contours, introducing new styles to mediate the boundaries between culture and nature. Lilies became a favored plant; dahlias grew fashionable. Peonies (from China) once were the rage. The inclusion of oriental species (e.g., Japanese camellias and hydrangeas; Chinese and Japanese chrysanthemums), some of which thrived and took off into the wild, left a strong imprint on the land. The popularity of petunias, discovered by the French in the warm valleys of Brazil in the 1830s, brought this plant to our shores. The Busy Lizzie came from Zanzibar around 1896, and modern sweet peas filled gardens at the turn of the twentieth century. Older flowering plants were seen as outmoded, as pansies, poppies, gillyflowers, larkspurs, hollyhocks, snapdragons, and thrift dropped in popularity.

What individuals sow today and perceive as ordinary garden plants draws on both the nineteenth-century heritage and the older legacy. The contents of Chesapeake gardens were never fully representative of plants grown worldwide for the plant exchange favored those that could be transported easily, such as bulbs, corms, tubers, and rhizomes, and those which had particularly sturdy seeds. Plant survival also showed climatic influences. Plants from regions with climates most like the Chesapeake had a better chance to live long enough to propagate. Those that liked the air in drafty Victorian houses and could stand the frost were better off than those that needed constant warmth and much moisture. The edible plants that people already knew how to cook (both African and European) were more likely to be incorporated into kitchen gardens. And if one could make money from a plant's sale, it was always grown. One result is that today a person can wander the paths of Maryland and Virginia nurseries and select plants from collections that are, in essence, multicultural. The all-American Baltimore or Richmond backyard garden typically displays global diversity, but native heirloom plants still grow in out-of-the-way places and in a few venerable gardens. Within the familiar countryside, from the Eastern Shore, across the Bay, and into the uplands—now inhabited by plants obtained worldwide—an astute observer can still see evidence of what once was but can never be again.

BIBLIOGRAPHY

Archaeological Studies

Kelso, W., and R. Most, eds. 1990. *Earth Patterns: Essays in Landscape Archaeology* (Charlottesville: University Press of Virginia).

Miller, N. F., and K. L. Gleason, eds. 1994. *The Archaeology of Garden and Field* (Philadelphia: University of Pennsylvania Press).

Noël Hume, A. 1974. *Archaeology and the Colonial Gardener.* Colonial Williamsburg Archaeology Series 7 (Williamsburg, Va.: Colonial Williamsburg Foundation).

Yamin, R., and K. Bescherer, eds. 1996. *Landscape Archaeology: Studies in Reading and Interpreting the Historic Landscape* (Knoxville: University of Tennessee Press).

Yentsch, A. E. 1994. *A Chesapeake Family and Their Slaves: A Study in Historical Archaeology* (Cambridge: Cambridge University Press).

The 1997 issue (vol. 17, no. 1) of the *Journal of Garden History, An International Quarterly* also contains articles on garden archaeology in the Chesapeake.

Botanical Studies

Bannister, J. 1680. A catalogue of plants observed in Virginia. Ms., on file at the Royal Society, London.

Berkeley, E., and D. S. Berkeley. 1963. *John Clayton, Pioneer of American Botany* (Chapel Hill: University of North Carolina Press).

Catesby, M. 1731–43. *The Natural History of Carolina, Florida, and the Bahama Islands* . . . (London: by the author).

Crosby, A. W. 1986. *Ecological Imperialism: The Biological Expansion of Europe, 900–1900* (Cambridge: Cambridge University Press).

Ewan, J., and N. Ewan. 1970. *John Bannister and His Natural History of Virginia, 1678–1692* (Urbana: University of Illinois Press).

Kunst, S. G., and A. O. Tucker. 1989. Where Have All the Flowers Gone? A preliminary listing of origination lists for ornamental plants. Association for Preservation Technology *Bulletin* 21(2): 43–50.

Parkinson, J. 1640. *Theatrum botanicum: The Theater of Plants, or an Herball of Large Extent . . . Distributed into Sundry Classes or Tribes, for the More Easier Knowledge of the Many Herbes of One Nature and Property . . . Collected . . . by John Parkinson Apothecary of London, and the Kings Herbarist* (London: Thomas Cotes).

Reveal, J. L. 1985. Colonial Maryland plants in D. C. Solander's "Descriptions of Plants from Various Parts of the World"—an unpublished 1767 manuscript. *Bartonia* 51: 80–92.

———. 1992. *Gentle Conquest: The Botanical Discovery of North America with Illustrations from the Library of Congress* (Washington, D.C.: Starwood).

Ross, P. L. 1984. *The John Tradescants: Gardeners to the Rose and Lily Queen* (London: Peter Owens).

Swem, E. G. 1957. *Brothers of the Spade: Correspondence of Peter Collinson of London and of John Custis, of Williamsburg, Virginia, 1734–1746* (Barre, Mass.: Barre Gazette).

Tradescant, J. 1656. Catalogus plantarum in horto Johannis Tredescanti [sic], nascentium. In J. Tradescant, *Musaeum tradescantianum or a collection of rarities preserved at South-Lambert neer [sic] London* (London: John Grismond).

Garden History

Favretti, R., and J. Favretti. 1990. *For Every House a Garden: A Guide for Reproducing Period Gardens* (Hanover, N.H.: University Press of New England).

Martin, P. 1991. *The Pleasure Gardens of Virginia from Jamestown to Jefferson* (Princeton: Princeton University Press).

McGuire, D. K. 1992. Early gardens along the Atlantic coast. In *Keeping Eden: A History of Gardening in America,* ed. W. T. Punch, for the Massachusetts Horticultural Society (Boston: Little, Brown), 13–29.

Sarudy, B. W. 1989. Eighteenth-century gardens of the Chesapeake. *Journal of Garden History* 9(3): 103–59.

———. 1998. *Gardens and Gardening in the Chesapeake, 1700–1805* (Baltimore: Johns Hopkins University Press).

Primary Historical Sources

Arby, E., ed. 1884. *Captain John Smith . . . Works.* Reprinted in 2 vols. with same pagination and with some corrections as *Travels and Works of Capt. John Smith* (1910), with an introduction by A. G. Bradley (New York: Burt Franklin, 1966).

Byrd, W., I. [1684] 1977. Correspondence. In *The Correspondence of the Three William Byrds of Westover, Virginia, 1684–1776,* vol. 1, ed. M. Trinley, with a foreword by L. B. Wright (Charlottesville: University Press of Virginia).

Calvert, C. [1633] 1925. *Instructions to the Colonists.* In *Narratives of Early Maryland, 1633–1684,* ed. C. C. Hall (New York: Scribners; facsimile, Bowie, Md.: Heritage Books, 1988).

Carter, L. [1752–1778] 1965. *The Diary of Colonel Landon Carter of Sabine Hall, 1752–1778,* ed. J. P. Green, 2 vols. (Charlottesville: University Press of Virginia).

Clayton, J. [1688] 1836–46. A Letter from Mr. John Clayton, Rector of Crofton at Wakefield in Yorkshire, to the Royal Society, May 12, 1688. Giving an Account of Several Observables in Virginia, and in his Voyage Thither, more particularly concerning the Air. In P. Force, *Tracts and Other Papers . . . ,* vol. 3 (Washington, D.C.: P. Force).

Jones, H. [1724] 1956. *The Present State of Virginia from whence is Inferred a Short*

View of Maryland and North Carolina, edited with an introduction by R. L. Morton (Chapel Hill: University of North Carolina Press).

Jordan, D. W. 1978. Maryland hoggs and Hyde Park dutchesses: A brief account of Maryland in 1697. *Maryland Historical Magazine* 73(1): 87–91.

Josselyn, J. 1672. *New-England Rarities Discovered: In Birds, Beasts, Fishes, Serpents, and Plants of That Country* (London: G. Widdowes).

Lawson, J. 1709. *A New Voyage to Carolina* (London: Printed for W. Taylor and F. Baker, 1714; facsimile, Ann Arbor: University Microfilms, 1966).

Quin, D., ed. 1967. *Observations Gathered Out of "A Discourse of the Plantation of the Southerne Colonie in Virginia by the English," 1606.* Written by that honorable gentleman, Master George Percy (Charlottesville: University Press of Virginia).

Reveal, J. L. 1983. Hugh Jones (1671–1702)—Calvert County naturalist. *Calvert History* 1(2): 1–11.

Sobel, M. 1987. *The World They Made Together: Black and White Values in Eighteenth-Century Virginia* (Princeton: Princeton University Press).

Stith, W. 1768. *The History of the First Discovery and Settlement of Virginia* (Williamsburg, Va.: William Parks; reprint, Spartanburg, S.C.: The Reprint Co., 1965).

CHAPTER THIRTEEN

Genteel Erosion

The Ecological Consequences of Agrarian Reform in the Chesapeake, 1730–1840

CARVILLE EARLE AND RONALD HOFFMAN

The Carroll family of Maryland provides a biography of change in attitude toward land use. Father and son stand on each side of the divide between empirical observation and simple technology in the final third of the eighteenth century and the complex technologies of plows, fertilizers, and "farming by the book" that brought the destructive practices of the next century.

An understanding of the Chesapeake and its ecological experience in the colonial and early national eras must begin with its economy. Situated within the trading networks of the Atlantic world, the Chesapeake's was a regional economy committed to producing staples, chiefly tobacco, for export, and it functioned within the context of the 45- to 60-year-long cycles or waves associated with capitalistic systems. Each of these waves consists of several distinguishable internal phases, specifically, depression, takeoff, acceleration, and deceleration. Although most often applied to industrial capitalism, this model works equally well in establishing a framework for agricultural growth and production (Earle, 1992a; Schumpeter, 1939; Goldstein, 1988; Egnal, 1998).

Depression, the most critical and interesting phase of the long wave, produces periods marked not only by great difficulty and hardship but also by creativity, experimentation, and innovation. Because depressions demand solutions, their adversities—falling prices, unemployment, and stagnation—invariably precipitate a range of creative reactions.[1] Historically, the creative responses that long-wave agricultural depressions initiate arise from either of two sources: the practical, empirical experience of individuals who might be called "folk capitalists," and the knowledge derived from theoretical scientific investigations and testimony. The environmental impacts of these agrarian innovations vary from benign to malignant, from constructive to destructive. The terms *constructive occupance* and its opposite *destructive occupance* effectively encapsulate these two very different outcomes. Although one of the principal aims of agricultural innovation is environmental improvement in the sense of restoring, maintaining, or improving soil fertility (the other, of course, is profit), the innovator's dilemma is that these consequences are usually unknowable in advance—that is, the environmental impacts of depression innovation tend to be discoverable only in retrospect. By the time environmental information becomes available, the destructive or constructive effects on the landscape have become virtually unstoppable.

In the American South during the seventeenth, eighteenth, and nineteenth centuries, the most ecologically sound agrarian innovations came, as the case study in this chapter demonstrates, from the practical advice of planters, while the innovations recommended by the testimony of science produced negative results such as soil exhaustion and accelerated erosion and sedimentation. Specifically, the agricultural innovations adopted during the two long-wave depressions that occurred in the colonial era initially produced innovations that were both profitable and environmentally constructive. However, the recurrence of a worsening economy in the 1780s forced the abandonment of these initiatives in favor of scientific reforms whose destructive practices had devastating environmental consequences.

To expand briefly, the steady drop in tobacco prices that began in the 1630s triggered a number of creative innovations such as tobacco topping (cutting off the top of the plant to prevent it from going to seed and to divert nutrients into its leaves) and housing of tobacco, procedures that vastly improved productivity at the expense of sharp increases in the demand for fresh cultivable fields. A half-century later, planters responded to the depression of the 1680s by introducing slavery and tenancy on a substantial scale to augment the labor force required to clear new lands and fallowed fields and by adopting a system of land rotation that combined tobacco with a diversified crop regime of corn, beans, peas, and small grains (Menard, 1980; Main, 1982; McCusker and Menard, 1985; Kulikoff, 1986; Carr, Menard, and Walsh, 1991). This new agricultural system, which proved to be profitable and ecologically efficient, constituted something of an ethnic amalgam—the land-rotation (or shifting cultivation) system seems to have been Indian in derivation, the system of bound labor rested primarily on Africans, and the notions of integrating diversified crops and tenancy drew upon English precedents.

Although land rotation created an unkempt and scraggly looking landscape as lands were cleared, cultivated, permitted to lie fallow, and, a quarter-century hence, cleared again, the recycling of land provided important benefits that included increased tobacco yields, a retardation of erosion, and the maintenance of soil fertility (Earle, 1975; Papenfuse, 1972). Tenant contracts accomplished the same ends by stipulating that renters must plant apple orchards, refrain from cutting timber except for making repairs to houses, barns, and fences, and employ no additional hands unless their wives and children were unable to work. Similarly, the transition from a labor system that relied primarily upon indentured servants to one that utilized a combination of slavery and tenancy offered an efficient means of meeting the requirements of an agricultural regimen that maintained soil fertility through the continual clearance of land. The change in the composition of the workforce appears to have resulted in part from a new regimen that required planters to amortize the costs of servant labor over a seven-year period—the amount of time that a field remained productive (its new useful life) before being recycled into its fallow period. A shorter useful life for fields meant that using servants to clear land became ever more expensive, while the cost of slave and tenant labor remained unchanged.

THIS CHAPTER CHRONICLES the history of two individuals, a father and a son, whose lives spanned two long waves. They were not the usual sorts of men encountered along the shores of Maryland's Chesapeake Bay or the banks of its tributaries. The father, Charles Carroll of Annapolis (1702–82), son of an Irish immigrant, defied the colony's intolerance toward his Catholic faith by building a fortune based on land, slaves, moneylending, and investment in an ironworks that made him one of British North America's richest men. Denied access to office holding and the franchise because of his Catholicism, Carroll of Annapolis provided his son, Charles Carroll of Carrollton (1737–1832), with the wealth and education that allowed him, during the American Revolution, to grasp political power and move into the highest echelon of the Maryland gentry (Hoffman, 2000).

Although their wealth and influence separated them from their less affluent neighbors, the Carrolls nevertheless shared with those folk a durable bond. The fortunes of all pivoted inexorably on land, agriculture, and the vicissitudes of Europe's markets for the great Chesapeake staples of tobacco and, later, wheat. Whether seeking a "rude" sufficiency, in the felicitous expression of Aubrey Land (1967: 474), or advancement into the ranks of the landed elite, these planters maneuvered between the fickle demands of a market economy and the frequently severe constraints that nature imposed on agrarian systems.

Fundamentally, the Carrolls and their neighbors sought to make as much money as they could—by maximizing the productivity of their staple crops—while preserving the soils upon which that productivity depended. Contrary to some versions of environmental history in which agrarian capitalists are imprisoned in a zero-sum game of economy and ecology, Chesapeake planters sought to ensure their profits in the former by maintaining their capital stock in the latter. Reducing their respective experiences to a morality play about capitalism's unsparing and relentless assault on the land (as environmental historians tend to do) runs the risk of misreading American history in the large and the wisdom and sagacity that occasionally informed the usage of American land and resources in the small (a promising step in the latter direction has been made in Taylor, 1998). At times capitalist planters in the Chesapeake alighted on innovations that maintained their lands and soils with a modicum of damage. To acknowledge these constructive phases of capitalist occupance is to acknowledge the kinds of precedents that inspire hopes for the eventual harmonization of ecological and economic imperatives, of nature and economy in the Chesapeake Bay watershed.

In deciding the best ways to manage the lands on which their fortunes depended, the Carroll father and son came to radically different conclusions. When summed up, their differences, which resembled those between hundreds of other fathers and sons, delineate a great divide in Chesapeake land use and management in the final third of the eighteenth century. On one side of that divide stood Charles Carroll of Annapolis, exemplar of an older agrarian system anchored in practical experience, close empirical observation of changes in the land and pragmatic responses to them, and the simple technologies of hoes, axes, and the rotation of "worn-down" acreage. On the other side stood Charles Carroll of Carrollton, practitioner of the new agrarian system known as "high farming" and premised on European agricultural theories, their American applications, and complex technologies of crop rotation, crop variety, fertilizers, and plows. Convinced that the new ways of agriculture surpassed the old, that theory and "farming by the book" would improve at once his lands and his purse, Charles Carroll of Carrollton and others of his generation strode confidently across this divide during the 1780s and 1790s. None of them imagined the damage that would follow in their wake nor the vast store of wisdom they had forsaken in leaving behind the practical agrarian world of their fathers. Ironically, capitalism was responsible for both worlds—for the ecologically constructive old system and for the ecologically destructive one that replaced it.

An examination of the experience of Charles Carroll of Annapolis reveals how one very successful agriculturalist managed to realize high annual incomes from slavery and tenancy while ensuring the continued productivity of his lands. Beginning in the early 1730s, Carroll implemented an aggressive strategy to bring large tracts of land into production. Already the owner of more than 120 slaves, he chose to continue an expansion of that labor force and the lands it cultivated while adding to his income by tenanting other of his properties, some 30,000 acres in all. Believing tenancy offered the quickest and cheapest means of making the land productive, he negotiated nearly 200 leases over the next 20 years, thereby bringing more than 19,000 acres into development.

Carroll began with a 10,000-acre frontier tract lying between the Potomac and Monocacy Rivers which he had patented as Carrollton in 1723. By December 1733, he was engaged in the process of drawing up leases for some 52 prospective tenants. Of these agreements, seven, all concluded on 25 March 1734, and recorded the following August, survive to document the arrangements he made with his original renters. Six of the lessees contracted for 100-acre tenements, while a seventh contracted for 200 acres. All received leases for 21 years, with rents, levied in cash, that doubled every 7 years, beginning with 8s 4d sterling a year for the initial period and rising to 16s 8d sterling for the second and £1.13.4 for the third, excluding quit rents.

The leases directed the tenants to pay their rents to Carroll at his house in Annapolis on a semiannual schedule based on traditional English quarter days—half on the feast of the Annunciation of the Virgin, or Lady Day, March 25, and the other half on the feast day of St. Michael Archangel, or Michaelmas, September 29—an arrangement that indicates he intended to oversee collections himself instead of engaging a steward. If a tenant fell more than 40 days behind in his payments, his delinquency entitled Carroll to distrain—that is, to sell the tenant's possessions and keep from the proceeds the amount due. Within six years of assuming a tenement, the leaseholder was required to plant an orchard of "100 good young apple trees," set 40 feet apart, and to always keep "the houses & plantation" in "good order & tenantable repair," lest Carroll levy a fine of £6 sterling against him for failing to comply with the terms of the lease. According to entries in Carroll's cash book for the 1730s, it was his practice to build his tenants' dwellings and tobacco barns, or pay the tenants to do so.

Having begun the process of settling his most remote property, Charles Carroll of Annapolis turned next to his holdings on the fertile soils of the Western Branch of the Patuxent River, in Prince George's County. By the end of 1734 he had rented 60% of the tract—800 acres of prime tobacco land—to eight tenants, all of whom received 21-year leases. As might be expected, the arrangements Carroll concluded with these leaseholders differed in some respects from those he made with the renters who undertook the more formidable task of carving farmsteads out of Carrollton's frontier wilderness. They faced similar restrictions with regard to cutting timber solely for repairs to tenement buildings and fences, using only the labor of wives and children with the caveat that "one able hand" could be taken on if the children could not work, and planting a 150-tree apple orchard within three years. But Carroll's tenants on the Western Branch of the Patuxent contracted to pay their rents in "clean and merchantable tobacco" rather than in cash. The rents varied from 800 to 1,200 pounds of tobacco per 100-acre tenement, and the staple had to be delivered "in a Convenient Cask" to the landing on the Eastern Branch of the Potomac River.

Through similar agreements made during the next four years Carroll brought tenants to three more Prince George's County tracts, placing renters on the 925-acre Clouen Couse in Eastern Branch Hundred in March 1735, the 1,700 acres called the Girles Portion in Rock Creek Hundred in 1736, and on a small tract called Darnalls Goodwill in 1738. These tenants, like those at Western Branch, held short-term leases of 21 years, except for two individuals on the Girles Portion who secured leases for three lives, generally those of a male leaseholder, his wife, and one child. All of these agreements levied rents in casked tobacco, at rates of 600–800 pounds of tobacco per 100 acres, to be delivered to the Eastern Branch landing every March 25, and included Carroll's usual stipulations regarding planting orchards, cutting timber, and hiring additional labor, with the exception that two Girles Portion tenants—one of them the sole female among Carroll's early renters—were allowed to hire two hands as long as neither tilled more than 15,000 plants, approximately three acres' worth.[2]

Carroll selected three Baltimore County properties to bring under cultivation through tenancy, beginning in 1742 by leasing 100-acre tenements on Clinmalira, a 5,000-acre tract that adjoined the proprietary manor Lord Baltimore's Gift, or My Lady's Manor, in Upper Gunpowder Hundred. The six recorded Clinmalira leases that have survived for 1742–45 differ from the agreements between Carroll and most of his other tenants in being long-term contracts for three lives.

The difference may be accounted for in several ways. Assuming that Clinmalira was a total wilderness, Carroll could well have turned to long-term leases as the most reliable means of securing a stable tenant population willing to undertake, in return for the equity such leases offered, the hard labor needed to make the land productive. Unlike short-term leases, which served the landlord's interests by guaranteeing opportunities to renew at higher rents as development progressed, long-term leases benefited the tenant by providing several important advantages, namely, "long term security, a salable asset, an inheritance for at least one child, and the same political privileges afforded to freeholders" (Walsh, 1985). However, Carroll had managed to attract tenants to his even more remote and equally undeveloped Carrollton acreage without locking himself into long-term leases, which makes other motives likely for his behavior on Clinmalira. In part, at least, he was probably reacting to competition from other landlords willing to adjust arrangements to accommodate prospective renters.

This difference aside, the Clinmalira contracts are similar in other respects, specifying tobacco rents of 600 pounds of tobacco per 100 acres, no destruction of timber, and the allowance of one additional hand, with a second permitted if the tenant's children could not work—plus additional specifications providing tenants access to any spring on the property and the use of a gristmill if Carroll decided to build one. Finally, each Clinmalira tenant was expected to supply Carroll with two capons annually, a stipulation he had also imposed upon his Carrollton tenants by the mid-1760s. In September 1744 he commenced renting the last two Baltimore County tracts he intended to develop—Ely O'Carroll, 1,000 acres, and the adjoining 1,130 acres of Litterluna, both lying on the west side of the Jones Falls in the area known today as the Green Spring Valley. The leases he executed in these tracts replicated those he had drawn for Clinmalira.

Because the records pertaining to the tracts Charles Carroll of Annapolis tenanted are fragmentary and incomplete, it is difficult to determine precisely the annual income his investment earned. From the leases and accounts available for analysis, rough estimates can be produced beginning in the 1750s. One hundred fifty of the 195 tenants Carroll had procured by midcentury owed rents directly to him, while the rents owed by the remainder were paid to the estates he managed for his deceased brother's children. Eighty-five of Carroll's leaseholders—specifically, those who rented his Prince George's County lands, excluding Carrollton, plus half the tenants on the tracts whose ownership he shared with his brother's heirs—were obligated to pay him tobacco rents at rates capable of generating at least 55,200 pounds of the staple a year. At the average mean tobacco price of 1.13 pence sterling for 1750–56, Carroll could expect an annual income of at least £260 from these properties.[3]

Carrollton's potential, on the other hand, did not begin to be realized fully until after 1755, when the expiration of the original leases allowed Carroll to switch his tenants from cash to tobacco rents at the rate of 1,000 pounds of tobacco per 100-acre tenement.[4] This change, which included all Carrollton leaseholders by 1762, considerably increased his revenue. For example, the £109 sterling in cash rents that his 65 Monocacy tenants owed in 1755 constituted less than a third of the gross annual return of £369,[5] the sum he could expect from all his agricultural rents. Nine years later, however, he reported with considerable satisfaction that Carrollton alone, with approximately half its acreage now "let to Tenants at Will," was currently "Producing Annually £250 Sterling," or £3.16.11 per tenant, a substantial increase over the £1.13.4 due each year of the last seven on most of the old cash rent leases (Charles Carroll of Annapolis [hereafter CCA] to Charles Carroll, 10 Apr. 1764, Carroll Papers, MS 206, Maryland Historical Society, Baltimore [hereafter MdHi]). With his total income from tenancy having climbed to over £500 sterling annually, he had reason to be pleased.[6]

What particularly concerns us here, however, is not so much Carroll's obvious determination to secure handsome profits but rather the equally decisive steps he took to limit tobacco production levels as a means of preventing abuse to his lands. Had he not established these restrictions, Carroll could have increased his rental income, at least in the short run, but he chose to institute a program of restraint that guaranteed the long-term productivity of his acreage. All of Carroll's leases, whether short term or long term, whether for tobacco or cash, contained provisions regarding the protection of timber resources, requirements for planting orchards, and, most significantly, limitations on the amount of labor that could be employed on the tenement. The lease he made with William Davis on 29 April 1749, provides a typical example. For 100 acres of land in Anne Arundel County, Davis agreed to pay a rent of "six hundred Pounds of clean Merchantable Tobacco clear of Trash and ground leaves" and further agreed not to sell or destroy any timber but to use it solely for necessary repairs to the premises. Promising to allow only his own wife and children to assist him in cultivating the land, he also pledged that "in case his children are uncapable to Work then only to take in one able Hand" and to plant within three years 100 "good apple trees in a Regular Orchard the trees at least 40 feet apart and fenced in." An analysis of the 63 leases officially entered in the county court records demonstrates Carroll's consistent, systematic use of such arrangements (Table 13.1).

Unfortunately, Carroll's accounts with more than 250 other tenants do not, as entered in his ledgers, detail his arrangements with them. Nonetheless, there is every reason to assume that those agreements embodied the same strict stipulations through which he meant to protect his land from overproduction and exploitation, thereby ensuring its long-term profitability. Moreover, Carroll enforced the thresholds imposed on cutting wood just as rigorously as the restrictions he placed on labor. Upon learning that several of his Carrollton tenants had ignored the provisions concerning timber and cleared their acreage extensively in order to plant more crops, he added 50 acres and an increased rent to the affected tenements with strict orders that trees could be taken only when needed for firewood and fencing. The explanation Carroll gave for his decision underscored his sensitive appreciation of both the land's capacity and its vulnerability: "Had [the transgressing tenants] acted honestly and used my land as they would have used it had it been their own, they would not have been under the necessity of adding 50 a[cres] to their old tenements, but as they destroyed my timber by deadening it and over running their tenements, common prudence obliges me not to suffer them to do so again." Indeed, when he was informed that some of the tenants found his policy harsh, he tersely averred that "I might with reason and justice have turned of[f] such tenants and made two tenements out of one by granting the 50 a of the cleared land to one tenant and 50 to another, adding to each 50 a of wood land, so that instead of complaining they ought to be thankful they are not turned of[f]" (CCA, Response to Notes on Tenants, c. Sept. 1766, Carroll Papers, MS 216, MdHi).

This highly productive method of agriculture began to come under attack before the Revolution, and the critique sharply intensified in the depression of the 1780s as reformers such as John Beale Bordley advocated abandonment of the land-rotation system in favor of modern "high farming" systems (Bordley, 1784, 1797; Nettles, 1962; Rossiter, 1976). Bordley and others argued that plows should replace hoes, fields should

Table 13.1 Charles Carroll of Annapolis, Recorded Leases, 1734–1775

County	Total no. of leases	Labor provisions				Timber usage	Orchards
		Wife, children, or if children cannot work, one hand	Wife, children, and one hand; two if children cannot work*	Wife, children, and two hands; three if children cannot work	Not to lease out any part or make more than one farm or plantation	No sale or destruction of timber; only necessary repairs allowed	100 to 150 young apple trees
Anne Arundel	6	6	0	0	0	6	6
Prince George's	47	29	2	0	16	47	47
Baltimore	9	7	1	1	9	9	9

*Lease with Thomas Harris contained stipulation limiting each hand to tending no more then 15,000 plants.

be totally cleared and clean-tilled, and a more continuous system of cultivation using fertilizers should replace the primitive system of land rotation (Walsh, 1985, 1989; Percy, 1992). Tragically, these enlightenment-inspired agricultural reforms had devastating consequences for the Chesapeake. Clean-tilled fields accelerated erosion, stripped away nutrients, and hastened soil exhaustion. Increased sediment flowed into the Bay's drainage system, a consequence supported by recent studies of sediment cores (Brush, 1984, 1986). As a result, twin damages occurred. First, siltation, with its attendant disruption of navigation, accelerated. Second, the nutrient depletion of Tidewater soils pushed many planters onto the fresh lands of the Piedmont, where the steeper slopes allowed destructive occupance to proceed with even greater intensity.[7]

In addition to his talent for uniting the pursuit of economic gain and wise environmental practices through ambitious rentier schemes capable of earning sizable income, ensuring capital improvements, and maintaining land and soil, the elder Carroll truly loved planting and on occasion waxed poetic in his responses to nature's seasonal round. He possessed a fine nose for natural cues, reporting to his son in the spring of 1772 that "it will not be mild & settled weather untill the wind goes Regularly Round from N to E, S & W" (Carroll Papers, MS 206). On April 1 of the following year, he recorded a similar observation, noting that "Our wheat & Rye looks charmingly & our Pastures begin to look green & we shall soon have plenty of Grass if we have not Pinching Cold Winds to check the Growth" (ibid.). Thoroughly attuned to what he called "all the Works of the Season," he worried over the selection and proper conservation of seeds for the next year's crops, and he gave directions for penning and slaughtering hogs (28 Oct. 1773, ibid.).

Above all, Carroll attacked any signs of incipient erosion, cataloguing the steps he was taking to guard against that possibility at Doughoregan Manor, the family's 12,500-acre dwelling plantation located just west of what today is Columbia, in Howard County. "I have," he wrote his son in late November 1773, "marked all the Places in the last Years wheat field where stops are to be made to fill up the Gullies in that field, which are very numerous & many of them deep, without this Precaution & trouble the field would be ruined" (26 Nov. 1773, ibid.). Such strictures not only attest to the trained eye of this practical farmer, they also underscore his appreciation of the dangers lurking in an emerging agroecological system increasingly reliant on broadcast small grains, plows, and clean-tillage practices. Although intrigued by agrarian innovation and reform and an occasional user of plow agriculture, Charles Carroll of Annapolis remained for the most part wedded to an older technology based on crops of tobacco and corn planted in hills and cultivated with hoes and axes. When the Carrolls hurried an order to England in January 1775, in hopes that a reconciliation between colonies and mother country would shortly lift the Continental Congress's recently imposed ban on importation, the list of items wanted included 48 large weeding hoes, 36 small ones, and 12 grubbing hoes.[8] Plows appeared nowhere among the varied requests, although ditching shovels did. Eventually, the younger Carroll would put both of these to much use.

Charles Carroll of Annapolis's death in 1782 placed control of the family's agricultural enterprises solely in the hands of his son, Charles Carroll of Carrollton. Unlike the father, the son, whose principal interests lay elsewhere, became a planter-farmer by default. Lacking the older man's sensitivity toward the land, he was far less practical in cultivating it. The difference is nicely illustrated by a letter the younger Carroll wrote to an English friend on 15 September 1765, soon after he had returned to America after 16 years of schooling on the Continent and in London. As much as he would have liked to pursue the life of an English gentleman, he lamented to his correspondent, that was simply not possible in Maryland. Income, he noted, was never certain, because "it depends upon the casual rise and fall in the price of Tobacco" and the rents received in that staple, realities he would later make it his business to change. He would therefore "confine [himself] to the improvement of [his] parental acres" and "live as becomes a gentleman." (Charles Carroll of Carrollton [hereafter CCC] to William Graves, 25 Sept. 1765, CCC, Letterbook 1765–1768, MdHi.) Most tellingly, in terms of his later actions, Charles Carroll of Carrollton declared that he would not hoard his annual income or spend it extravagantly, but he explicitly refused to accept his friend's advice that he reinvest all of his returns in improving his property, thereby opening a door to partial disengagement from the farming life that so absorbed his father. Although he never walked all of the way through that opening, he expended the greatest portion of his time and effort in nonagricultural pursuits—provincial and national politics, business and industry, and urban realty.

Since the generational transition in agrarian interests between father and son played out primarily on the acreage of Doughoregan Manor, a word or two about the estate's geography would perhaps be helpful. Initially acquired in 1702 by Charles Carroll of Annapolis's father, the manor's lands boxed the Middle Branch of the Patuxent River, whose northwest-to-southeast course bisected the tract's roughly rectangular shape. The property's southern boundary crossed the river just north of Simpsonville; its northern border, slightly southeast of West Friendship. In the river's four-mile traverse across the manor, it fell more than 60 feet in elevation, with the gradient steepest in the most northern and southern portions—some 20 feet per mile—and least in the middle two miles—only about 10 feet per mile. Highly valued bottom lands broadened out along the east side of the river's middle portion as well as along the river's eastern tributaries, which drained the relatively gentle side slopes of the Piedmont and the descending uplands beyond the manor house. The terrain on the west side of "middle river," as the Carrolls called it, bore a quite different aspect, with steep slopes that came down nearly to the river's edge and far less bottom land.[9]

The dramatically different ways in which the practical farmer and his well-educated, more genteel son managed this agrarian fundament had quite different environmental consequences. Under the elder Carroll's command during the 1760s and 1770s, the prevailing farming system accented planted or row crops (tobacco for the market and corn for slave and livestock subsistence), hilling, with its emphasis on partial tillage of the surface (a kind of early modern no-till), and hand-held grubbing and weeding hoes and axes. These methods of cultivation generally produced from 90 to 100 hogsheads of tobacco, which required an equal number of acres, and between 1,500 and 2,000 barrels of corn for feeding slaves and an assortment of cattle, hogs, and horses. Much less time and acreage were devoted to small grains, including wheat and rye, although the records do not provide detailed and precise accounts. By the 1770s, Charles Carroll of Annapolis had 330 slaves living on the ten quarters into which he had organized Doughoregan. In addition, he leased out, with his usual strictures, ten tenements for farming and planting and at least one mill site.[10]

Not long after he assumed full responsibility for the manor in May 1782, Charles Carroll of Carrollton introduced a number of agrarian changes that resounded across the Piedmont landscape. The farm journal, which he kept intermittently between 1792 and 1802, records the following. First, in the mid-1790s, he trimmed his slave labor force to 260. Second, he reduced tobacco production from 90–100 hogsheads to 50 in the 1791 crop year, and five years later he contemplated forsaking the tobacco staple entirely during the next growing season. Third, he increased the number of manor tenants from 10 to 16 and insisted on receiving their rents in wheat. Fourth, in 1802, he considered feeding his hogs and horses rye in order to reduce corn production sharply. Fifth, his journal entries never mention hoes and axes but dwell instead on plows, which seem to have been deployed for row crops as well as for grains. He admired, for example, the practice in Frederick County, where "the wheat & rye Sowed in the corn fields was really put in, & the fields as clean as fallows." Finally, he was, in consequence of the preceding methods, increasingly preoccupied by the problems of poor drainage and flooding in the manor's bottom lands.[11]

By the time Charles Carroll of Carrollton's farm journal ends in 1802, the trends resulting from his system of management are obvious. As the estate inventory made at his death three decades later confirms, he had carried post-Revolutionary agricultural reforms to their logical conclusion. For example, tobacco did not appear on the list of Doughoregan products. Ranked by value, that catalog began with hogs and hog meat ($2,498), followed by wheat ($2,368), corn ($2,280), cattle ($1,448), and modest amounts of rye, oats, potatoes, and hay. The inventory also shows that Carroll had succeeded in reducing his slave population, though not to the level of 160 as he had proposed in the 1790s but rather to 222, slightly more than 100 fewer than his father had owned. All of this meant lower population densities and fewer acres under cultivation. On the eve of his death, Carroll could well have reflected with pride on the distance he had traveled from his father's "old-fashioned" ways of farming and on the extent of his restructuring of Doughoregan's agricultural economy through the introduction of scientific agrarian reforms.

At the same time, it is hard not to imagine the father's ghost hovering reproachfully over the son's ambitious projects. "But at what cost to the lands and soils of our ancestral manor have you brought these changes?" the specter might inquire. Judging from Charles Carroll of Carrollton's farm journal, the "improved" methods inflicted considerable environmental damage over a span in which neither precipitation nor runoff seems to have increased appreciably (Landsberg, Yu, and Huang, 1968). Carroll's decade-long daily account is littered with references to dramatic changes in the fluvial landscape of "middle river" and its tributaries—changes that are traceable to accelerated upland erosion in association with plowing and clean tillage.[12] The journal speaks repeatedly about the need to ditch the low grounds; straighten or foreshorten the river, its tributaries, and drainage ditches and canals; level off stream banks, the natural levees from overbank flow; mason in caving banks; and fill holes in the valuable bottom lands. This was an environment *in extremis.*

For 1793, a year in which Carroll spent some 174 days at Doughoregan, notes on ditching and ditchers appear in 59 of his daily entries. In contrast to his father, who apparently managed with just three slave ditchers, Charles Carroll of Carrollton employed ditchers in gangs consisting of ten or more slaves and Irish hirelings, assisted frequently by masons and carpenters. Much of their time and energy was directed toward the bottom lands, ditching the 14 meadows their employer noted in his journal and, to less avail, those portions of the floodplain used for tobacco and wheat. Of the 59 ditcher entries between April and October 1793, 29 appear in the month and a half between July 26 and September 9, precisely when the uplands were being plowed in preparation for sowing winter wheat. Preoccupied, if not consumed, by problems of hydraulic management, Carroll sought ways to protect the manor's valuable lowlands against flooding, wetland expansion, and the eventual swamping of the low ground.[13] Also unlike the father, however, the son failed to get at the heart of the problem. Tinkering with water management in the bottom lands was unlikely to resolve difficulties that originated in the uplands—that is, in the agrarian reforms that had accelerated erosion on the upland surfaces of Doughoregan.

Viewing the landscape in cross section clarifies the inextricable linkages between upland land use, erosion, and deposition in the fluvial system. (For more general discussions of upland-lowland interactions in fluvial systems, see Trimble, 1992, 1974; Knox, 1977; and Happ, Rittenhouse, and Dobson, 1940.) Fig. 13.1 depicts the river, the floodplain, and the adjoining side slopes of Doughoregan at two instants: the situation prior to accelerated upland erosion is portrayed in (a), while (b) is after erosion has accelerated. Note that the sediment accumulated in the river channel elevates the river bed (a special problem in the low-gradient portion of "middle river") and causes more frequent overflows. Overflows deposit heavier sediments in a gentle down-gradient across the floodplain and away from the river. In time, this leveeing process traps the overbank flow and, in the absence of hydraulic management, wetlands form and "back swamping" occurs on the back edge of the floodplain. A combination of waterlogging and hydrostatic pressures further destabilizes bottom-land soils by producing surface "holes." Higher base levels in the main channel of the river also back up tributary streams, canals, and ditches, causing them to meander across the floodplain. Efforts to straighten or foreshorten these streams and ditches shift base levels, causing erosional knickpoints to migrate headward into the uplands (Happ, Rittenhouse, and Dobson, 1940; Kesel and Yodis, 1992; Yodis and Kesel, 1993).

Father and son approached their geographic dilemma in radically

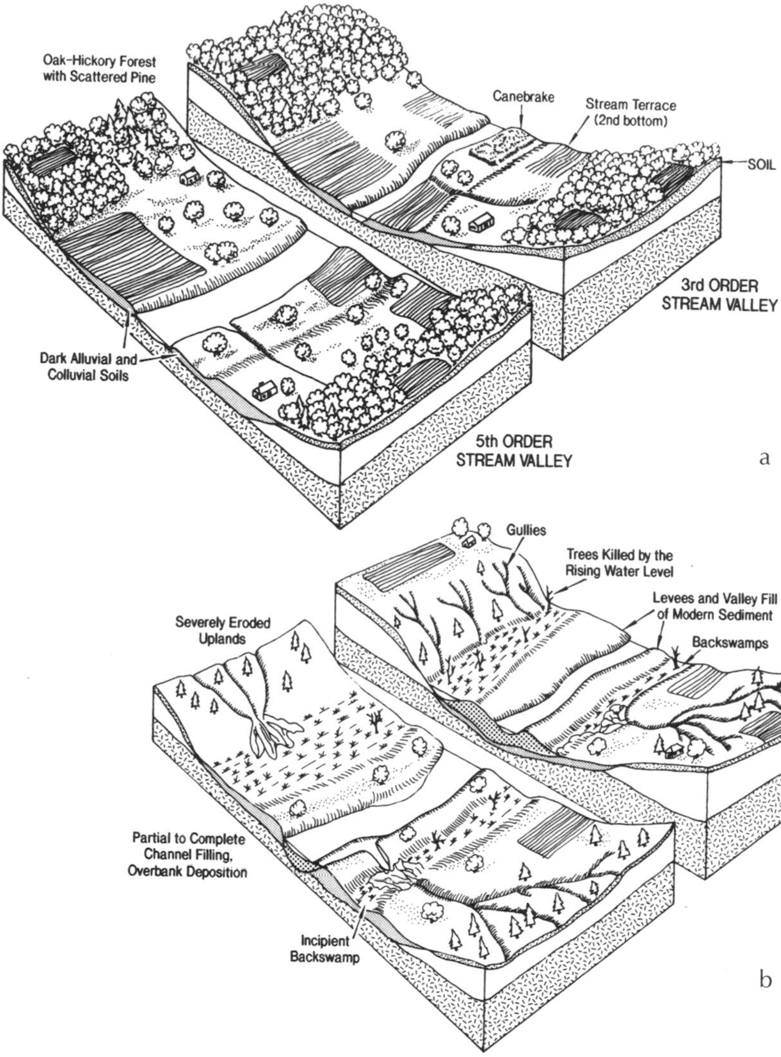

Fig. 13.1. Agrarian practice and landscape change in the Maryland Piedmont, 1700–1840. The figures depict generalized landscapes in branches and streams (third-order stream valleys) and the main river (fifth-order stream valleys) under regimes of (a) land rotation, long fallows, hilling, and hoe-and-axe technology, 1700–1770s, and (b) agrarian reform, small grains, clear tillage, and plow technology, 1770s–1890s (adapted from Trimble, 1974: 177).

different ways, with the former relying upon careful management of technology, crops, and soils in the uplands—the source of the problem—and the latter upon increasingly ornate hydraulic management strategies in the lowlands—the locus of the erosional consequences. Charles Carroll of Carrollton's farm journal offers painful reminders of his miscalculation. Following the flood of 28 May 1793, he reported that it "has done much damage to low grounds; the floods have washed away all of the Tobo hills, in the new meadow at Jacob's parallel with the race from the fish pond: also in the new ground cleared by Mr. Busey, also at the Folly—also in Valentine's meadow." The work of repairing the various water-gates and repairing and cleaning out the ditches that drained the low grounds kept the ditchers busy indeed, and soon after the flood, Carroll recorded that the grass (for hay) in one of his bottom-land meadows was much damaged by lodging and sotting. A month later, he embarked on a major hydraulic effort to recover some of the low grounds. As he explained in his entry for June 26,

> In order to prevent the overflowing & washing of Jacobs lower meadow, it will be necessary to dig a large ditch under the hill below the race: this race should be at least 8 feet wide at the bottom and 12 wide at top: the earth should all be thrown on the west or meadow side of the ditch: a large ditch should be dug to convey Reed's branch into the said ditch under the Hill: this later ditch should be made as wide as the former, the earth should be all cast up on the south side of it: other mounds must be made to convey the water in floods into the first ditch to prevent its spreading over the lowland and washing it.

Carroll's instructions to his ditchers flew thick and fast in the months and years ahead. On 21 August 1793 he laid out their schedule for the ensuing fall and winter: cut up driftwood on the river, dig up large weeds in the pool meadow, level the banks of ditches (the natural levees), fill with stone two rivulets at the lower end of the pool meadow, and cleanse the ditches in the Great Savannah. Five days later he added the tasks of filling "sunken places" in the meadows with earth and cleaning out ditches and leveling their banks. The following spring, he ordered new ditches "as are needed to drain the grounds wh[ich] will be tended this summer" (CCC, Journal, 1 Apr. 1794).

Efforts at hydraulic engineering intensified after 1793 when Carroll turned his attention to projects to straighten or foreshorten ditches and tributaries and later the "middle river" itself. On 1 October 1801 he ordered a Mr. Henry to contract with hirelings "to remove the ground in the channel of the river, & cast off projecting points of land, wh impede & retard the flowing of the water," noting that Henry thought that by straightening the river, "a fall may be obtained fully sufficient to dray the bottoms above mentioned" (CCC, Journal). In other words, the fall created by straightening out the river would scour out the sediment deposited on the river's elevated channel and lower its base level, thereby scouring out and draining the tributary ditches in the low grounds.

Judging from the rectilinear patterning of fluvial courses on the 1834 map of Carroll's estate, the strategy of straightening (foreshortening) rivers, branches, ditches, and canals must have been pursued with some vigor during the first third of the nineteenth century. The problem with this strategy was that neither Carroll nor Mr. Henry anticipated the acceleration of erosion in the upland waters of these tributaries. Their clever hydraulic schemes for rescuing the bottom lands therefore offered only a momentary reprieve before a new cycle of upland erosion and bottom-land deposition commenced (Yodis and Kesel, 1993).

Although there is a gap of three decades between 1802, when the farm journal entries stop, and Carroll's death in 1832, we can easily predict the continuing frustrations Carroll experienced during those years as he wrestled with the challenges posed by his ancestral estate. As confirmed by the inventory of his possessions in 1833, he never returned to his father's practical strategies of agrarian management. Livestock, which do reasonably well on wet bottom lands, and wheat had replaced tobacco and, to some extent, corn. The old technology of hilling row crops and cultivating with hand hoes and axes had given way to the supposedly more advanced technology of plowing and clean tillage. Moreover, reading Carroll's farm journal for 1792–1802 leaves little doubt that these "bookish" agrarian reforms, however modest, were etched into the landscape through the fluvial dynamics of his "middle river" and its tributaries. Ironically, between the lines of a decade-long diary kept by a man who considered himself an enlightened agriculturalist there lies an unintended case study of post-Revolutionary agrarian reform and destructive occupance.

As the experience of Charles Carroll of Carrollton illustrates, advocates of high farming failed to anticipate the corrosive consequences of their well-intentioned programs. In many instances, farmers able to adopt the plow did not bother, or could not afford, to introduce the right rotations of clover, turnips, and small grains needed to build up the soil to counteract runoff and erosion. The newly clean plowed fields—although far more attractive in appearance than the disheveled landscape of long-fallow shifting cultivation—were susceptible to sheet erosion, rilling (the first stage of linear erosion), and gullying. Nor did agrarian reformers allow for climatic differences between Western Europe and the United States. In the maritime climate of Western Europe, gentle rainfalls over long periods of time penetrate the soil instead of running off over it. By contrast, the violently intense storms of short duration characteristic of the continental climates of the United States often produce volumes of rainfall too great to be absorbed. Combined with the increasingly clean-plowed landscape advocated by agrarian reformers in the late eighteenth and early nineteenth centuries and the unfortunate coincidence of late-summer plowing and intense convectional rainstorms, such storms simply intensified runoff and accelerated soil erosion (Sauer, 1941; Percy, 1992). Because high-farming reforms produced ruinous cumulative effects, it is this agrarian strategy, rather than the Chesapeake's slave and tenant labor systems, that must be held responsible for the destructive occupance of the landscape that occurred from 1780 to 1840.

For all his tenacity and wit in trying to salvage Doughoregan's bottom lands, Charles Carroll of Carrollton's bookish understanding of farming in Maryland's Piedmont paled in comparison to the wisdom of his father's practiced eye. Instead of addressing the problem at its source by stopping up rills and gullies in the uplands, as the elder Carroll would have done, the younger tried to deal with the hydraulic consequences as they played out in the manor's bottom lands. Thus did the genteel and "enlightened" patriarch of one Chesapeake family in the late 1700s and early 1800s—joined, in all probability, by many of his planting neighbors—stride confidently across the divide that separated practical farming from farming by the book. For all of them, nature had much in store.

NOTES

1. It is possible to identify an impressive range of agricultural innovations within the depression phases of the eight long waves that occurred in England and the American South between 1580 and 1980 (Earle, 1988, 1992b, 1992c). Here is a brief chronological summary of these developments. *1580s–1630s:* Improvements in English wheat production double per-acre yields. *1630s–1680s:* Chesapeake tobacco yields per worker triple as a consequence of innovations in topping and housing the plants. *1680s–1730s:* The English incorporate clover and turnips into crop rotation; Chesapeake agriculturalists adopt land rotation in tobacco production, centralize and expand production through tenancy, and begin

to use slave labor to cultivate a variety of crops besides tobacco. *1740s–1780s:* Several agricultural policy innovations take place, most notably the Chesapeake system of tobacco inspection. *1790–1840s:* The cotton gin is adopted, internal improvements facilitate steamboat and barge traffic, and "high farming" reform begins. *1840–1880s:* Use of the reaper becomes common, railroads develop, and the southern crop-rotation system (cotton, corn, and cowpeas) is introduced. *1880s–1920s:* Commercial fertilizers, tractors, and threshers appear. *1930s–1970s:* Hybrid corn, cotton harvesters, and English combine threshers make their appearance.

2. The foregoing chronology of development is based upon entries in Charles Carroll of Annapolis (CCA), "An Old Cash Book and Accounts: Raised in 1749 with People indebted to John Digges whose Bonds are assigned to C: Carroll," fol. 48 (Library of Congress, Washington, D.C.), combined with recorded leases found in the county land records. For recorded leases on Clouen Couse, the Girles Portion, and Darnalls Goodwill, see Prince George's County, Land Records, Liber Y., fols. 139–40, 196–202, 214–15, 260–61, 398–400, 592–94, 602–3, 669–70; Liber E.E., fol. 53; and Liber B.B. no. 1, fols. 318–19, Maryland State Archives, Annapolis (MdAA).

3. Tobacco prices were computed from U.S. Bureau of the Census (1976: 1198).

4. The account of Thomas Brashears for 1754 provides the first evidence of the change from cash to tobacco rents on Carrollton. Brashears's rent is stated as 1,000 pounds of tobacco and two capons. The last recorded lease for Carrollton, executed in 1751, specified a cash rent for a term of 11½ years. CCA, Account Book and Index, 1754–84, fol. 64, Maryland Historical Society, Baltimore (MdHi); Frederick County, Land Records, Liber B., fols. 181–84, MdAA.

5. The 13 Carrollton cash rent leases recorded between 1742 and 1751 show that tenures decreased from a maximum of 19 years in 1742 to 11½ years in 1751. All of these contracts would have expired by 1762, with the exception of CCA's 1749 agreement with John Darnall, Esq., for 240 acres at 24s sterling a year for three lives. Darnall's son took over 100 acres of the tenement upon his father's death in 1768 and paid his rent in tobacco. Prince George's County, Land Records, Liber B.B. no. 1, fols. 179–90, 196–97, 264–65, 379; Frederick County, Land Records, Liber B., fols. 91–92, 181–84; John Darnall, 1768, will, Frederick County Wills, box 4, folder 4, MdAA; account of John Darnall, CCA, Account Book and Index, 1754–84, fol. 5, MdHi.

6. For an analysis of CCA's agricultural income based on a different reading of the data, see Papenfuse (1975: 48).

7. A number of scholars seem to have misread the erosional evidence from the late eighteenth century by erroneously assuming it applies to the whole colonial period, when most of the damage seems to have occurred in the post-Rev-

olutionary era with the ascendance of agrarian reform. For examples of these misreadings, see Craven (1965) and Gottschalk (1945).

8. Charles Carroll of Carrollton, hereafter CCC, to Wallace, Davidson, and Johnson, 8 Jan. 1775, CCC, Letterbook, 1771–1833, Arents Collections, New York Public Library. Although the order came from CCC, the agricultural tools and implements listed therein reflect CCA's agrarian priorities and strategies.

9. This summary of the physiography and hydrology of Doughoregan Manor is based on the following maps: "Map of Doughoregan Manor Enlarged: The Property of Charles Carroll Esqr," Sorbench and Draton, [1834], Prints and Photographs Department, MdHi; "Maryland Ellicott Quadrangle, 1:62,500, Contour Interval 20 feet," State of Maryland and U.S. Geological Survey, 1906 (based on 1902 and 1904 survey); "Maryland Laurel Quadrangle, 1:62,500, Contour Interval 20 feet," Department of the Interior, U.S. Geological Survey and State of Maryland Geological Survey, 1926 (based on surveys in 1890, 1900, 1904–6, and 1913–15); and "Clarksville Quadrangle, Maryland, 1:24,000, Contour Interval 20 feet," 7.5 Minute Series (Topographic), U.S. Department of the Interior, Geological Survey (1957 survey, photorevised in 1979).

10. Under CCA, Doughoregan Manor typically produced about 2,000 barrels of corn, on perhaps 600 to 1,000 acres (CCA to CCC, 26 Aug. 1773 and 18 Nov. 1773, Carroll Papers, MS 206); about 2,000 bushels of wheat, on perhaps 130 to 200 acres (24 June 1773, ibid.). CCA also noted 182 sheep and 99 lambs on Doughoregan, while giving lesser notice to cattle and hogs and hay, rye, oats, and potatoes (20 June 1775, ibid., no. 296). On slaveholdings, see Deans (1990) and "A List of Negroes—on Doohoregan Manor—Taken in Familys with their Ages Dec. 1, 1773," Carroll-McTavish Papers, MdHi. CCA divided his 330 Doughoregan Manor slaves among ten quarters. Between 19 and 27 slaves lived on each of seven quarters, while the number of residents at two of the remaining three locations varied widely from 8 to 37, with 130 at Riggs's, or the "home quarter."

11. CCC, Journal Kept at Doughoregan Manor, 1 June 1792 to 1 Oct. 1802, MdHi. On the reduction in numbers of slaves, see entry for 24 Aug. 1792. A year later CCC speculated on lowering his slave population to 160 (entry for 19 May 1794). Between Aug. 1792 and Oct. 1794, he sold off 40 of his 222 slaves (entry for 28 Sept. 1794). On CCC's reduction of the tobacco crop, see entries for 18 and 20 Sept. 1792 (based on production in the 1791 crop year), and for 7 Sept. 1796, which notes "no tobo to be made next year." On tenants and rentals on Doughoregan Manor, see entry for 30 Oct. 1792 and CCC, Inventory of Real Estate, 1833, MdHi. On CCC's shift from corn to rye for his horses and hogs, see the end of the 1802 journal entries. On plows and their usage even for corn, see entries for 11 Aug. 1795 and 11 Aug. 1796.

12. CCC, Journal. The landscape transformation of Doughoregan Manor at the end of the eighteenth century closely resembles erosional processes in the upland south at a later date. See Trimble (1992).

13. Owing to the fertility of their alluvial soils (at least prior to flooding and sanding over), "bottom" lands commanded higher rentals than uplands. CCA charged rents of 1,000 pounds of tobacco per 100 acres for "Bottom tenants" compared to 800 pounds of tobacco for "Upland Tenants." CCA to CCC, 11 Apr. 1780, Carroll Papers, MS. 206. Deteriorating drainage conditions in and the sanding over of lowland floodplains no doubt reversed this rental structure.

BIBLIOGRAPHY

This chapter is based in substantial measure on the family papers of Charles Carroll of Carrollton. These documents include correspondence, financial and land records, account books, probate materials, estate inventories, farm journals, and other sources that detail commercial and agricultural transactions. For a listing of the range of topics encompassed by the Carroll manuscripts, see R. Hoffman and E. S. Darcy, eds., *The Charles Carroll of Carrollton Family Papers: A Guide to a Microfilm Edition of Documents Located at the Maryland Historical Society* (Baltimore: Electronic Publication, Maryland Historical Society, 1996). Other primary sources held by the Maryland State Archives have been consulted, especially land patents, deeds, and surveys. We also used map publications of the United States Geological Survey prepared under the auspices of the Department of the Interior.

Barnes, R., and D. F. Thomas. 1978. *The Green Spring Valley: Its History and Heritage* (Baltimore: Maryland Historical Society).

Bliss, W. F. 1950. The rise of tenancy in Virginia. *Virginia Magazine of History and Biography* 58 (Oct.): 427–41.

Bordley, J. B. 1784. *A Summary View of the Courses of Crops, in the Husbandry of England & Maryland, with a Comparison of Their Products, and a System of Improved Courses, Proposed for Farms in America* (Philadelphia: Charles Cist).

———. 1797. *Sketches on Rotations of Crops, and Other Rural Matters. To Which Are Annexed Intimations On Manufactures; On the Fruits of Agriculture; and on New Sources of Trade, Interfering with Products of the United States of America in Foreign Markets* (Philadelphia: Charles Cist).

Brush, G. S. 1984. Patterns of recent sediment accumulation in Chesapeake Bay (Virginia and Maryland, U.S.A.) tributaries. *Chemical Geology* 44: 227–42.

———. 1986. Geology and paleoecology of the Chesapeake Bay: A long-term monitoring tool for management. *Journal of the Washington Academy of Sciences* 6: 146–60.

Carr, L. G. 1968. County government in Maryland, 1689–1709. Ph.D. diss., Harvard University.

Carr, L. G., R. R. Menard, and L. S. Walsh. 1991. *Robert Cole's World: Agriculture and Society in Early Maryland* (Chapel Hill: University of North Carolina Press, for the Institute of Early American History and Culture).

Craven, A. O. 1965 [1926]. *Soil Exhaustion as a Factor in the Agricultural History of Virginia and Maryland, 1606–1860* (Gloucester, Mass.: Peter Smith).

Deans, D. B. 1990. The Carroll family slaves, 1721–1833: Genealogies. Ms., University of Maryland.

Earle, C. 1975. *The Evolution of a Tidewater Settlement System: All Hallows' Parish, Maryland, 1650–1783*. University of Chicago Department of Geography Research Papers, no. 170.

———. 1988. The myth of the southern soil miner: Macrohistory, agricultural innovation, and environmental change. In *The Ends of the Earth: Perspectives on Modern Environmental History,* ed. D. Worster (Cambridge: Cambridge University Press), 175–210.

———. 1992a. *Geographical Inquiry and American Historical Problems* (Stanford: Stanford University Press).

———. 1992b. The price of precocity: Technical choice and ecological constraint in the cotton South, 1840–1890. *Agricultural History* 66 (summer): 25–60.

———. 1992c. Into the abyss . . . again: Technical change and destructive occupance in the American cotton belt, 1870–1930. In *The American Environment: Interpretations of Past Geographies,* ed. L. M. Dilsaver and C. E. Colten (Lanham, Md.: Rowman & Littlefield), 53–88.

Egnal, M. 1998. *New World Economies: The Growth of the Thirteen Colonies and Early Canada* (New York: Oxford University Press).

Goldstein, J. S. 1988. *Long Cycles: Prosperity and War in the Modern Age* (New Haven, Conn.: Yale University Press).

Gottschalk, L. C. 1945. Effects of soil erosion on navigation in upper Chesapeake Bay. *Geographical Review* 35: 219–38.

Happ, S. C., G. C. Rittenhouse, and G. D. Dobson. 1940. *Some Principles of Accelerated Stream and Valley Sedimentation.* U.S. Department of Agriculture, Technical Bulletin No. 695 (Washington, D.C.: Government Printing Office).

Hoffman, R. 2000. *Princes of Ireland, Planters of Maryland: A Carroll Saga, 1500–1782* (Chapel Hill: University of North Carolina Press, for the Omohundro Institute of Early American History and Culture).

Kesel, R. H., and E. C. Yodis. 1992. Some effects of human modification on sand-bed rivers in southwestern Mississippi. *Environmental Geology and Water Science* 20: 93–104.

Knox, J. C. 1977. Human impacts on Wisconsin stream channels. *Annals of the Association of American Geographers* 67 (Sept.): 323–42.

Kulikoff, A. 1986. *Tobacco and Slaves: The Development of Southern Cultures in the Chesapeake, 1680–1800* (Chapel Hill: University of North Carolina Press, for the Institute of Early American History and Culture).

Land, A. C. 1967. Economic behavior in a planting society: The eighteenth-century Chesapeake. *Journal of Southern History* 33: 469–85.

Landsberg, H. E., C. S. Yu, and L. Huang. 1968. Preliminary reconstruction of a long time series of climatic data for the eastern United States. University of Maryland Institute of Fluid Dynamics and Applied Mathematics Technical Note, BN-571 (College Park, Md.).

Main, G. L. 1982. *Tobacco Colony: Life in Early Maryland, 1650–1720* (Princeton, N.J.: Princeton University Press).

McCusker, J. J., and R. R. Menard. 1985. *The Economy of British America, 1607–1789* (Chapel Hill: University of North Carolina Press, for the Institute of Early American History and Culture).

Menard, R. R. 1980. The tobacco industry in the Chesapeake colonies, 1617–1730: An interpretation. *Research in Economic History* 5: 109–77.

Nettles, C. P. 1962. *The Emergence of a National Economy, 1775–1815* (New York: Holt, Rinehart and Winston).

Papenfuse, E. C., Jr. 1972. Planter behavior and economic opportunity in a staple economy. *Agricultural History* 46 (Apr.): 297–312.

———. 1975. English aristocrat in an American setting. In *"Anywhere So Long As There Be Freedom": Charles Carroll of Carrollton, His Family and His Maryland*, ed. A. C. Van Devanter (Baltimore: Baltimore Museum of Art).

Percy, D. O. 1992. Ax or plow?: Significant colonial landscape alteration rates in the Maryland and Virginia tidewater. *Agricultural History* 66 (spring): 66–74.

Rossiter, M. W. 1976. The organization of agricultural improvement in the United States, 1785–1865. In *The Pursuit of Knowledge in the Early American Republic: American Scientific and Learned Societies from Colonial Times to the Civil War*, eds. A. Oleson and S. C. Brown (Baltimore: Johns Hopkins University Press), 279–98.

Sauer, C. 1941. The settlement of the humid East. In *Climate and Man*. Yearbook of Agriculture (Washington, D.C.: U.S. Department of Agriculture).

Schumpeter, J. 1939. *Business Cycles*. 2 vols. (New York: McGraw-Hill).

Stiverson, G. A. 1977. *Poverty in a Land of Plenty: Tenancy in Eighteenth-Century Maryland* (Baltimore: Johns Hopkins University Press).

Taylor, A. 1998. "Wasty ways": Stories of American settlement. *Environmental History* 3 (July): 291–310.

Trimble, S. 1974. *Man Induced Soil Erosion on the Southern Piedmont, 1700–1970* (Soil Conservation Society of America).

———. 1992. The Alcovy River swamps: The result of culturally-accelerated sedimentation. In *The American Environment: Interpretations of Past Geographies*, ed. L. M. Dilsaver and C. E. Colten (Lanham, Md.: Rowman & Littlefield), 21–32.

U.S. Bureau of the Census. 1976. Prices of Maryland tobacco: 1711–1775. In *The Statistical History of the United States, Colonial Times to 1970*, ed. B. J. Wattenberg (New York: Basic Books), ser. Z 578–82.

Walsh, L. S. 1985. Land, landlord, and leaseholder: Estate management and tenant fortunes in southern Maryland, 1642–1820. *Agricultural History* 59 (July): 373–96.

———. 1989. Plantation management in the Chesapeake, 1620–1820. *Journal of Economic History* 49 (June): 393–406.

———. Forthcoming. *"To Labour for Profit": Plantation Management in the Chesapeake, 1620–1820* (Charlottesville: University Press of Virginia).

Yodis, E. C., and R. H. Kesel. 1993. The effects and implications of base-level changes to Mississippi River tributaries. *Zeitschrift für Geomorphologie* n.s. Bd. 37: 385–402.

CHAPTER FOURTEEN

Farming, Disease, and Change in the Chesapeake Ecosystem

G. TERRY SHARRER

As a borderland temperate area, the watershed is a place where northern and southern plant and animal life overlap. Plant and animal diseases overlap too and they have played a disastrous role in the developing agriculture of the watershed. Each case involved a confluence of destructive pathogen, such as stem rust of wheat, fusarium wilt of tobacco, hog cholera or equine fever, with a susceptible host, under disease-conductive circumstances that arose from economic developments or war.

Iron-age farming in the Chesapeake ecosystem has just plowed into its fourth century. What the future holds can only be guessed, but agricultural change is inevitable—and perhaps dramatically so. Currently, there are about 60,000 farms in Maryland and Virginia, occupying some 11 million acres, or 35% of the total land area. Farmers in the two states sell around $3.5 billion worth of crops and livestock: Maryland's most valuable commodities are broilers, greenhouse and nursery plants, and dairy products; Virginia's are cattle, broilers, and dairy. Yields, in most cases, are nearly the same as or slightly below national averages. The population (including the District of Columbia; 1991) is 11,744,511, at a density of 237.5 people per square mile (three and a half times that at the beginning of the twentieth century), and growing 1.6% annually. Population pressure bids up land values, increases taxes, and encourages farmers to sell. Agriculture may therefore decline over time because of its competitive disadvantage for space.

On the other hand, farmers could improve yields or raise more valuable commodities. Biotechnology already holds some startling prospects.

For example, it normally takes 55 days to raise a marketable broiler. Experimentally, researchers have reduced that growth time to 25 days. Let's suppose 21 days might be possible. If that were commercially feasible, productivity with poultry could double. Increased twinning in beef cattle, improved feed conversions for dairy herds, enhanced photosynthetic efficiency in corn, and greater disease resistance in all plants and animals are some, among many, biological strategies that could make farming much more profitable. It also may become possible to produce the world's most valuable substances—human-protein pharmaceuticals—in transgenic plants or animals. A company in Blacksburg, Virginia, now raises transgenic sheep and extracts two human protein pharmaceuticals from their milk. Ironically, recombinant tobacco may have an especially bright future in "pharming." It is not far-fetched to suppose that the principal crops and livestock Chesapeake farmers will raise over this next century do not exist today as "species" and may hold in store changes that could shock all other agricultural policy issues.

Farming, of one sort or another, likely will continue to be a major use of land in Maryland and Virginia well into the foreseeable future. Profitability will determine how well farmers cope with collateral concerns like soil conservation, groundwater contamination, and maintaining wildlife habitat. Over the past 400 years, agriculture's most nagging problem, and its most observable impact on the Chesapeake ecosystem, has been soil erosion. In simple terms, people—with axes and hoes at first and then with increasingly sophisticated machines—opened more and more land to the scarifying force of rainwater running downhill from the Appalachians to the Atlantic. Today, even after more than 50 years of organized soil-conservation efforts, a river such as the 110-mile-long Patuxent can carry away from its drainage area over a million tons of topsoil per year.

Land use in farming, of course, involves far more than tools. It is largely the outcome of ideas about how things grow, practical means for growing them, and market-related motivations. Of these, the market forces may be the best known, and the Chesapeake's economic development has drawn constant attention since the beginning of modern American historiography, including the work of several distinguished authors in this book. In recent years, historical farming practices have been studied in much finer detail, perhaps with no better example than *Robert Cole's World: Agriculture and Society in Early Maryland* by Lois Carr, Russell Menard, and Lorena Walsh (1991). However, biological thought in agricultural history—that is, how farmers understood and dealt with breeding, nutrition, and diseases—is relatively unexplored. Such ideas may provide useful insights about the way people used the land and brought about change in the ecosystem.

This chapter delves into the role of plant and animal diseases in the Chesapeake's agricultural history. Specifically, it is a story about six major agricultural diseases—stem rust of wheat, fusarium wilt of tobacco, yellows virus of the peach, cattle fever, hog cholera, and equine glanders—that wreaked havoc in Maryland and Virginia, mainly during the nineteenth century. Each case involves the confluence of a destructive pathogen, a susceptible host, and disease-conducive circumstances that arose from economic developments or war. Together, they help explain why land was exploited so extensively. The late nineteenth century also was the time when farmers first learned, as one skeptic put it, about the "'malarial germs,' 'microscopic animalcules,' and imaginary 'fungi,' which we are told are doing [so much] damage."[1] In agriculture, as in medicine, the germ theory of disease made "health" a more achievable goal and thereby conditioned all other developments.

Disease, of course, challenges farmers everywhere. What makes the Chesapeake unusual in this regard is its borderland character; it is a place where typically northern and typically southern phenomena overlap. Bald cypress, for example, grows from the Texas coast north to Calvert County, Maryland, while black ash ranges from Quebec Province south to Baltimore County, Maryland. The Chesapeake is a little too warm for tamarack and a little too cool for live oak, but middling enough for loblollies and shortleaf pine, red cedar and red maple, white ash, black oak, and yellow poplar. Temperate conditions also meant that farmers had to fight northern rusts and southern wilts, northern lung plague and southern tick fever. Ubelaker and Curtin (Chap. 7) observe this environmental "temperament" as a factor in human diseases. Similarly, the Chesapeake has been a meeting ground for northern and southern social, economic, and political traditions. The national capital took root on the Potomac because it was a place between north and south, and the nation split apart there during the Civil War. With the Chesapeake region not entirely a part of the north or the south, its experience actually is relevant to both.

Chesapeake Phytopathology

Before the nineteenth century, crop epidemics seem to have been a minor concern in Maryland and Virginia, probably for several reasons.

First, the early colonists escaped Old World disease cycles for a while, and the long distance across the ocean provided a barrier that delayed pathogenic introductions. Further, most accounts indicate that early farmers worked the ground with hoes rather than plows and raised nearly all their crops in rows of hills—even wheat, barley, and oats—sometimes intercropping. Landon Carter, in one example just before the Revolution, sowed peas between alternating hill rows of barley and spelt (a relative of wheat). This conserved seed, while creating a sort of checkerboard environment that controlled the growth of pathogenic organisms. Most important, human habitation grew slowly. After a century of colonization, Maryland and Virginia together had, perhaps, no more than 100,000 settlers (87,192 by actual count in 1701). Not until enough farmers raised enough of any crop could disease-causing organisms become established. In 1800, with over a million inhabitants, most of them farmers, pathogenic conditions had emerged, and the chances for epidemics increased along with population growth.

As the nineteenth century began, the wars of the French Revolution and Napoleon brought an agricultural bonanza to the Chesapeake region. Wheat sold for over two dollars a bushel in 1801 (a level it would reach in only seven other years of the century). Contemplating the state of affairs abroad, Thomas Jefferson wryly surmised: "This [war] we cannot help, and therefore we must console ourselves with the good price of wheat which it will bring us. Since it is so decreed by fate, we have only to pray that their souldiers [sic] may eat a great deal" (Sears, 1927: 16). Indeed, their soldiers did eat a great deal, but really luxurious prices came from smuggling breadstuffs to the combatant's West Indian colonies. A barrel of flour that sold for $7.50 in Maryland could bring $20 to $40 in Havana. Baltimore developed as a leading flour milling and marketing center; automated mills, where a miller could dump wheat into one chute and fill barrels from another, without human labor in between, became its wheels of fortune.

In part, though, the Chesapeake's wheat bonanza arose because a fungal disease devastated northern farmers. Stem rust wiped out wheat-growing in New England before the Revolution and gradually worked its way south. The pathogen, *Puccinia graminis,* had a heteroecious life cycle, meaning that its winter and summer spores survived on different hosts. When the Puritans introduced the European barberry, *Berberis vulgaris,* for hedges, along with wheat culture, the disease became established. *P. graminis* also found the native barberry, *B. canadensis,* which flourished in the deciduous woods from New England to Virginia (and generally above 37° N, west to Missouri), an equally suitable winter host. Spores spread on the wind, and the disease migrated when southerners raised enough wheat to establish a pathogenic threshold.

When stem rust first appeared in the Chesapeake is uncertain. It may have arrived before the Revolution. The English traveler John Smyth mentioned in 1784 that Virginians were growing an early maturing "Sicilian" variety because it somewhat escaped rust damage. But from the 1790s onward, as wheat growing expanded over plowed fields, the disease became endemic and was the subject of constant discussion in the agricultural societies and journals.

Grain suffering from stem rust looked drought-stricken or deprived of sufficient nutrients—both of which were the case, as the parasite stole what it needed. Kernels shriveled and the straw turned brown and brittle; losses ranged up to 90%. Weather influenced the incidence, with the worse damage occurring after a mild winter, a damp, early spring, and a hot, dry, late spring. Attacks came suddenly, either locally or over a wide area, and could last for a single season or several. Farmers suspected barberry was involved, but they could do little about it. They also realized that changing their planting date let their crops escape rust injury to some degree, but it also exposed the crops to other problems—most fearfully, the terribly destructive gall midge, *Mayetiola destructor,* popularly known as the "Hessian fly" (thought to have been introduced in the straw bed mats of Hessian soldiers during the Revolution). Rust and the fly became true horns of a dilemma.

Although some of the disease's circumstances were completely mysterious, the starved appearance of the plants conformed with the general understanding of how any crop ailment arose. Before the discovery of microorganisms, farmers commonly supposed that all aspects of a plant's growth and development were functions of "nutrition." It was a concept that had ancient roots. In the fifth century B.C., the Greek philosopher Empedocles proposed that all substances were composed of four elements—earth, air, fire, and water. Each element represented one of four qualities—heat, cold, dryness, and wetness, which corresponded to the four seasons. Aristotle grafted onto this hypothesis the notion that "life" originated with the ability to absorb nutriment. This explained why seeds swelled in germinating and even how insects generated "spontaneously" from the soil. Plants developed roots to soak up earth and water, while stems and leaves absorbed air and sunlight. "Fertility," in this scheme, originated in the soil (rather than in the plant); "fertilizers" stimulated absorption. Deprived of proper nutriment, cultivated plants were thought to degenerate into weeds (wheat devolving into chess brome, for example). Diseases had the same causation. Thus, stem rust, which may have involved certain proximate factors, ultimately emanated from a problem with soil fertility, or so people thought. Over time, "natural philosophers," from Pliny the Elder to Walter of Henley, Gervase Markham, Jan Ingen-housz, Albrecht Daniel Thaer, and Justus von Liebig (a representative few), refined the details but refashioned the general theory very little. Indeed, nutrition provided a systematic way of understanding growth and thus directed practices of how to grow things for about two dozen centuries.

The wheat bonanza ended in 1815: the European war concluded, and Britain passed prohibitive tariffs against American grain imports. Wheat fetched two dollars a bushel in 1817 because of the "year without a summer" (in 1816, following the explosion a year earlier of Tambora, a volcano in the East Indies, dust in the atmosphere altered weather conditions worldwide). But thereafter prices plummeted, reaching a low point in 1843, just as stem rust seemed on the verge of dooming wheat growers in Maryland and Virginia, as it had in New England. In that year, however, the first commercial shipment of Peruvian guano arrived in Baltimore, which farmers immediately hailed as a salvation—not surprising, considering their understanding of the problem. Later, a traveler who took the steamer *Virginia,* which was carrying a load of fertilizer from Baltimore to Fredericksburg, supposed "a guano epidemic [had broken out] among the planters of the Old Dominion." "I shall never forget the bouquet," he wrote, "it was inhaled by me with every thing I ate; indeed, after attaining my journey's end, it was a long time before I forgot the rich Islands of Peru."[2] When farmers, then and later, became disappointed with the fertilizer's results, they widely supposed unscrupulous dealers had swindled them. Enough dishonesty probably existed that any exaggeration became believable—even to the extent that fertilizer fraud, as one of many alleged ways northerners exploited the South, figured in the rationale for secession. Not coincidentally, the chief proponent of such ridiculous hyperbole—"the father of soil chemistry in America," Virginia's Edmund Ruffin—had the "honor" of firing the first shot on Fort Sumter.

Contemporary descriptions that mention wheat yields per acre, indicating prevailing conditions, vary considerably. As one historian put it, speaking about the lower Shenandoah Valley in 1850: "In wheat, 5 to 10 bushels was the usual range, with an average of maybe 7 bushels an acre. Farmers could produce 25 bushels an acre in wheat . . . , but almost never did so. When they did, they dashed off a letter to the editor to impress their neighbors and befuddle historians" (Schlebecker, 1971: 468). William Strickland, writing in 1796, and Joseph Scott, in 1807, both supposed average yields in the Chesapeake states to be between five and seven bushels per acre—the same level the U.S. Commissioner of Agriculture reported (for Virginia) in 1873. Apparently, normal yields changed very little over three-quarters of the century, even though new wheat varieties, marling, guano fertilizer, iron plows, drill planters, McCormick's and Hussey's reapers, and other "improvements" offered greater productivity. New techniques had to be reckoned against the dominating threat from stem rust. Chesapeake farmers could raise a few bushels of wheat per acre at an acceptable level of disease risk, or two or three times as much if they were willing to change methods and bet against the odds. Apparently, they largely chose the former, and in so doing, extended wheat husbandry in Maryland and Virginia over some 2.7 million acres for the record harvest of 1860 (19.2 million bushels, the largest amount of any census year during the nineteenth century), all the while exposing an enormous area to soil erosion.

Although wheat became an increasingly difficult crop, tobacco had a well-deserved reputation for being troublesome. Two observers, among many, 75 years apart, told the same story. Isaac Weld, traveling through Prince George's County, Maryland, in 1795, wrote:

> The country is flat and sandy, wearing a most dreary aspect. Nothing is to be seen for miles together, but extensive plains that have been worn out by the culture of tobacco, overgrown with yellow sedge and interspersed with groves of pine and cedar trees. . . . In the midst of these plains are the remains of several good homes which show that the country was once very different from what it is now. These houses . . . have been suffered to go to decay as the land around them is worn out and the people find it more to their advantage to remove to another part of the country and clear a piece of rich land than to attempt to reclaim these exhausted plains. (Craven, 1926: 156)

Another traveler, describing Virginia's southern Piedmont in 1870, reported: "[Pittsylvania] . . . is one of the leading tobacco counties, and the system of cultivation pursued ends in total or partial exhaustion in five or six years." In neighboring Patrick County, he found "the system of farming is rude and primitive; hence no attention is given to alternation of crops; the term is hardly appreciated."[3]

The long-standing consensus held that tobacco was a "heavy feeder" and, as such, could deplete soil fertility (i.e., nutrients) quite quickly, especially if growers practiced continuous cropping. Farmers could see that successive plantings resulted in stunted growth and diminishing returns. Once the land became exhausted, the restorative alternatives were fertilizer or "resting" (i.e., keeping the field out of production for a number of years, permitting fertility to revive). Yet in spite of these problems, tobacco remained almost irresistibly attractive as a cash earner.

To be sure, tobacco or any other plant required nutrients from the soil for growth, but "feeding" also depended upon root health. Unknown to farmers until the late nineteenth century, Maryland- and Virginia-grown tobacco faced formidable subterranean adversaries—principally, two genera of nematodes (root knot, *Meloidogyne* spp.; and tobacco cyst, *Globodera* spp.) and three fungal organisms (*Sclerotium rolfsii,* southern blight; *Thielaviopsis basicola,* black root rot; and *Fusarium oxysporum,* tobacco wilt). Generally, all of these organisms existed wherever tobacco was grown in the eastern United States, though they were sensitive to weather, soil types, and farming practices. Nematodes preferred lighter soil and warmer temperatures; hard winters could improve the following tobacco harvests for that reason. They disseminated by any means that moved soil—transplanted seedlings, dirt on implements, even the washing action of rainwater. The root knot species, more commonly found in the southern Piedmont, hatched from eggs in the ground, swam to the roots when it rained, and began to feed. Crop rotations—even a single season of wheat or oats, or two years of bare fallow, where less susceptible native weeds grew—checked the nematode population, but continuous cropping allowed their numbers to explode. In the topsoil space allotted to a single tobacco plant, roughly one square meter, the population of nematodes may have been one million to several million. With life cycles as short as a few days and reproductive ratios around 1:500, they awaited the farmer's cropping decisions.

Nibbling on roots, nematodes could be quite devastating by themselves, but they also opened the way for the more serious fungal infections. Temperature influenced where and when these diseases became most severe. Black root rot peaked at a relatively cool 17–23°C and mainly affected seedlings. Southern blight and tobacco wilt worsened at similarly warm temperatures, but differed enough (30–35°C and 28–31°C, respectively) that blight posed somewhat more of a threat in the eastern Piedmont and Tidewater, while wilt increased toward the Blue Ridge. *Fusarium oxysporum* survived in the soil and infected residue. It spread on seeds and by any means that moved soil. Once the pathogen entered a plant's roots, it blocked the xylem pathway and produced enzymes and toxins that broke down cell walls. Yellowing started at the top of the plant and moved down one side. Generally, wilt disease damaged an entire field.

Unlike the nematode problem, fungal diseases were less responsive to control through crop rotations. The black root rot organism long survived in the ground and infected several plants unrelated to tobacco. When farmers let a field "rest" for many years, it was the pathogens' slow subsidence rather than a restoration of nutrients that eventually made the field "fertile" again. Fusarium wilt had a wide range of alternative hosts, including cotton, flax, and several legumes; thus, rotating tobacco with red clover or beans potentially worsened the incidence of the disease. Perhaps this explains why tobacco farmers in particular had so little faith in alternating crops.

After the Civil War, tobacco growing fell on hard times, especially in Virginia's southern Piedmont, ostensibly because of soil exhaustion. Blame for poor conditions often fell on the freedmen, a claim that was largely racially motivated and mistaken. After all, the former slaves knew better than anyone how to bring a crop to market. Instead, and far more likely, tobacco growers experienced a severe and persistent wilt epidemic. The counties where crop rotations were described as being "hardly appreciated" also had been the principal flax-growing area of the Confederacy (Patrick, Franklin, Bedford, and Campbell Counties). Flax was extremely susceptible to fungal wilt and, because of its denser spacing, built up an enormous *Fusarium* load in the soil that was capable of injuring subsequent tobacco crops for many years. After the war, desperately cash-poor Virginians turned to tobacco, and it is exceedingly likely that wilt disease produced the conditions that, at the time, could be understood only as "soil exhaustion."

The traditional solution to "worn out" land simply was to abandon it and clear more, relying on slave labor. Virginia, the leading slaveholding state in 1860, had nearly twice as much unimproved land as land improved in farms, which had an average size of 324 acres. In Maryland, average farm size was 212 acres. The end of slavery made land clearing more expensive, partly explaining why average farm size in both states fell to 167 acres in 1870. Smaller farms and more expensive labor necessitated agricultural diversification, and increasingly, Chesapeake farmers turned to growing fruits and vegetables, creating new opportunities but also bringing new problems.

Baltimore was the leading canning center in the nation by 1860, and the Civil War increased demands for its most renowned seasonal specialties, winter oysters and summer peaches. When Eastern Shore farmers shifted out of wheat growing, some switched to raising peaches, and as one account described it in 1865, "peaches were planted all along the water courses in the first district of Cecil county, in all of Kent county, and in parts of Queen Anne county, until there was a continuous forest of peach trees all along the water courses and extending from one to two miles back from the water" (Gould, 1901: 130). Postwar railroad construction and Andrew Shriver's invention of the steam retort (a pressure cooker) in 1874 helped Baltimore pack about one-third of all the canned foods in the nation by 1880.

As peach growing came into "full blossom" during the mid-1880s, a disease called "yellows" suddenly and completely collapsed the business. It had appeared earlier—first, near Philadelphia in 1806; from there, it migrated across the north from Connecticut to Michigan by the 1860s. When peach culture extended south, so did the disease—through Maryland and Virginia to Arkansas and Texas by the end of the century. Orchardmen had no idea of its cause or how it spread, though they suspected it involved a fertilizer deficiency (nitrogen could restore a green color to the leaves). Fertilizer, however, only masked the disease; once infected, the trees died. After the discovery of tobacco mosaic virus in 1892, plant pathologists categorized yellows as a viral disease, though they reclassified the pathogen as a mycoplasmalike organism in 1967. It was disseminated by leafhoppers and also infected native wild cherry—explaining how peach yellows, which did not exist in the Old World, originated in America: the combination of an indigenous pathogen and introduced cultivars.

The collapse of peach growing obviously hurt Tidewater orchardmen, but it also injured the Baltimore canners, who apparently tried to make up in oysters the income they lost from fruit. It hardly seems coincidental that the record oyster harvest of Maryland—15 million bushels in 1884–85 (representing about one-third of all commercial oyster fishing in the world at the time)—occurred in the midst of the peach yellows disaster. Agriculture's problems, in short order, could translate into the exploitation of other resources.

Actually, peach yellows was only one of several devastating diseases that struck fruit growers in Maryland and Virginia almost simultaneously. A substantial pear industry had grown up in Virginia's Southside (the Tidewater counties south of the James River) in the early 1870s. Then, suddenly, fire blight swept through the groves. The unknown cause was the rodlike bacterium *Erwinia amylovora*, which lived symbiotically in the digestive tracts of honeybees. When the bees drank from spring blossoms, the pathogen entered the tree through the nectar glands. Extensive infection left the branches looking as black as though they had been burned. About a decade after they had been started, the pear orchards of Surry, Isle of Wight, and Prince George Counties were obliterated.

Meanwhile, in the Shenandoah Valley, growers of red apples encountered a disease identified later as cedar-apple rust. As in the stem rust of wheat, the agent responsible for the disease was a heteroecious fungus, one which completed its life cycle only on red apple trees and the native red cedar (*Juniperus virginiana*). Honeybees distributed its spores. Young trees died, while others became weakened against opportunistic pathogens. Sharing the misery, wine-grape growers in Albemarle and several adjoining counties encountered black rot (caused by *Guignardia bidwellii*), which wrecked the area's vineyards and led the editor of the *Monticello Farmer and Grape Grower* in October 1883 to conclude: "Very many appear to be discouraged at the outlook for the grape industry. . . . The question comes home to every one who has suffered from the rotting of the fruit on the vines, as to how long we can stand such losses. We are all in the dark as to the cause" (25).

Crop diseases became pervasive in Chesapeake agriculture during the nineteenth century, completely destroying some enterprises and locking others into low-yield circumstances that carried with them an extensive exploitation of the land. In either case, farmers had to accept their losses, as there was very little they could do about them without knowing the cause.

Animal Plagues and the Civil War

Unlike the crop diseases that seem to have accumulated over the century, numerous livestock epidemics struck all at once, during the Civil War. Along with the animal supply for the cavalry, for transport, and for food of the opposing armies, almost every major livestock disease in the nation came to Maryland and Virginia. Nothing like this had happened before, nor would it happen again, though once established, glanders, tick fever, hog cholera, pleuropneumonia, and numerous other diseases proved exceedingly hard to suppress. Farm animals, like farm boys, went off to war relatively unexposed to communicable diseases and so had little immunity. Poor sanitation, exposure, stress, and accidents triggered outbreaks. Figures do not exist for animals lost to disease as they do for soldiers (414,000 humans died of dysentery, typhoid, measles, scarlet fever, etc., compared to 204,000 killed in battle or dead from wounds), but some indicators suggest enormous losses.

Early in the war, neither side realized how large the struggle would become. The Confederate commissary general, Lt. Col. Frank Ruffin, supposed he could supply meat to the forces in Virginia from within the state. He built one packing plant at Richmond and another—the largest in the South—at Thoroughfare Gap (Prince William County), sandwiched between Virginia's two wealthiest livestock-raising counties, Fauquier and Loudoun. By the time Gen. Joseph Johnston abandoned Manassas Junction in March 1862, Ruffin had managed to process and stockpile more than two million pounds of pork and beef at Thoroughfare Gap—perhaps the only time in the war the Confederate commissary truly got ahead of the game. When the packing house went up in flames, one retreating soldier wrote that he could smell bacon frying for 20 miles.

The government abattoir at Richmond continued operating until the end of the war. Ruffin also set up several subsidiary depots, at Danville, Lynchburg, Dublin (Pulaski County), Boykins (Southampton County), and Milford (Caroline County), and at Charlottesville and Staunton. At these locations, on railroad lines, commissary officers received cattle and hog shipments, forwarding them to the troops as needed. Unlike the Union commissary operating in hostile territory, Ruffin's office contracted with local farmers to feed the animals in these holding areas, either around the depots or close to the front. A typical contract specified that J. W. Curtis, of Williamsburg, agreed to keep 500 head of cattle at his farm on Skiffs Creek. Such arrangements exposed local livestock to any disease the transported animals brought into Virginia.[4]

As the war dragged on, the Confederate commissary drew meat animals from greater distances—from Tennessee, via the Virginia & Tennessee Railroad, and from the Carolinas and Georgia, over the Richmond & Danville. These supply lines ran to the central breeding areas of the South's major livestock diseases, hog cholera and cattle fever. And, presumably, as Gen. Robert E. Lee's situation deteriorated, animals in almost any condition were considered better than none at all.

The full extent of the problem surfaced after the war, when the Department of Agriculture's statistician in Washington, D.C., J. R. Dodge, gathered data on the conditions of crops and livestock in the restored Union. Hog cholera caused widespread losses, but in Nelson County,

Virginia, fully a quarter of all the swine died from the disease, and in Craig County, half. Worst of all, farmers in Floyd County, near the Confederate depot at Dublin, lost three-fourths of their hogs—the highest incidence of hog cholera reported to Dodge from any county in the nation. There were no confirmed cases in Virginia before the war.

Cattle fever had appeared before 1860, but only in isolated instances. During the late 1860s, however, it struck often, in every part of Virginia, and with great severity. A physician, G. M. Brown, in Cumberland County, thought the "Carolina distemper" so terrible in 1866 that it might be rinderpest. Stockmen, particularly in southwestern Virginia, supposed cattle fever had permanently infected their area, while others in the state worried about hog diphtheria, blind staggers, hog measles, and blackleg, which also seemed to be legacies of the war.

Of all the diseases stemming from the Civil War, probably only glanders could have been anticipated; it had been the bane of cavalries since antiquity. Still, no one fully realized how horrible it could become—partly because no one ever had seen a horse-handling facility like the U.S. Cavalry Bureau's immense depot in Washington, D.C., at Giesboro Point (now the location of Boling Air Force Base). Built between June and August 1863, the post covered 625 acres and could accommodate about 30,000 horses (one entire remount for the army of the Potomac). Its barns sheltered 6,000 head, but most of the horses mingled in several open corrals, each covering 45 acres. With 1,500 workers, typically feeding 142 tons of hay every day, Giesboro's operating expenses averaged $1 million a day from 12 August 1863 until it closed on 30 June 1866. At times, its daily expenses even reached $4 million.

Glanders appeared as soon as the depot opened. The disease, caused by the bacterium *Pseudomonas mallei,* had chronic and acute forms. Sick horses showed a high fever and a thick nasal discharge. The acute form killed an animal a few days after the symptoms appeared, though an incubation period of about two weeks preceded the onset of signs. Horses with chronic glanders lived for years, all the while disseminating it, mostly by rubbing noses. In the open corrals, where the horses were free to move around and drink from common water troughs, the disease spread relentlessly. In all, 24,321 horses died at Giesboro, and 47,721 more were sold at public auction for being unfit, quite likely with chronic, communicable symptoms. The *American Agriculturist* warned its readers in September 1865 that "the Government might better have shot every horse, than to have them spread contagion and death (for the disease is utterly incurable) among the stables of the country, far and near" (269).

Ironically, even acts of kindness contributed to the spread of disease. At Appomattox, Gen. Ulysses S. Grant allowed Lee's soldiers to keep their horses because, he said, "they will need them in putting in their spring crops." After the surrender, the federal quartermasters turned over "many broken down horses and mules" to the local people. Glanders raged among the southerners' horses as well, emanating from the Confederates' central horse depot at Lynchburg. The release of these animals, perhaps done with the best intentions, opened a floodgate of disease upon the countryside.

After the war, glanders eventually subsided. Cattle fever continued to cause problems for another generation. Hog cholera, a highly contagious virus (a single-stranded RNA, similar to viruses responsible for yellow fever and hepatitis C in humans), raged on. Farmers who moved their swine during the disease's 5- to 10-day incubation period, when the animals still appeared healthy, left trails of infection. The virus survived in processed pork, axle grease, and soap. If a dog dug up the bone of a hog that had died of cholera, the pathogen found a new opportunity. When a government inspector investigated an outbreak along the Potomac in the summer of 1884, he saw an enormous number of carcasses floating and piled up in the river at Point of Rocks, between Loudoun County, Virginia, and Frederick County, Maryland. He described the Chesapeake and Ohio Canal as a stinking sewer of dead hogs.[5]

Livestock diseases inhibited Chesapeake farmers from climbing the so-called agricultural ladder. Commercial animal husbandry, besides being a generally higher value business, lent itself to crop rotations, a "homegrown" source of fertilizer, and other practices that allowed farming to be carried out more profitably on less land. Until healthy animals became a more achievable goal, farmers found it difficult to escape from the extensive land-exploiting circumstances that had prevailed for generations.

Germ Theory in Agricultural Practice

The agricultural disease crises that converged around the Civil War also coincided with the beginning of germ theory in Europe. In April 1864, French scientist Louis Pasteur presented a lecture before an audience of dignitaries at the Sorbonne which challenged a central concept in classical biological thought—the notion of spontaneous generation. In arguing that living organisms were themselves responsible for decomposition and putrefaction, Pasteur undermined the humoral theory of disease that had been the intellectual fortress of medicine and veterinary medicine since antiquity. About the same time, German botanist Anton De Bary demonstrated that the late blight of potato (the disease that caused the great famine in Ireland during the 1840s) was caused by a fungus rather than by excessive absorption of water. Without the living pathogen, no blight appeared—much as Pasteur contended. In 1865, De Bary discovered the spore stage that connected the heteroecious life cycle of the stem rust organism. Quickly thereafter, botanists discovered the cause of other fungal diseases of plants, as well as the role fungi played in forming humus in the soil and in germinating certain higher plants.

Nearly all the crucial arguments and basic research that founded germ theory took place in Europe between 1860 and 1885. Americans observed skeptically at first, but with rising concern over one catastrophic livestock or crop disease after another. Gradually, sentiment mounted that the Department of Agriculture should do more than simply report disease outbreaks. In 1884, several cattlemen's associations successfully lobbied Congress to create within the department a Bureau of Animal Industry (BAI), specifying that the bureau would begin with efforts to control bovine pleuropneumonia. Two years later, at the urging of the American Association for the Advancement of Science, Congress also authorized the department to establish in the Division of Botany a Section of Mycology, where federal research into plant diseases began. Both new entities set up research facilities in Washington, D.C.—the BAI operated its Veterinary Experiment Station on Benning Road, and the Mycology Section used the department's research plots adjacent to the Smithsonian Institution—and conducted field studies at nearby farms in Maryland and Virginia.

The BAI certainly was well located to fight pleuropneumonia. The disease first appeared among cattle imported from England to Boston in 1843, but it spread to Maryland and northern Virginia during the Civil War. In 1875, USDA statistician J. R. Dodge speculated that the disease prevailed in Maryland more than anywhere else. BAI investigators built their epidemiology from Robert Koch's studies of tuberculosis in Germany (ironically, both worked off the mistaken assumption that the responsible pathogens were bacteria; they were later shown to be mycoplasmas). Armed with regulatory authority to control the railroad shipment of sick animals and with funding for a slaughter-indemnity program, BAI inspectors and state health officials confined pleuropneumonia to Baltimore, New York, Brooklyn, and one county in New Jersey by 1888. Four years later, the secretary of agriculture announced that pleuropneumonia had been eliminated entirely from the United States—the first livestock disease eradication.

Almost simultaneously, the mycologists proclaimed a parallel victory. In 1891, Beverly T. Galloway, then head of the Section of Vegetable Pathology, told a Farmers' Institute meeting in Charlottesville that the black rot disease of grapes, which devastated Virginia's wine industry almost 20 years earlier, was controllable with the copper-sulfate fungicide (Bordeaux mixture) discovered in France in 1885. Farmers soon began using fungicides and oil sprays against many crop pests. With the creation of the agricultural experiment stations in 1887, ideas about plant and animal pathology that began in Europe with De Bary, Pasteur, Koch, and others, and which Department of Agriculture scientists demonstrated, disseminated throughout the nation.

Germ theory redefined disease and allowed researchers to begin sorting out specific etiologies and clinical findings. "Cholera" had once applied to almost any contagious disease of the hog, but the discovery of different organisms eventually segregated salmonellosis, erysipelas, swine influenza, vesicular disease, and others, including viral hog cholera. Similarly, the problem of "soil exhaustion" broke into several components, involving many types of pathogens, mineral nutrition, and soil structure. Disease specificity allowed countermeasures to be developed—antitoxins, bacterins, hyperimmune serums, antiseptics, insecticide dips, and vaccines for livestock; fungicides, insecticides, dormant sprays, and fumigants for crops. Knowledge of microorganisms also supported laws for milk pasteurization, the removal of crop pests, plant and animal quarantine, livestock fencing, and compensation for slaughter in controlling epidemics. At least indirectly, germ theory paved the way for Mendelian genetics. As BAI Chief Daniel Salmon told a meeting of the U.S. Veterinary Medical Association in 1898: "We must . . . accept the fact that . . . there is a world of life that the microscope is powerless to reveal, just as we have long known of a world that our unaided vision could not detect."[6] Botanists "rediscovered" Mendel's work in 1900, and thereafter disease resistance became the most consistent goal of all plant and animal breeding.

For the farmer, improved plant and animal health allows more intensive husbandry with higher yields. In 1990, Maryland and Virginia wheat growers raised about the same size crop as they did in 1860, though using roughly one-seventh the acreage. Resistant cultivars and removal of the common barberry have minimized the threat of stem rust. Certified disease-free seed, systemic, foliar, and seed-protectant fungicides, and crop rotations control many other problems. Dramatically higher yields, in virtually all crops and livestock, is agriculture's parallel to medicine's remarkable success in doubling the life expectancy of Americans since the beginning of the twentieth century. In neither medicine nor agriculture, however, has disease really become less complex or less critical. If anything, the threat of epidemics is greater (if less frequent) as increasing densities (plants, animals, and people) create novel circumstances in which new pathogens can emerge. Certainly, the poultry industry in Maryland and Virginia seem particularly vulnerable in this regard. It is impossible to predict future crop and livestock plagues, but historical experience shows that agriculture in the Chesapeake and its embracing ecosystem are wedded, for richer or poorer.

NOTES

1. *Journal of the Virginia State Agricultural Society* 1 (April 1879): 115.
2. *American Farmer* 9 (May 1854): 347.
3. *Report of the Commissioner of Agriculture for the Year 1870* (1871): 276.
4. Undated contract, probably winter 1861 or early 1862, between J. W. Curtis of Williamsburg and Capt. John Henry Wayt (CSA), Letters, Telegrams and Orders Received and Sent by Capt. J. H. Wayt, National Archives Record Group 109.
5. U.S. Bureau of Animal Industry, *First Annual Report of the Bureau of Animal Industry for the Year 1884* (1885): 446.
6. *American Veterinary Medical Journal* 22 (October 1898): 448.

BIBLIOGRAPHY

Bierer, W. A. 1955. *A History of Veterinary Medicine in America* (Baltimore: the author).

Carr, L. C., R. Menard, and L. S. Walsh. 1991. *Robert Cole's World: Agriculture and Society in Early Maryland* (Chapel Hill: University of North Carolina Press).

Craven, A. O. 1926. *Soil Exhaustion as a Factor in the Agricultural History of Virginia and Maryland, 1606–1860* (Gloucester, Mass.: Peter Smith; 1965 reprint of 1926 ed.).

Crozier, A. A. 1882. *Popular Errors About Plants* (Washington, D.C.: Rural Publishing Co.).

Fussell, G. E. 1970. *Crop Nutrition: Science and Practice before Liebig* (Lawrence, Kans.: Coronado Press).

Gould, H. P. 1901. *Observations on Peach Growing in Maryland* (College Park: The Maryland Agricultural Experiment Station Bulletin No. 72, March 1901).

Gray, L. C. 1933. *History of Agriculture in the Southern United States to 1860* (Baton Rouge: Louisiana State University Press).

Johnson, P. J., ed. 1988. *Working the Water: The Commercial Fisheries of Maryland's Patuxent River* (Charlottesville: University Press of Virginia for the Calvert Marine Museum).

Russell, J. E. 1966. *A History of Agricultural Science in Great Britain, 1620–1954* (London: George Allen & Unwin).

Schlebecker, J. T. 1971. Farmers in the Lower Shenandoah Valley, 1850. *Virginia Magazine of History and Biography* 79, 4 (1971): 462–76.

Schumann, G. L. 1991. *Plant Diseases: Their Biology and Social Impact* (St. Paul, Minn.: American Phytopathological Society Press).

Sears, L. M. 1927. *Jefferson and the Embargo* (Durham: Duke University Press).

Stalheim, O. H. V. 1994. *The Winning of Animal Health* (Ames: Iowa State University Press).

Wiser, V., L. Mark, and H. G. Purchase. 1987. *100 Years of Animal Health* (Beltsville, Md.: Associates of the National Agricultural Library).

CHAPTER FIFTEEN

Bird Populations of the Chesapeake Bay Region

350 Years of Change

JAMES F. LYNCH

The distribution and abundance of birds in the Chesapeake watershed in the last 350 years has altered according to species-specific responses to environmental changes occurring since European settlement. Deforestation, hunting, loss of submersed aquatic vegetation in the Bay, introduction of exotic species, and environmental contamination have had major effects on birds. The future of birds and other wildlife depends on measures taken to protect their habitat in the watershed and estuary.

As seventeenth-century European colonists transformed the landscape of the Chesapeake Bay area, many bird species retreated westward. A few disappeared completely from the region. Other species remained widespread but in lesser numbers. In contrast, some birds profited from the man-made changes in their environment, and their numbers and geographic ranges increased. Losses of native bird species were balanced, numerically if not ecologically, by the arrival of exotic species that successfully established themselves in the wild. Not all historical changes in bird populations can be linked directly to human causes, but no bird species could have been unaffected by the tide of environmental change that swept over the Chesapeake Bay region during the past 350 years.

In this chapter I identify the historical factors that seem to have had the greatest effect on birds in the Chesapeake Bay watershed. I describe a few examples in detail but do not attempt to trace population trends for all of the hundreds of bird species that are found, or formerly were found, in the Bay region. I emphasize Maryland and the District of Columbia, where birds have been more intensively studied than elsewhere. My focus is on two broad ecological groupings of birds: landbirds that inhabit the forests and fields of the watershed, and waterbirds that make use of the Bay, its tributaries, and fringing wetlands.

The Historical Record

The fossil history of birds in the Chesapeake Bay region has been discussed by Steadman (Chap. 5), who also addresses the zoogeographic affinities of the local avifauna. Stewart and Robbins have summarized the history of ornithological work in Maryland and the District of Columbia. Quantitative historical information on the Bay region's birds is practically nonexistent until well into the nineteenth century, by which time major changes in abundance and distribution had already occurred. Earlier descriptive accounts mainly concern species that were heavily hunted for food, such as ducks, geese, swans, wild turkeys, and passenger pigeons.

Two centuries after the arrival of the first English colonists, Warden in 1816 listed only 32 bird species known in the District of Columbia and adjacent Maryland. By the time of the Civil War, the District's bird list had grown to 226 species; by 1918 the number had reached 286. These figures reflect an increase in the number of competent observers, as well as the steady accumulation of records on rare or sporadically seen species. Generally speaking, the statement by a reliable observer that a bird species occurred at a given place and time constitutes strong evidence of the species' presence, but a lack of comment does not mean that species was absent. Thus, Warden's failure to mention a number of species common at present, like the song sparrow or the Carolina wren, does not mean that those species were absent in 1816, but the inclusion on his list of both the passenger pigeon, now globally extinct, and the loggerhead shrike, now locally extirpated, is evidence that they were still around in the early nineteenth century.

The dearth of early historical information forces us to infer the eighteenth- and nineteenth-century status of most bird species by combining knowledge of their current ecology with information about historical land-use changes. Given that about 90% of the Chesapeake Bay watershed was forested at the time of European contact, most forest-dependent birds must have been more abundant and widely distributed then than they were to become centuries later, when more than 75% of

Fig. 15.1. Audubon painting of passenger pigeons (Special Collections, Milton S. Eisenhower Library of the Johns Hopkins University)

the original forest would give way to cropland and pastures. Conversely, birds whose preferred habitat is open grassland or savannah were presumably scarce or absent in presettlement times and have subsequently spread throughout the region as the forest was cleared. Some population changes are not simple functions of habitat availability; important non-habitat factors include overhunting, pesticide contamination, and competition with exotic species.

After the mid-nineteenth century, qualitative accounts of the Bay region's birds become more numerous, and from the 1930s onward periodic quantitative status reports are available for waterfowl and some birds of prey. Since the 1950s, reasonably detailed distributional information has been available for a wide range of species. The *Atlas of the Breeding Birds of Maryland and the District of Columbia,* edited by C. S. Robbins (1996), provides an invaluable baseline for assessments of future changes in distribution and abundance.

In the following discussion I adopt the common usage of the term *extinct* to refer to species that no longer exist, either in nature or in captivity, and *extirpated* to refer to species that have disappeared from a particular place, such as the Chesapeake Bay watershed, but survive elsewhere.

The Significance of Bird Migration

Interpreting historical changes in the Chesapeake region's bird populations is complicated because so many "local" species spend much of the year elsewhere. The vast majority of ducks, geese, and swans that overwinter on the Bay's waters abandon the region every spring and return to their breeding grounds in the north-central United States, Canada, and Alaska. Some of the most familiar songbirds that frequent Bay area backyards and bird feeders during winter, such as the white-throated sparrow and dark-eyed junco, also abandon the area each spring to return to their northern and western breeding grounds. A different pattern of seasonal migration is shown by most of the hummingbirds, cuckoos, flycatchers, swallows, thrushes, vireos, warblers, and tanagers that enliven the region's forests during summer. Most species in these groups spend five to seven months of the nonbreeding season at tropical or subtropical latitudes in the West Indies and Mesoamerica. A significant number of the birds that breed in the Chesapeake Bay region, including such familiar species as the osprey, purple martin, red-eyed vireo, and scarlet tanager, range south to Brazil and Peru during winter. All told, probably 80–90% of the individual birds that bred in the presettlement forest were neotropical migrants.

Seasonal movements of "short-distance" migrants tend to be less clear-cut than those of "long-distance" neotropical migrants. After the breeding season, large numbers of blue jays, American robins, gray catbirds, and northern flickers leave the Chesapeake region for the southeastern United States and northern Caribbean, but some individuals of these species remain in the Bay area over the winter. Yet another group of species, including many shorebirds and wood warblers, breeds north of the Bay region but passes through the area briefly each fall and spring on its way to and from its tropical wintering grounds. Finally, a sizable number of nonmigratory species remain in the Bay area year-round. Familiar examples include the Carolina chickadee, tufted titmouse, northern mockingbird, northern cardinal, and most species of woodpeckers.

Changes in the numbers of the Bay region's migratory birds may be caused by habitat disruptions that occur thousands of miles away. Tropical deforestation and the draining of prairie potholes for agricultural development are but two examples of remote habitat changes that may influence how many birds we observe in the Bay region. Not all members of the local breeding population of a given species will overwinter in the same part of the tropics, and not all individuals of an overwintering population necessarily share the same breeding grounds. Thus, the effects of a localized disturbance of breeding or wintering habitat may be diffused across the entire range of a species at other times of the year, making it very difficult to ascribe specific causes to population changes.

Still another source of variation in the numbers of migratory birds is that some species will modify their migration pattern in response to changes in weather, food availability, or other factors. Historical rises or falls in the numbers of some species within the Bay region, such as those of the Canada goose, snow goose, and redhead, may reflect migratory shifts, not overall population changes.

The Presettlement Landscape and Its Birds

Except for tidal marshes, almost the entire watershed was forested at the time of European contact. The slash-and-burn style of maize cultivation practiced by local Indians in precontact times created relatively small patches of cleared ground interspersed with numerous patches of successional regrowth, all within an extensive matrix of forest. Lacking metal implements or draft animals, Native Americans typically prepared land for agriculture by killing the trees by girdling with stone axes and burning off the undergrowth. The land was not "cleared" in the modern sense of killing and removing all native vegetation, nor was the soil tilled. A given plot could be cultivated for only two or three years before declining soil fertility and the vigorous regrowth of weedy vegetation forced

Fig. 15.2. Audubon painting of a wild turkey (Special Collections, Milton S. Eisenhower Library of the Johns Hopkins University)

its abandonment. This style of agriculture, which today remains widespread in the wet tropics, was adopted by the English settlers for cultivation of maize and tobacco.

Natural openings in the presettlement forest were constantly being created by the death of trees, from causes including old age, disease, insect damage, uprooting by wind, lightning strikes, and flooding from beaver ponds. The most extensive gaps in the presettlement forests of the Chesapeake Bay region were grassy or shrub-covered "barrens," some of which covered thousands of hectares in the Piedmont region of present-day Pennsylvania, Maryland, and Virginia. Some of these unforested clearings may have been natural plant communities associated with unusually dry or sterile substrates, like serpentine-derived soils. Most barrens, however, were intentionally created and maintained by Indians, who repeatedly burned large tracts in order to improve hunting. Possibly the combination of natural conditions and sustained human intervention was necessary for barrens vegetation to develop. In any event, once the European settlers had suppressed this early form of prescribed burning, all but a few barrens quickly reverted to forest or were converted to productive farmland.

Barrens may have been the original strongholds for a number of non-forest bird species. Some of these species could not adapt to a cultivated landscape and are now rare or absent in the Bay region, such as the extirpated greater prairie chicken. Other open-country species successfully colonized the hayfields, pastures, and croplands created by eighteenth- and nineteenth-century settlers but have been much less successful at coping with modern-day agricultural practices. Examples include the upland sandpiper, loggerhead shrike, lark sparrow, Henslow's sparrow, savannah sparrow, Bachman's sparrow, and dickcissel.

The original forest cover of the Bay region was diverse at various spatial scales. Within a local tract of forest, the presence of watercourses and the constant creation and filling-in of canopy gaps resulted in a heterogeneous mosaic of vegetation. At a larger spatial scale, the floristic composition and structure of the forest varied with elevation, latitude, soil type, and available moisture. At a still broader regional scale, extensive portions of the Piedmont and northern Coastal Plain were cloaked with deciduous forest that, when mature, contained few or no coniferous trees, but conifers were major components of forests in the southern Coastal Plain and the western highlands. Swamps and flooded forest were extensive in two disjunct areas, the tidewater fringes of the Bay and the freshwater bogs of the Appalachian highlands. Today, the distributions of several bird species associated either with conifers (e.g., pine warblers) or wetlands (e.g., sedge wrens) mirror the discontinuous distribution of their respective habitats. Another group of water-associated species (e.g., osprey, American black duck, blue-winged teal, king rail, willet, least tern, marsh wren, and prothonotary warbler) are essentially restricted to the extensive marshes or riparian forests of the Coastal Plain. Many forest bird species that have a western, upland distribution in the Chesapeake watershed are outliers of the boreal avifauna of New England and southern Canada, while those with an exclusively eastern, lowland distribution tend to be related to the south.

Throughout the Bay region, the presettlement forest contained numerous large, nut-bearing trees, mainly oaks, hickories, American beech (*Fagus grandifolia*), black walnut (*Juglans nigra*), and American chestnut (*Castanea dentata*). Acorns and other edible nuts were a particularly important food source for the wild turkey and the passenger pigeon, which abounded at the time of European colonization. The large trees and dead snags of mature forests would have provided optimal feeding and nesting habitat for species such as the northern goshawk, pileated woodpecker, hairy woodpecker, hooded merganser, wood duck, purple martin, winter wren, cerulean warbler, and Blackburnian warbler.

One can only speculate about the detailed composition of the avifauna associated with the presettlement forest, especially because contemporary studies provide little information on larger and rarer species, many of which either have been locally extirpated or were too rare to be regularly encountered. Recent studies of Maryland bird communities in the forest area may give a fairly reliable indication of the smaller species that would have occurred there historically (Table 15.1).

The Presettlement Estuary and Its Birds

The Chesapeake Bay and its myriad tributaries form a huge and complex array of habitats for large numbers of waterfowl (ducks, geese, and swans), waders (herons, egrets, and bitterns), shorebirds (sandpipers and plovers), as well as rails, grebes, gulls, terns, loons, cormorants, and other waterbirds. In presettlement times the Bay and its birds were relatively little influenced by the indigenous human population. The surrounding blanket of forest blunted the destructive erosional force of storms and absorbed large amounts of rainwater. This moisture was only gradually released into the groundwater, and the waters of the Bay and its tributaries were much clearer than at present. Much of the presettlement Bay's primary productivity was channeled through "meadows" of subtidal rooted macrophytic plants, which are collectively termed *submersed aquatic vegetation* (SAV). These plants removed excess nutrients from the Bay, trapped suspended sediments, provided hiding places and breeding habi-

Table 15.1 Sensitivity to forest fragmentation

Species	Migratory status	SSF	Species	Migratory status	SSF
European starling	R	0	Wood thrush	N	3
Downy woodpecker	R	0.5	Hairy woodpecker	R	3.5
Red-bellied woodpecker	R	1	Acadian flycatcher	N	4
Blue jay	R	1	Red-eyed vireo	N	4
American crow	R	1	Scarlet tanager	N	4
Carolina chickadee	R	1	Blue-gray gnatcatcher	N	4.5
Tufted titmouse	R	1	American redstart	N	5
Northern cardinal	R	2	Northern parula warbler	N	5
Eastern wood pewee	N	2	Yellow-throated vireo	N	5.5
White-breasted nuthatch	R	2.5	Ovenbird	N	6
Great crested flycatcher	N	2.5	Black-and-white warbler	N	6.5
Carolina wren	R	3	Worm-eating warbler	N	7
Pileated woodpecker	R	3	Kentucky warbler	N	7
Yellow-billed cuckoo	N	3	Hooded warbler	N	7

Source: Whitcomb et al. (1981).

Note: Migratory status: R = year-round resident; N = neotropical resident. SSF: Scores for sensitivity to forest fragmentation, on scale of 0 (no negative response to fragmentation) to 8 (highly negative response). Scores are based on territory size, nest construction, nest placement, degree of habitat specialization, avoidance of "edge" habitat, reproductive potential, existence of ecologically competing species, and ability to colonize new areas of habitat. All species with SSF scores of less than 4 are neotropical migrants.

tat for fish and invertebrates, and served as food for waterfowl and other wildlife. The seemingly limitless beds of oysters also played an important role in maintaining water clarity, by filtering sediments and removing nutrients from the estuary. Water clarity was a critical characteristic of the presettlement Bay, for the plant species that constitute SAV require unusually high intensities of light for their survival.

By today's standards, seventeenth-century populations of waterfowl were staggering. In c. 1670, an observer in Cecil County, Maryland, at the head of the Bay, wrote: "The screeching of the wild geese and other wild fowl in the creek before the door prevented us from having a good sleep . . . the water was so black with [birds] that it seemed [like] a mass of filth or turf, and when they flew up there was a rushing and vibration

Fig. 15.3. Audubon painting of canvasbacks in Baltimore harbor (Special Collections, Milton S. Eisenhower Library of the Johns Hopkins University)

of the air like a great storm coming through the trees while the sky over the whole creek was filled with them like a cloud" (quoted in Semmes, 1937).

Although we lack detailed information on the species composition of the waterfowl flocks that used to mass on the Bay, we do know that canvasbacks, redheads, and American black ducks were vastly more abundant than at present. Tundra swans abounded; the local Indian name for the Potomac River was Cohonguroton (river of swans), and the now-extirpated trumpeter swan and whooping crane were both common. Conversely, the Canada goose and the mallard, the most abundant waterfowl species in the Bay region today, were relatively—though perhaps not absolutely—less common in the seventeenth century. Contemporary accounts make it clear that the bald eagle and the osprey were more widespread and abundant in early colonial times than at present.

Historical Changes in Bird Populations

Destruction and Degradation of the Forest

Early English settlements in the Bay region were clustered within a few hundred yards of navigable waterways. Corn and tobacco were cultivated in small temporary plots that were not plowed or completely cleared of native vegetation. This agricultural system, adopted by the settlers from their Indian predecessors, resulted in a landscape that retained extensive native vegetation, even in settled areas. Human population growth was slow through the seventeenth century, and much of the original forest remained uncleared. Even as late as 1720, only 2% of the land in southern Maryland was under cultivation. During the first century of settlement, probably the only upland birds that felt the impact of the human presence were a few heavily hunted species such as the wild turkey. Hunters probably decimated this large, conspicuous species well beyond the immediate vicinity of the tidewater settlements. At the regional level, no bird species are known to have been extirpated during this early phase of settlement, and species associated with forest margins or early successional scrub (white-eyed vireo, common yellow throat, blue-winged warbler, yellow-breasted chat, and indigo bunting) undoubtedly increased in number.

It was not until the mid-eighteenth century that the interior portions of the coastal counties of Maryland were finally claimed for settlement, and it was still later before these areas were actually occupied and cleared. Much of the remaining forests, however, had been disturbed by selective logging and the introduction of free-ranging pigs, cattle, and horses. The resulting compaction of forest soils and damage to understory vegetation would have been detrimental to the many species of birds that nest or feed on the forest floor.

By the 1700s, farming began to shift away from the long-fallow, slash-and-burn method. Plows and fertilizer were used to grow a variety of crops in permanent fields, often with the labor of African slaves. Although the depletion of certain valuable timber species had been noted at least since the early eighteenth century, it was not until after the American Revolution that the combination of rapid population growth, more intensive agricultural practices, and the development of an interior transportation network began the ecological transformation of the entire Chesapeake watershed. This is also the period when deforestation, soil erosion, deteriorating water quality, and siltation of waterways first became widespread problems. By 1820, the proportion of cultivated land in southern Maryland had reached 40%; by the late nineteenth century, 70–80% of the Bay's watershed was under cultivation or being grazed. Even the forest that did survive was hardly pristine, having been heavily exploited for lumber, charcoal, and firewood. The nineteenth century is the historical period that witnessed the extirpation of many species of birds from the Coastal Plain and Piedmont. Two formerly abundant species, the passenger pigeon and the Carolina parakeet, eventually declined to extinction.

Over the past century, large-scale farm abandonment allowed regeneration of forest throughout much of eastern North America. At present, about 40% of the Bay region is forested, roughly twice the amount of forest that was present a century ago. In recent decades, however, the trend toward reforestation has been reversed in the Coastal Plain and parts of the Piedmont, owing to an unprecedented increase in residential development.

There are important qualitative as well as quantitative differences between the presettlement forests and today's reduced woodlands. Because most of today's forests are less than 50 years old, trees are smaller in stature and there are fewer dead snags. The original forest was dominated by oaks and other mast-bearing species, whereas contemporary forests often are composed predominately of non-masting trees, such as Virginia pine (*Pinus virginiana*), red maple (*Acer rubrum*), sweet gum (*Liquidambar styraciflua*), and tulip poplar (*Liriodendron tulipifera*). Especially in the heavily populated Piedmont and Coastal Plains, present-day forests tend to exist as small fragments that are partially or completely isolated from other forested areas by farmland or residential areas.

The degree of forest fragmentation, as well as the overall reduction of forest, appears to influence the suitability of forests as habitat for many bird species. One would not expect large, wide-ranging bird species to make use of forest patches too small to contain their home ranges. What is surprising is that small, highly migratory species with a body mass less than 75 gm, which prefer the interior of forested tracts, are also sensitive to forest fragmentation (see Table 15.1). A number of these species tend to be absent from otherwise suitable patches of forest large enough to accommodate many breeding pairs.

Why certain bird species are sensitive to forest fragmentation is poorly understood. Some species simply avoid tracts that are smaller than some critical size. Others may attempt to breed in small forest patches but then abandon them after suffering inordinately high nest predation. In much of North America a major cause of reproductive failure in fragmented forests is nest parasitism by the brown-headed cowbird. This savannah-adapted midwestern species invaded the Chesapeake Bay region during historic times as the original forest was cleared.

As forest tracts have become smaller, more isolated, and more subject to nest predators and nest parasites, like cowbirds, local populations of many forest-associated bird species have declined. Whether this is a direct cause-and-effect relationship remains controversial because tropical deforestation and other factors may also play a role in the decline of some migratory species. It seems significant, however, that many of the same species whose abundance has declined in forest fragments are holding their own in large expanses of forest.

Hunting of Waterfowl

The initial response of European settlers to the Bay's bounty of waterfowl was sadly predictable. Slaughter of the seemingly limitless flocks of ducks, geese, and swans began almost immediately, and early accounts detail the number of birds killed with a single shot or in one hunting season. The number of hunters, however, was relatively small for the first century after settlement, and the early colonists lacked the weapons and technical expertise to wreak havoc on a grand scale. Uncontrolled hunting had nevertheless already caused major reductions in waterfowl on the Bay by the time of the naturalists Alexander Wilson's and John James Audubon's observations in the early nineteenth century. With the introduction of better equipment and more effective hunting techniques, and with a seemingly insatiable market clamoring for edible waterfowl, the rate of killing accelerated even as the great flocks of waterfowl declined. In 1925, J. C. Phillips cited the example of one hunter who killed more than 7,000 canvasbacks during the 1846–47 season, and individual daily bags exceeding 100 birds were by no means unknown during this period.

Until well into the twentieth century, most waterfowl were killed by commercial hunters, not amateur sportsmen. Indeed, the Bay's waterfowl fueled what has been termed the largest market hunting business known to man. At the height of commercial waterfowling, c. 1870 to 1910, ducks and geese from the Chesapeake region were shipped all over eastern North America and even to Europe. Uncontrolled commercial hunting continued until 1918, when the Migratory Bird Treaty Act was enacted by the U.S. Congress. By that time, the great flocks of waterfowl had dwindled to mere remnants. With the protective hunting regulations, populations began to rebound, but several factors prevented numbers from approaching their former levels. Continent-wide drought severely limited the breeding habitat for waterfowl during the 1930s. Even worse was the permanent draining of prairie potholes for agricultural development, which accelerated from the 1950s onward. Finally, the widespread die-off of SAV since the 1960s eliminated the main food supply of many

Fig. 15.4. Old-time market gunners aboard the *Sherwood* proudly display one weekend's bag of waterfowl (Dr. Harry Walsh Collection, Chesapeake Bay Maritime Museum)

of the Bay's waterfowl, including its "flagship" duck species, the canvasback.

Hunting of Upland Birds

The passenger pigeon, at one time possibly the most abundant bird species on earth, was the only upland bird species in the Bay region that supported a commercial hunting industry. Huge nomadic flocks of these pigeons once roamed the deciduous forests of eastern North America during fall and winter, feeding voraciously on acorns and other mast. Their conspicuous communal roosts were easy targets for "hunters," who clubbed to death the sleeping birds and salted them down by the barrelful. At times passenger pigeons were such a glut on the market that their carcasses were fed to hogs.

In the eighteenth and early nineteenth centuries flocks of passenger pigeons numbering in the thousands were commonly observed as far east as Baltimore and Washington, D.C. By the mid-nineteenth century, the combined effects of overharvesting and deforestation had pushed the breeding range of the species west to the Appalachian highlands; by 1880, the huge flocks were a thing of the past. The last reliable sighting of passenger pigeons in Maryland occurred about 1900, and the world's last-known passenger pigeon died in captivity in 1918.

The wild turkey and the ruffed grouse are two other avidly hunted forest birds that once ranged across the entire Chesapeake Bay watershed. As late as 1836 a hunter was reported to have shot 9 turkeys from a flock of 20–30 birds that appeared in the District of Columbia, but by the mid-nineteenth century both species had been extirpated from the Coastal Plain and Piedmont. Today the ruffed grouse remains restricted to the western highlands, but the wild turkey has responded spectacularly to reintroduction programs by state game departments. In the 1930s the remnant Appalachian population of turkeys in the Virginia-Maryland-Pennsylvania region was at most a few thousand. By 1960 the population had increased to more than 60,000 birds, and by 1990 an estimated 260,000 wild turkeys roamed the forests of the three-state area, from the Appalachian highlands to the Delmarva Peninsula.

Loss of Submersed Aquatic Vegetation (SAV)

The link between the abundance of waterfowl in the Bay and the presence of extensive beds of SAV was recognized by Audubon in 1840: "The Chesapeake with its tributary streams has from its discovery been known as the greatest resort of waterfowl in the United States. This has depended on the profusion of their food, which is accessible on the immense flats . . . near the mouth of the Susquehanna, along the entire length of the North-east and Elk Rivers, and . . . as far south as the York and James Rivers."

Despite the effects of earlier uncontrolled market hunting, the Susquehanna Flats supported impressive numbers of wintering waterfowl as recently as the 1930s. In November 1937, the Maryland Game and Fish Commission estimated that just one section of the upper Bay between Baltimore and the Susquehanna Flats held 500,000 canvasbacks. To put this number into perspective, it is equal to today's entire North American canvasback population.

Although the original areal coverage of SAV was never accurately measured, it probably exceeded a half-million acres. SAV was still widespread in the Bay at the time of the 1937 waterfowl survey. In recent decades, however, concentrations of key nutrients (nitrogen, phosphorus, silicon) in the Bay have risen owing to an increased influx of human sewage, animal waste, and agricultural runoff. Algae and other phytoplankton adapted to high nutrient concentrations have come to dominate the Bay's primary productivity, and its once-clear waters have become increasingly turbid. Light penetration has been reduced to such an extent that many species of SAV, which are some of the most light-demanding plants known, are unable to survive.

Loss of SAV was first noticed in the 1930s and 1940s in the oligohaline sections of the Potomac estuary below Washington, those which have less than 5 parts salt per thousand. The decline there accelerated in the 1960s; by the mid-1970s, SAV had been extirpated from the upper-tidal portion of the river. In the main stem of the Chesapeake Bay, SAV loss became evident around 1960. By 1990, SAV was absent from nearly 80% of the area where it had formerly occurred in the upper Bay, traditionally the most important overwintering area for waterfowl. SAV loss was also severe in the headwaters of the Bay's major tidal tributaries. In 1991, aerial surveys indicated a total SAV coverage of just over 50,000 acres, or approximately 10% of the estimated historical level.

The various species of waterfowl have responded very differently to the decline of SAV in the Chesapeake Bay. Overwintering Canada geese and snow geese, two herbivorous species that historically fed both on terrestrial vegetation and SAV, now feed almost entirely in fallow agricultural fields. This shift has been favored by the substitution of machines for human corn pickers on the Delmarva Peninsula's farms. Mechanical harvesters leave relatively large amounts of waste corn in the fields each fall. For Canada geese and snow geese, this rich, easily obtained source of food has proved to be a bonanza that more than compensates for the loss of SAV. Bay area populations of both species have skyrocketed in recent decades, to the point where the Chesapeake region has become the main overwintering destination for the entire Atlantic flyway population of snow geese and Canada geese. In a classic example of birds shifting their migratory pattern to take advantage of changing conditions, both species of geese have all but abandoned their traditional wintering grounds in the Carolinas.

A dramatically different response to declining SAV is that of the brant, a small herbivorous goose that inhabits saline waters of the lower Bay and Atlantic coast. The brant feeds almost entirely on SAV, and it has been unable to substitute terrestrial plants or grain. In the 1930s a disease killed off most local beds of eelgrass (*Zostera marina*), the brant's preferred food plant. The population of brant subsequently plummeted and has remained low ever since.

Among the Bay's ducks, the canvasback was historically one of the most dependent on SAV. Indeed, the scientific name of the canvasback (*Aythya valisineria*) refers to wild celery (*Vallisneria americana*), its preferred food plant. Over the past half-century, canvasbacks in the Chesapeake Bay have made a dramatic shift from an SAV-dominated diet to one composed mainly of clams and other aquatic animal food. Whereas plants constituted 70–80% of the canvasback's diet prior to 1930, the percentage of plant food dropped to 50% by the 1960s, and by the mid-1970s plants made up only a trivial proportion of the diet. The redhead, a close taxonomic relative of the canvasback, has not shown the same dietary flexibility. Unable to subsist on animal food, the redhead has declined precipitously in the Bay area. Unlike the Canada goose and the snow goose, which abandoned their traditional southern wintering grounds in favor of the Chesapeake Bay region, the redhead and the canvasback shifted their winter ranges southward to the Carolinas and Gulf Coast, where SAV is still abundant. The total North American population of canvasbacks has declined significantly over the past few decades, but the redhead's continental population has remained stable, despite declining numbers in the Bay region. Again, these patterns emphasize the need to consider the entire distribution of migratory species when interpreting local changes in abundance.

Introduction of Exotic Species

The most clear-cut examples of invasion of the Chesapeake region by exotic birds involve Old World species that were either intentionally introduced into the wild in North America or escaped from captivity. Examples include the domestic pigeon brought to the region from England in earliest colonial times; the ring-necked pheasant, an Asian species repeatedly introduced for hunting since the early nineteenth century; the house sparrow, a European species intentionally introduced into Washington, D.C., and Baltimore in the 1870s; the European starling, intentionally introduced into New York in 1890 and first reported in Maryland in 1906; and the mute swan, a European species that escaped from captivity and became established on the Bay in the 1950s. The house finch, a native of western North America, was introduced into New York in the 1950s and subsequently spread up and down the East Coast. This finch has been widespread and common in the Bay area since the 1970s. The only Old World species that has become well-established in North America without direct human intervention is the cattle egret. This small terrestrial heron spontaneously colonized northern South America from Africa in the 1870s and rapidly expanded its New World distribution. The first stragglers reached Maryland in the late 1950s, and the cattle egret is now a common breeding summer resident throughout the Bay area.

Although these exotic species are a diverse group taxonomically, they share a propensity for highly disturbed habitats created by humans, including cities and stockyards (domestic pigeon, house sparrow, European starling); agricultural fields (ring-necked pheasant, European starling), pastures (cattle egret, European starling); suburban and rural residential developments (house finch, house sparrow, European starling); and disturbed shorelines (mute swan). Although most exotic species tend to avoid pristine native vegetation, some introduced species may compete with native birds for nest sites, food, or other critical resources. Certain cavity-nesting native species associated with semi-open habitats (red-headed woodpecker, eastern bluebird, tree swallow, purple martin) are subject to intense nest-site competition from European starlings and house sparrows. It seems ironic that these native species, which must have been scarce or localized in the region prior to European settlement, initially proliferated as the forest was cleared, only to be displaced later by aggressive Old World exotics that had adapted to human-modified landscapes over many centuries.

A good example of a native bird that has suffered from intense competition with exotic species is the purple martin. This large swallow is thought to have bred in emergent dead snags in the presettlement deciduous forest. During that period the purple martin would have been widespread but not particularly abundant because suitable nest sites were quite limited. As European settlers began cutting the forest, they created the large clearings preferred by purple martins for foraging. Like the Native Americans before them, the new human population welcomed the purple martin's mosquito-eating habits and provided apartment-style birdhouses for martin colonies. An almost limitless number of additional nesting sites was unintentionally provided in the form of cornices, drains, and eaves on city buildings. At some point, the purple martin ceased breeding in the forest and became essentially a commensal of human settlements. During the eighteenth and nineteenth centuries, the species became abundant even in cities, where thousands nested in buildings. This "golden age" for purple martins came to an end in the late nineteenth and early twentieth centuries, with the spread of the introduced house sparrow throughout North America. The house sparrow, and later the European starling, completely coopted the purple martin's urban breeding niche and made strong inroads even in rural areas. Largely as a result of nest-site competition with house sparrows, the purple martin population is estimated to have declined by as much as 90% over the twentieth century.

Detrimental Effects of Some Native Species

The handful of native species that have proliferated in response to historical land-use changes may pose a threat to other native species whose ecological or behavioral characteristics make them less tolerant of disturbance. Thus, aggressive captive-reared mallards appear to displace both American black ducks and wild mallards. Populations of the common grackle and American crow have thrived in the fragmented landscape of the Chesapeake watershed, and both species are frequently mentioned as nest predators of other native birds. The overall impact of crows and grackles on other native birds is questionable, although egg predation by common crows and fish crows on the regionally declining American black duck sometimes reaches high levels.

One native species that clearly has had a detrimental effect on other native birds is the brown-headed cowbird. Females of this species are nest parasites, which lay their eggs in the nests of other species. There are as many as 50 known host species for the brown-headed cowbird in Maryland and the District of Columbia alone. The host parents generally feed the large, aggressive cowbird chick at the expense of their own young, and in many instances only the cowbird survives to fledge. A female cowbird can produce scores of eggs over the course of one nesting season, so the local impact of even a few cowbirds on vulnerable host populations can be large. Cowbird numbers skyrocketed in the twentieth century, and its host species suffered accordingly.

The brown-headed cowbird evolved in the ecotone between the Great Plains and the eastern forests, but it became widespread in the East in the wake of deforestation. The species feeds on grain in open fields and has benefited from the conversion of forests to pastures and croplands. Female cowbirds, moreover, will fly hundreds or even thousands of yards into forest in search of host nests; only very large tracts of contiguous forest are immune to their depredation. Unfortunately, forest-interior species have not had time to evolve behavioral safeguards (e.g., removal of cowbird eggs, abandonment of parasitized nests) that protect some potential host species elsewhere in the range of the brown-headed cowbird.

Environmental Contamination

M. C. Perry, G. H. Heinz, and others have studied environmental contamination of habitats in the Chesapeake Bay region, where birds associated with wetlands and agricultural fields have been most severely affected. Toxic substances can induce direct mortality of adult birds as well as cause reproductive failure. A growing body of evidence, moreover, suggests that sublethal toxicity, often the result of several contaminants, poses a greater threat to wildlife than the direct effect of any one toxin. Environmental toxins that have been implicated as causes of mortality or reduced viability in Bay area birds include organochlorine pesticides (especially DDT, DDE, dieldrin, and kepone), organophosphorus and carbamate insecticides (e.g., Abate, Furaden), various herbicides, polychlorinated biphenyls (PCBs), heavy metals (especially lead and cadmium), and spilled petroleum distillates.

Dieldrin has been implicated as the cause of adult mortality in bald eagles, cattle egrets, and great blue herons. Carbofuran, an insecticide commonly applied in cornfields, is reported to have caused adult mortality in bald eagles, red-tailed hawks, American kestrels, and various songbird species. The most widespread and insidious effect of pesticides, however, has been reproductive failure caused by organochlorine poisoning. Common manifestations of DDT or the related DDE poisoning include collapse of eggs owing to thinning of their shells and the reduced viability of embryos. These organochlorines are persistent pesticides and tend to accumulate as they move up the food chain. Thus, top predators in the Bay region's ecosystems are especially prone to organochlorine poisoning; the bald eagle and osprey have been especially hard hit.

Heavy metals, especially lead, cadmium, and mercury, enter the Bay by way of industrial pollution. Metallic lead is also injected into game birds in the form of shotgun pellets or ingested when feeding waterfowl accidentally swallow spent shot. Cadmium levels are highest in diving ducks, such as the canvasback, oldsquaw, and white-winged scoter, but lead levels are highest in dabbling ducks, which frequent the shallow waters where the concentration of spent shot tends to be highest. The wood duck, a permanently resident dabbling species, shows the highest levels of lead contamination of any waterfowl species in the Chesapeake Bay region. Bald eagles have died from lead poisoning in the Bay area, probably from eating waterfowl crippled or killed by hunters. Fortunately, the recent banning of lead shot pellets in the Bay region has significantly reduced the incidence of lead poisoning.

Recent Trends and Future Prospects

Waterfowl Populations

For the past half-century the combined winter population of ducks, geese, and swans in the Chesapeake Bay region has remained relatively stable at about one million birds, but the species mix has changed dramatically. The Canada goose increased from an average of about 150,000 birds in the 1950s to around 600,000 in the 1970s and 1980s and is now the most abundant waterfowl species in the region. Traditionally, the Canada goose has been a winter visitor to the Delmarva Peninsula, but thousands of birds now overwinter on the western shore of Chesapeake Bay and a growing breeding population inhabits the Bay area year-round. The snow goose has shown an even more impressive increase, with the average winter population exploding from a mere 1,000 birds in the early 1950s to about 120,000 in the 1980s. The only duck whose local numbers have increased markedly since the 1940s is the bufflehead, a small diving species that is rarely hunted and does not depend on SAV for food. One of the few major waterfowl species whose numbers are stable in the Chesapeake region is the tundra swan, whose winter population has fluctuated around 35,000 birds in recent decades. Although tundra swans traditionally restricted their feeding to SAV, they have begun to feed in fallow agricultural fields as well.

In stark contrast to the reasonably favorable outlook for a few waterfowl species, alarming declines have occurred in Bay populations of dabbling ducks as a group, diving ducks as a group, mergansers (two species), scaup (two species), ruddy duck, redhead, American wigeon, American black duck, northern pintail, and canvasback. An especially alarming decrease has affected the American black duck, one of the few waterfowl species that both breeds and overwinters in the Bay area. In Maryland the average number of wintering black ducks plummeted by about 85%, from around 130,000 birds in the 1950s to 18,000 in 1980. During this same period, about 150,000 pen-raised mallards were released around the Bay each year by fish and game agencies and private hunting clubs. It is likely that at least part of the decline of the American black duck can be attributed to competition and hybridization with these behaviorally aggressive, semidomesticated mallards.

The effect of a massive influx of captive-raised mallards on wild mallards is difficult to assess, but it may be considerable. Wild mallards did not traditionally breed in Maryland, but a resident breeding population of mallards has arisen in recent decades. Many of these sedentary mallards are noticeably tame, and many show morphological evidence of introgression with domestic birds. It is likely that such birds are captive-reared, or at least descendants of captive-reared birds. Winter counts of mallards, wild and captive-reared birds combined, rose in the early 1950s, declined sharply in the late 1950s and 1960s, then began a moderate increase during the 1970s. What proportion of the recent increase is attributable to nonwild birds is an important question, because the total North American population of wild mallards is at an all-time low.

Local Return of SAV to the Estuary

About 1980, the exotic SAV species *Hydrilla verticillata* appeared in the Potomac River near Washington, D.C., having been inadvertently introduced by a local SAV revegetation project. Since 1980 *Hydrilla* has rapidly expanded into oligohaline sections of the Potomac subestuary, possibly in response to the introduction of improved tertiary sewage treatment in Washington. The areal coverage of *Hydrilla* and minor amounts of native SAV species in the upper-tidal Potomac increased from zero in 1979–81 to more than 5,000 acres in 1992. In 1990, *Hydrilla* first appeared in the Susquehanna Flats. At present, *Hydrilla*-dominated beds account for about 10% of all SAV in Chesapeake Bay.

Observers have noted an increase in waterfowl and fish numbers in areas where *Hydrilla* has taken hold, but this one exotic plant species is no panacea for the Bay-wide problem of declining SAV. One problem is *Hydrilla's* low salt tolerance, which prevents it from colonizing reaches of the Bay where salinity exceeds 5 parts per thousand. Nor can one introduced species be expected to substitute functionally for a native SAV community that includes more than 20 species. Finally, the twentieth-century history of the Potomac subestuary, and of the Bay generally, was one of recurring "boom and bust" cycles of invasion and decline by various exotic SAV species. Local fishermen report that even as *Hydrilla* continued to spread down the Potomac, it began to die back in the upper reaches where it first appeared.

Populations of Forest and Nonforest Landbirds

At both the continental and regional levels, populations of most small birds that breed in forests, including neotropical migrants, remained fairly stable or increased from the mid-1960s to about 1980. Since that time, a number of species have declined, prompting concern that tropical deforestation rates have finally reached the critical point. However, it is also possible that these declines are short-term responses to unfavorable weather patterns in the breeding grounds rather than an irreversible downward trend.

Most attention has been focused on declining populations of forest-associated species that migrate to the neotropics, like the wood thrush and cerulean warbler, and a controversy has developed over the relative importance of tropical versus temperate-zone forest degradation as the major cause of observed declines. However, a number of shrub-associated species (yellow-breasted chat, rufous-sided towhee) and grassland species (bobolink, eastern meadowlark), most of which are not long-distance migrants, are also declining. None of the Bay region's landbirds are yet in danger of global extinction, but increasing rates of conversion of forest and other bird habitats, both locally and in the tropics, could severely affect many species in coming decades.

Population Increases by Previously Declining Species

The bald eagle and the osprey, two of the Chesapeake Bay's "flagship" species, provide especially good examples of birds whose populations recovered once human society ceased certain destructive practices. The original bald eagle population around Chesapeake Bay probably numbered at least several thousand pairs. Following European settlement, bald eagles were reduced by shooting, and they also suffered from loss of habitat and reductions in their principal food supply (carcasses of large fish, mammals, and birds). It is estimated that 3,000 eagles survived in the region in 1900. In 1936, the first accurate survey gave 1,800–4,000 individuals, so it would appear that the population had stabilized in the early decades of the twentieth century. In the 1940s, when the use of DDT for mosquito control became widespread, bald eagle populations plummeted throughout the United States and Canada. In the Chesapeake region, DDT-caused reproductive failure as well as adult mortality induced by dieldrin and PCBs have been implicated in the decline. The Bay's eagle population never completely died out, but by 1962 had reached an all-time low of 50–60 breeding pairs. In that year only 20 eaglets successfully fledged in the entire region. This dismal situation quickly improved after DDT was banned in 1972. Fourteen years later, 133 breeding pairs of eagles successfully fledged 188 young.

The osprey's decline was driven by the same factors, but losses were less severe. The Chesapeake Bay osprey population probably never sank below 50% of its pre–World War II level, and numbers rebounded rapidly following the banning of organochlorine pesticides. The Chesapeake Bay now supports more than 1,500 breeding pairs of ospreys, approximately 20% of the entire U.S. population. The Bay's osprey population appears to be at or near carrying capacity, given the current reduced status of the Bay's fish populations.

Among upland birds, the wild turkey is perhaps the best example of a species that was hunted nearly to extinction, only to recover once societal attitudes changed. Successful transplantation of wild turkeys by public agencies has been followed by spontaneous range expansion of the species into most areas where suitable forested habitat still exists.

The pileated woodpecker is North America's largest extant woodpecker; only the extinct ivory-billed woodpecker was larger. The pileated woodpecker has traditionally been associated with large trees in extensive tracts of forest. In the nineteenth century the species steadily retreated westward in the face of deforestation and was treated as no more than a "timber pest" until the 1940s. By the mid-1950s it had disappeared from the Maryland Coastal Plain, except along the forested floodplains of the largest rivers and swamps. For reasons that are not entirely clear, the species began a strong comeback in the 1970s and today is common and widespread in wooded sections of the Coastal Plain and interior. One intriguing, but unproven, possibility is that behavioral adaptations enabled the pileated woodpecker to cope more effectively with a fragmented landscape. In any event, these impressive woodpeckers are now commonly observed flying across wide expanses of open country as they move from one forest patch to another.

Species That Require Active Intervention

Unfortunately, not all declining bird species become abundant if persecution simply ceases. For some species, modern land-use practices are incompatible with their fundamental needs for food or nest sites. In the Bay region, this would appear to be the dilemma faced by several species of grassland birds—such as the upland sandpiper, loggerhead shrike,

Henslow's sparrow, Bachman's sparrow, dickcissel, and bobolink—that profited from the initial clearing of the presettlement forest but cannot cope with modern farming and grazing practices.

Other species might have the potential to thrive in the modern landscape, but predation or competition for food or nest sites keeps their numbers low. Examples include the purple martin, eastern bluebird, and tree swallow, hole-nesting species that are adversely affected by competition with the exotic house sparrow and European starling. Concerted efforts by the bird-loving public have provided tens of thousands of specially designed nest boxes for purple martins and eastern bluebirds. The bluebird boxes are also extensively used by tree swallows. This intervention may have halted the population declines of bluebirds and martins, but a return to their previous abundance probably is not possible except in the unlikely event that the numbers of house sparrows and European starlings are drastically reduced. The wood duck is another native species that has benefited from a widespread campaign to provide artificial nest boxes. For this duck, the main problem was lack of suitable large, natural tree hollows, as well as depredation of natural nests by mammalian predators.

Cowbird control has proven effective in managing localized populations of seriously threatened host species, but the enormous size and great mobility of the cowbird population means that broad-scale regional control would be prohibitively expensive—and probably would be ethically unacceptable to the American public.

Apparently Spontaneous Range Expansions

In recent decades a few native bird species have expanded their ranges into the Bay region for reasons that are not obviously tied to human activities. One prominent example, the black vulture, is a southeastern species that was a rare straggler to the Washington, D.C., area early in the twentieth century. Up until the 1950s, Washington remained the northernmost breeding locality for the species. Today, black vultures breed north to Pennsylvania and New Jersey, and stragglers regularly reach southern New England. This may be just one of many instances of a southern species expanding its range northward as the climate has warmed. Other species that have spread northward into New England in recent decades, such as the northern mockingbird, northern cardinal, tufted titmouse, and Carolina wren, already existed in the Chesapeake Bay region before their expansion farther north.

Not all range expansions have been from south to north. The opposite pattern is exemplified by the double-crested cormorant, a northern species that used to occur in the Bay region only as a nonbreeding transient. In recent decades the double-crested cormorant has become a regular breeder in the area.

Future Prospects

The Chesapeake Bay region's avifauna will continue to be a kaleidoscopic mix of changing populations. Some species will steadily increase or decrease, others will fluctuate erratically from year to year, still others will maintain relatively stable numbers. In most instances, the reasons for population changes or stability will not be evident. Even in those few cases where we understand the causes of population changes, we may lack the economic means or the political will to reverse undesirable trends. There have been some heartening exceptions to this generalization: the rapid recovery of the osprey and the bald eagle after chlorinated hydrocarbon pesticides were banned; the spectacular increase in the range and numbers of the wild turkey following its protection and reintroduction; the positive responses of wood ducks, purple martins, and eastern bluebirds to the provision of artificial nest boxes. But what are we to do about the burgeoning populations of brown-headed cowbirds, house sparrows, and European starlings which negatively affect so many native species? Are we as a society willing to endorse control programs that not only cost millions of dollars but also result in the deaths of millions of "undesirable" birds? And no degree of effective conservation action in the Chesapeake Bay region will save a duck species whose breeding sites in Canada have been converted to farmland, or a warbler whose winter habitat in Latin America is being turned into overgrazed cattle pastures.

To re-create the Bay area's presettlement avifauna would be impossible; the passenger pigeon, once the most abundant of all species, is gone forever, and massive transformations of the landscape have occurred over the past 350 years. Nor would we necessarily want to return to pristine conditions, even in the unlikely case that this were economically feasible. If the Bay's watershed were to revert to a fully forested condition, many presently common species would become much less abundant and some would likely disappear. We might not mourn the passing of the brown-headed cowbird under such a scenario, but what about the many grassland and savannah species whose precarious status is currently of concern to conservationists? A more realistic long-term conservation goal for the birds of the Chesapeake Bay watershed would be to maintain viable populations of all existing native species, with special emphasis on (1) forest-associated species that are sensitive to fragmentation and other forms of habitat disturbance; (2) shrub and grassland species that require fallow fields, successional scrub, or other nonforest upland habitat; and (3) waterfowl and other aquatic birds that depend on a healthy estuarine environment.

The growing public commitment to reversing the Bay's deterioration is encouraging, and legislative measures to clean up the Bay are beginning to bear fruit. The banning of some of the most potent environmental toxins has spurred the recovery of the osprey and bald eagle. The recent return of SAV to the upper-tidal reaches of the Potomac estuary is a positive development, even if an exotic SAV species is the main colonist. SAV provides food and cover for birds and other wildlife, helps to purify the Bay's waters, and serves as an overall indicator of the Bay's health. Although fish and waterfowl numbers have begun to increase in the upper-tidal Potomac over the past decade, the battle to restore the Bay is nowhere near won, and we cannot be certain that all of the important problems, much less their solutions, have yet been recognized.

The often subtle effects of land-use practices on wildlife have barely begun to penetrate public awareness. Although state and federal laws protect almost all Chesapeake region bird species from wanton killing, there has been little regulatory attention directed at the fundamental cause of population declines: loss of appropriate habitat. Thus, a citizen who would be subject to a severe fine for illegally killing a single bird can destroy the habitat of that bird, and of many others, with impunity. Maryland's progressive Chesapeake Bay Critical Area legislation is an important exception to the lack of adequate safeguards for bird habitat. Among other provisions, these laws place significant controls on the cutting of trees within a 300-yard buffer zone around the Bay and its tidal tributaries. Several local ornithologists were invited to participate in drafting sections of this legislation which explicitly aim to protect fragmentation-sensitive forest birds and their breeding habitat.

Preserving the Bay area's rich avifauna for future generations will require a growing human population to regulate its use of the landscape and estuary in ways that will not always be politically popular. Important commitments to bird conservation would include (1) protection of large contiguous tracts of forest from development; (2) use of active interventions such as tree planting alongside such passive measures as allowing secondary succession to expand and link forest patches; (3) use of tax benefits, conservation easements, and other incentives to encourage landowners to manage open fields and successional habitats in ways that benefit nonforest birds; and (4) continued pressure to enact local, state, and regional legislation that reduces the Bay's load of excess nutrients and other pollutants. Our long-term goal should be to eliminate wasteful and unnecessary destruction of critical wildlife habitat and to minimize the negative effects of necessary human activities.

APPENDIX: SCIENTIFIC NAMES OF BIRDS MENTIONED IN THE TEXT AND TABLE

Names and sequence of families follows American Ornithologists' Union (1983).

Pelicans and cormorants: double-crested cormorant (*Phalacrocorax auritus*). **Herons and egrets:** great blue heron (*Ardea herodias*), cattle egret (*Bubulcus ibis*), American bittern (*Botaurus lentiginosus*). **Swans, geese, and ducks:** tundra swan (*Cygnus columbianus*), trumpeter swan (*C. buccinator*), mute swan (*C. olor*), snow goose (*Chen caerulescens*), brant (*Branta bernicla*), Canada goose (*B. canadensis*), wood duck (*Aix sponsa*), American black duck (*Anas rubripes*), mallard (*A. plutyrhynchos*), northern pintail (*A. acuta*), American wigeon (*A. americana*), blue-winged teal (*A. discors*), canvasback (*Aythya valisineria*), redhead (*A. americana*), oldsquaw (*Clangula hyemalis*), white-winged scoter (*Melanitta fusca*), bufflehead (*Bucephala albeola*), hooded merganser (*Lophodytes cucullatus*), ruddy duck (*Oxyura jamaicensis*). **Vultures, osprey, eagles, hawks, and falcons:** black vulture (*Coragyps atratus*), osprey (*Pandion haliaetus*), bald eagle (*Haliaeetus leucocephalus*), northern goshawk (*Accipiter gentilis*), red-tailed hawk (*Buteo jamaicensis*), American kestrel (*Falco sparverius*), peregrine falcon (*Falco peregrinus*). **Grouse, turkeys, and quail:** ring-necked pheasant (*Phasianus colchicus*), ruffed grouse (*Bonasa umbellus*), greater prairie chicken (*Tympanuchus cupido*), wild turkey (*Meleagris gallopavo*). **Rails:** king rail (*Rallus elegans*). **Cranes:** whooping crane (*Grus americana*). **Sandpipers and plovers:** willet (*Catoptrophorus semipalmatus*), upland sandpiper (*Bartramia longicauda*). **Gulls and terns:** least tern (*Sterna antillarum*). **Pigeons and doves:** domestic pigeon (*Columba livia*), passenger pigeon (*Ectopistes migratorius*). **Cuckoos:** yellow-billed cuckoo (*Coccyzus americanus*). **Parrots:** Carolina parakeet (*Conuropsis carolinensis*). **Woodpeckers:** downy woodpecker (*Picoides pubescens*), hairy woodpecker (*P. villosus*), pileated woodpecker (*Dryocopus pileatus*), ivory-billed woodpecker (*Campephilus principalis*), red-bellied woodpecker (*Melanerpes carolinus*), red-headed woodpecker (*M. erythrocephalus*), northern flicker (*Colaptes auratus*). **Fly-**

catchers: eastern wood pewee (*Contopus virens*), Acadian flycatcher (*Empidonax virescens*), great crested flycatcher (*Myiarchus crinitus*). **Swallows:** purple martin (*Progne subis*), tree swallow (*Iridoprocne bicolor*), cliff swallow (*Hirundo pyrrhonota*). **Jays and crows:** blue jay (*Cyanocitta cristata*), American crow (*Corvus brachyrhynchos*), fish crow (*C. ossifragus*). **Titmice:** Carolina chickadee (*Parus carolinensis*), tufted titmouse (*Parus bicolor*). **Nuthatches:** white-breasted nuthatch (*Sitta carolinensis*). **Wrens:** winter wren (*Troglodytes troglodytes*), marsh wren (*Cistothorus palustris*), sedge wren (*C. platensis*), Carolina wren (*Thryothorus ludovicianus*). **Gnatcatchers:** blue-gray gnatcatcher (*Polioptila caerulea*). **Thrushes:** eastern bluebird (*Sialia sialis*), American robin (*Turdus migratorius*), wood thrush (*Hylocichla mustelina*). **Mockingbirds and thrashers:** northern mockingbird (*Mimus polyglottos*). **Shrikes:** loggerhead shrike (*Lanius ludovicianus*). **Starlings:** European starling (*Sturnus vulgaris*). **Vireos:** red-eyed vireo (*Vireo olivaceus*), white-eyed vireo (*V. griseus*), yellow-throated vireo (*V. flavifrons*). **Wood warblers:** Kentucky warbler (*Oporornis formosus*), northern parula warbler (*Parula americana*), ovenbird (*Seiurus aurocapillus*), Blackburnian warbler (*Dendroica fusca*), pine warbler (*D. pinus*), cerulean warbler (*D. cerulea*), hooded warbler (*Wilsonia citrina*), blue-winged warbler (*Vermivora pinus*), common yellow throat (*Geothlypis trichas*), black-and-white warbler (*Mniotilta varia*), yellow-breasted chat (*Icteria virens*), prothonotary warbler (*Protonotaria citrea*), worm-eating warbler (*Helmitheros vermivorus*), American redstart (*Setophaga ruticilla*). **Tanagers:** scarlet tanager (*Piranga olivaceas*). **Sparrows and allies:** dickcissel (*Spiza americana*), lark sparrow (*Chondestes grammacus*), swamp sparrow (*Melospiza georgiana*), song sparrow (*M. melodia*), northern cardinal (*Cardinalis cardinalis*), Savannah sparrow (*Passerculus sandwichensis*), Bachman's sparrow (*Aimophila aestivalis*), Henslow's sparrow (*Passerherbulus henslowii*), indigo bunting (*Passerina cyanea*). **Blackbirds and allies:** bobolink (*Dolichonyx oryzivorus*), common grackle (*Quiscalus quiscula*), red-winged blackbird (*Agelaius phoenicius*), brown-headed cowbird (*Molothrus ater*). **Cardueline finches:** house finch (*Carpodacus mexicanus*). **Weaver finches:** house sparrow (*Passer domesticus*).

ACKNOWLEDGMENTS

I extend special thanks to Chan Robbins, dean of Maryland ornithologists, for allowing me prepublication access to his *Atlas of the Breeding Birds of Maryland and the District of Columbia*. Gene Morton informed me on the plights of the American black duck, purple martin, and eastern bluebird, species with which he has been deeply involved for decades. Alan Poole shared his expert insights on the state of the Bay's osprey population. Dan Higman provided several important historical references.

BIBLIOGRAPHY

American Ornithologists' Union. 1983. *Checklist of North American Birds,* 6th ed. (Lawrence, Kans.: Allen Press).

Askins, R. A., J. F. Lynch, and R. Greenberg. 1990. Population declines in migratory birds in eastern North America. *Current Ornithology* 7: 1–57.

Audubon, J. J., and J. B. Chevalier. 1840. *Birds of America,* 7 vols. (Philadelphia: published by the authors).

Bollinger, E. K., and T. A. Gavin. 1992. Eastern bobolink populations: Ecology and conservation in an agricultural landscape. In *Ecology and Conservation of Neotropical Migrant Landbirds,* ed. J. M. Hagan and D. W. Johnston (Washington, D.C.: Smithsonian Institution Press), 497–506.

Brittingham, M. C., and S. A. Temple. 1983. Have cowbirds caused forest song birds to decline? *Bioscience* 33: 31–35.

Carter, V., and G. M. Haramis. 1980. Distribution and abundance of submersed aquatic vegetation in the tidal Potomac River—implications for waterfowl. *Atlantic Naturalist* 33: 14–19.

Cooke, M. T. 1921. Birds of the Washington area. *Proc. Biol. Soc. Washington* 34: 1–22.

Correll, D. L. 1987. Nutrients in Chesapeake Bay. In *Contaminant Problems and Management of Living Chesapeake Bay Resources,* ed. S. K. Majumdar, L. W. Hall, and H. M. Austin (Philadelphia: Pennsylvania Academy of Sciences), 298–320.

Davis, F. W. 1985. Historical changes in submerged macrophyte communities of upper Chesapeake Bay. *Ecology* 66: 981–93.

Dennison, W. C., R. J. Orth, K. A. Moore, J. C. Stevenson, V. Carter, S. Kollar, P. Bergstrom, and R. A. Batiuk. 1993. Assessing water quality with submersed aquatic vegetation. *Bioscience* 43: 86–94.

Dickson, J. G., ed. *The Wild Turkey* (Harrisburg, Pa.: Stackpole Books).

Freemark, K., and B. Collins. 1992. Landscape ecology of bird breeding in temperate forest fragments. In *Ecology and Conservation of Neotropical Migrant Landbirds,* ed. J. M. Hagan and D. W. Johnston (Washington, D.C.: Smithsonian Institution Press), 443–54.

Frieswyk, T. S., and D. DiGiovanni. 1988. *Forest Statistics for Maryland, 1976 and 1986* (Broomall, Pa.: Northeastern Forest Experiment Station).

Gerrard, J. M., and G. R. Bortolotti. 1988. *The Bald Eagle* (Washington, D.C.: Smithsonian Institution Press).

Heinz, G. H., S. N. Wiemeyer, D. R. Clark, P. Albers, P. Henry, and R. A. Batiuk. 1992. Status and assessment of Chesapeake Bay wildlife contamination. Basinwide Toxics Reduction Strategy Reevaluation Report. CBP/TRS 80/92 (Washington, D.C.: U.S. Environmental Protection Agency).

Johnston, D. W., and J. M. Hagan. 1992. An analysis of long-term breeding censuses from eastern deciduous forests. In *Ecology and Conservation of Neotropical Migrant Landbirds*, ed. J. M. Hagan and D. W. Johnston (Washington, D.C.: Smithsonian Institution Press), 75–84.

Kirkwood, F. C. 1895. A list of the birds of Maryland. Transactions of the *Maryland Academy of Sciences* 2: 241–382.

Krementz, D. G., V. D. Stotts, D. B. Stotts, J. E. Hines, and S. L. Funderburk. 1991. Historical changes in laying date, clutch size, and nest success of American black ducks. *Journal of Wildlife Management* 462–66.

Lynch, J. F., and D. F. Whigham. 1984. Effects of forest fragmentation on breeding bird communities in Maryland. *Biological Conservation* 28: 287–324.

Lynch, J. F., and R. F. Whitcomb. 1978. Effects of the insularization of the eastern deciduous forest on avifaunal diversity and turnover. In *Classification, Inventory, and Analysis of Fish and Wildlife Habitat: A Symposium,* ed. A. Marmelstein (Washington, D.C.: U.S. Fish and Wildlife Service, OBS-78/76), 461–89.

Marye, W. B. 1955. The great Maryland barrens. *Maryland Historical Magazine* 50: 11–23, 120–42, 234–54.

Martin, T. E. 1992. Breeding productivity considerations: What are the appropriate habitat features for management? In *Ecology and Conservation of Neotropical Migrant Landbirds,* ed. J. M. Hagan and D. W. Johnston (Washington, D.C.: Smithsonian Institution Press), 455–73.

Mayfield, H. 1965. The brown-headed cowbird, with old and new hosts. *Living Bird* 4: 13–28.

McAtee, W. L. 1918. A sketch of the natural history of the District of Columbia. *Bulletin of the Biological Society of Washington* 1: 1–142.

Meanley, B. 1982. *Waterfowl of the Chesapeake Bay Country* (Centreville, Md.: Tidewater Publishers).

Morton, E. S. 1988a. Swansong of the purple martin. *Atlantic Naturalist* 38: 38–48.

———. 1988b. The truly long term trend in populations of the purple martin. *Purple Martin Update* 1: 20–22.

Orth, R. J., and K. A. Moore. 1983. Chesapeake Bay: An unprecedented decline in submerged aquatic vegetation. *Science* 222: 51–53.

———. 1988. Submerged aquatic vegetation in Chesapeake Bay: A barometer of Bay health. In *Understanding the Estuary: Advances in Chesapeake Bay Research,* ed. M. P. Lynch and E. C. Krome. Chesapeake Research Consortium Publication No. 129 (Solomons, Md.), 619–29.

Orth, R. J., J. F. Nowak, G. F. Anderson, K. P. Kiley, and J. R. Whiting. 1992.

Distribution of Submerged Aquatic Vegetation in the Chesapeake Bay and Tributaries and Chincoteague Bay–1991. Final report to U.S. Environmental Protection Agency (Annapolis, Md.).

Perry, M. C. 1987. Waterfowl of Chesapeake Bay. *Contaminant Problems and Management of Living Chesapeake Bay Resources,* ed. S. K. Majumdar, L. W. Hall, and H. M. Austin (Philadelphia: Pennsylvania Academy of Sciences), 94–115.

Phillips, J. C. 1922–26. *A Natural History of the Ducks,* 4 vols. (Boston: Houghton Mifflin).

Poole, A. F. 1989. *Ospreys: A Natural and Unnatural History* (Cambridge: Cambridge University Press).

Robbins, C. S., ed. 1996. *Atlas of the Breeding Birds of Maryland and the District of Columbia* (Pittsburgh: University of Pittsburgh Press).

Robbins, C. S., D. K. Dawson, and B. A. Dowell. 1989. Habitat area requirements for breeding forest birds of the Middle Atlantic states. *Wildlife Monograph* No. 103.

Robinson, S. K. 1996. Population dynamics of breeding neotropical migrants in a fragmented landscape. *Atlas of the Breeding Birds of Maryland and the District of Columbia,* ed. C. S. Robbins (Pittsburgh: University of Pittsburgh Press), 408–18.

Semmes, R. 1937. *Captains and Mariners of Early Maryland* (Baltimore: The Johns Hopkins Press).

Stevenson, J. C., and N. M. Confer. 1978. Summary of available information on Chesapeake Bay submerged vegetation. U.S. Fish and Wildlife Service OBS-78/66.

Stewart, R. E., and C. S. Robbins. 1958. *Birds of Maryland and the District of Columbia* North American Fauna No. 62 (Washington, D.C.: U.S. Fish and Wildlife Service Publication).

Terborgh, J. 1989. *Where Have All the Birds Gone?* (Princeton: Princeton University Press).

Walkinshaw, L. H. 1983. *Kirtland's Warbler: The Natural History of an Endangered Species* (Bloomfield Hills, Mich.: Cranbrook Institute of Science).

Warden, D. B. 1816. *A chorographical and statistical description of the District of Columbia* (Paris: printed by Smith).

Whitcomb, R. F., C. S. Robbins, J. F. Lynch, B. L. Whitcomb, M. K. Klimkiewicz, and D. Bystrak. 1981. Effects of forest fragmentation on avifauna of the eastern deciduous forest. In *Forest Island Dynamic in Man-dominated Landscapes,* ed. R. L. Burgess and D. M. Sharpe (New York: Springer-Verlag), 125–205.

Wilcove, D. S. 1985. Nest predation in forest tracts and the decline of migratory songbirds. *Ecology* 66: 1211–14.

———. 1988. Changes in the avifauna of the Great Smoky Mountains: 1947–1983. *Wilson Bulletin* 100: 256–71.

COMMENTARY

Reading the Palimpsest

WILLIAM CRONON

Over and over again, the chapters in this book make the point clear: the Chesapeake is like a palimpsest, one of those ancient medieval manuscripts whose authors were so concerned to reuse the expensive material on which they wrote that they scraped away previous texts, leaving beneath the visible text faint layers of ink which can be read if only one knows how to look at them aright. At times, the simile could hardly be more literal. Watch Grace Brush extract her pollen cores from the silt of the Bay, and you find yourself reading minute historical documents of the trees and other plants that have been sending their clouds of silica-encrusted germ plasm out across the waters for millennia. On top, in the muds that squish between your toes when you walk the estuary, are the pollens that record the landscape we know today: forests of oak and hickory and pine, fields of corn and grass and ragweed. Scrape away just a little of that top layer and you read the text of a different manuscript: a landscape in which oaks shared the forest canopy with chestnuts, and elms offered their lovely shade to suburban lawns and rural streambanks. Both the chestnut and the elm have succumbed in this century to blights from overseas, and their buried pollen in the estuary bears ghostly witness to their absence from today's countryside.

Scrape away more layers of the muddy ink, and the ragweed pollen that is so abundant up above begins to diminish until it nearly disappears. The plant had been present in this watershed long before the arrival of European settlers, but its opportunities for population growth dramatically increased as the colonists' plows and grazing animals proliferated across the landscape. By the nineteenth century, the lands of the Chesapeake had become ragweed heaven. As you go farther down through the mud, scraping away the nineteenth century, scraping away the cattle and pigs and pastures, scraping away the colonists and their clearings, you finally come to a layer in which the ragweed had not yet begun its rise to glory. This is Indian Country. Here the mud shows signs of occasional burning and other modest evidence of human occupation, but the big patterns are driven more by climate than anything else. Two thousand years down, the oaks and chestnuts give way to hemlocks and other species betokening a moister climate. Nearly eight thousand years down, these give way to other needle-bearing conifers, pines and firs, and finally, ten thousand years into the past, the spruce trees that invaded this land as the great ice sheets retreated to the north. Here at last the pollen-bearing muds give out altogether. We stand now at the very edge of the glaciers themselves, and the palimpsest at the bottom of the Bay must be read in different ways if it is to tell of that frozen world.

Yet despite these similarities in the ways they can be read, a landscape is far richer than a medieval manuscript. Its record of the past is written in many different languages with many different kinds of ink, each of which requires a different set of techniques to be properly understood. One has only to look at the disciplinary affiliations of this book's contributors to get a sense of how many kinds of readers are necessary to try to understand a document as complicated as the Chesapeake Bay. Not just pollen scientists, but historians, archaeologists, geographers, geologists, climatologists, ecologists, marine biologists, ornithologists, botanists, epidemiologists, and others have all aided in interpreting how the Bay and its watershed have changed over the course of postglacial history.

Some, such as the pollen scientists and archaeologists, work with stratified evidence very much like the layers of ink that medievalists try to decipher when they read an actual palimpsest. Others use evidence of different sorts. Geologists and climatologists may look at ongoing natural processes to see how phenomena we observe today might have manifested themselves under the altered conditions of the past. Historians and geographers look at written records, contrasting their descriptions of the Chesapeake country with the modern landscape. Evidence of this sort is sometimes less like a palimpsest than like Sherlock Holmes's dog that didn't bark. The absence of certain things in the historical record—failure on the part of travelers to mention starlings and house sparrows, for instance, or the eventual disappearance of references to passenger pigeons, which once numbered in the millions—can say as much about the changing landscape as the details that actually get recorded. No one person could hope to have all the skills necessary for these different kinds of readings, which is an important reason why this book is a collection of essays and not a unified narrative by a single author.

To think of the Chesapeake as a palimpsest is to realize that landscapes exist not just in three dimensions but in four. In addition to the present landscape which we see, touch, smell, and move through, there is also the nearly infinite series of past landscapes that preceded the present one *in time*—landscapes that in a very real sense now depend for their existence on our ability to see them in our mind's eye. To understand why the environment around us has the shape it does, why the plants and animals and people who inhabit it live here as they do, we must connect the present of this place to its past. Once we know where to look, we see the signs and artifacts of that past almost everywhere; indeed, they are the very stuff of which the present landscape is made. Thus, as we begin learning to read the palimpsest, we may feel more and more that to dwell with open eyes on the land and water is to inhabit the pages of an extraordinarily rich and fascinating history book. The manuscript may be incomplete and many of its pages torn or missing, but what remains is more than enough to carry us backward to discover vanished lives and places which we cannot hope to know in any other way.

Still, there is a problem here. As the chapters in this volume suggest, there are nearly as many ways of reconstructing the past landscapes of the Chesapeake as there are disciplines and scholars to study them. Each technique for interpreting the palimpsest, each filter through which to read the landscape, tells a slightly different story. A pollen scientist's narrative may have much to say about hickories and chestnuts and ragweed while remaining utterly silent about passenger pigeons. An agricultural historian's account may detail for us how various generations of farmers went about their work and yet say nothing about how farming on the land might have been connected to fishing in the Bay. Each narrative of a past landscape may be valuable in itself and entirely accurate but still resemble the old story of the blind men and the elephant. As usually happens, we have a better sense of the parts than we do of the whole. To say this is not to criticize these fine essays or their authors; it is rather to admit the obvious, which is that the task of synthesis is almost always harder than the task of analysis. The very complexity which makes the landscape such a rich and tantalizing document also makes it hard to see whole.

So perhaps the question that needs to be asked here at the end of this book is simply this: When we step back and consider the Chesapeake whole, trying to see it as a four-dimensional landscape existing as much in time as in space, what can we say of it? In the wonderfully broadened vision of the past that these authors have assembled for us—in these histories that tell tales not just about human beings but about oysters and oak trees, tobacco fields and tidal flows, cowbirds and cattle fevers—can we still identify a central theme or plot line? Can the Chesapeake past be narrated as One Big Story, or should we more accurately think of it as many little stories, sometimes flowing together and sometimes flowing apart? What's the Big Picture here?

We could probably all agree that the past 500 years have seen human interventions in the Chesapeake ecosystem which become ever more massive and intricate the nearer we approach the present. In this, the Chesapeake is not much different from most other corners of the planet, so that it might be tempting to attribute its changing environment to one or another of the Big Causes that have so often been blamed for ecological damage and destruction in the modern era. Why has the Chesapeake experienced ecological change? The possible answers are pretty familiar. Human overpopulation: too many people. Human greed: the disastrous environmental effects of capitalist markets and economic growth. Human technology: people's growing ability to ravage land and water alike with the high-energy tools of a petrochemical society. Human values: the Judeo-Christian impulse toward dominance supposedly taught by the Book of Genesis. Human stupidity: a phenomenon that needs no elaboration. Any of these can serve as One Big Cause capable of turning the tale of the Chesapeake into One Big Story of a landscape and an ecosystem changing inexorably for the worse as a product of human foolishness and misdeeds.

And yet even to list Big Causes in this way is to call their accuracy into question. That is one problem with synthesis, which tempts us to resort to these unitary explanations of past change as a device for imposing some all-encompassing order on an otherwise messy reality. As soon as we recognize more than one synthesizing explanation, the adequacy of any one of them becomes suspect. Satisfying as unitary causes may be in reducing an excessively complicated reality into a more manageable morality tale, they obscure as much as they reveal in their efforts to make sense of the past. Much of what is most fascinating about the ecological changes narrated in this book is their unexpectedness: the surprising feedback loops and quirky interconnections that somehow tie the behavior of nonhuman systems and species to the activities of people on the land. One could, for instance, assert that the remarkable expansion of the nest-parasitizing cowbird over the course of the twentieth century is a product of human overpopulation, greed, and technological dominance. One might even be correct in so doing. But it is not clear how much one would actually learn from such a claim, or what it would teach us about cowbirds, human beings, or the problem of living responsibly in the Chesapeake landscape. There may also be a certain hubris in ascribing the cowbird's ascent to one or another of these Big Causes, because doing so reflects an inclination to see human beings as the only agents shaping the modern landscape. People surely bear great responsibility for the ecological dynamics of the modern Chesapeake, and we should never flinch from acknowledging those responsibilities. But cowbirds deserve some credit too. Nature is still at work here.

Better than One Big Cause as the explanation of ecological change in the Chesapeake country, then, is a synthesizing approach that recognizes from the start the many overlapping processes that have left their inscriptions on this palimpsest. One of the great contributions of this book is to suggest in considerable detail just how complex and interacting those processes are. Some, like the flow dynamics of the estuary itself, have almost nothing to do with human influence and yet drive the entire system. By squinting at the Bay and seeing its movements in the fullness of time, we become aware of the elaborate nested cycles on which all else depends. The daily ebb and flow of the tides, which reflect the intricate dance of sun and moon and earth, could hardly be more crucial to everything that makes up the Chesapeake, now and in the past: the shifting boundary of salt water and fresh, the flow of nutrients and pollutants through the estuary, the diverse habitats that render the place hospitable to the creatures who live there. None of these are static. All depend on the oscillations of the sea.

Superimposed upon the tides are many other cycles. Night and day alternate in their 24-hour rhythm to demarcate the different worlds of light and dark which punctuate the lives of plants and animals who schedule their daily rounds by the celestial clock. The seasons alternate as the earth makes its journey around the sun, defining the flux of solar energy and thereby governing the great cycles of hot and cold, dry and wet, plenty and want. The annual outpouring of the spring meltwaters from the Susquehanna, the migrations of birds along the coast, the spawning of fish and shellfish in the Bay, the leafing of trees and other plants on the land: all take their cue from the sun's migration up and down across the lines of latitude. And beyond the familiar time scales of days and seasons and years (familiar because we measure our own human lives by their rhythms) are other cycles operating on scales far beyond our reckoning. The wobbling of the planet on its axis, the tug of war between sun and earth as the latter moves closer and farther away in its orbit, the inner dynamics of the sun itself: all have their effects on the lands and waters of the Chesapeake, and in fact are ultimately responsible for the existence of the Bay. The coming and going of the ice sheets, the depression of the land beneath the enormous weight of the glaciers, the flooding and drying of the great valley as the seawater rises and falls: all of these originate not as a result of any human agency but ultimately from the earth below and the heavens above.

These large natural cycles have operated on the Chesapeake from the beginning and will presumably continue to do so as long as there is water in the Bay to respond to their effects. The bulk of this book, however, deals not with the deep past of geological time but rather with the past few hundred years, during which the Chesapeake has experienced dramatic ecological change linked in one way or another to human activity. Here the processes whose stories we need to narrate seem to become less cyclical and more linear: they seem not to return repeatedly to their beginnings but to make a definite passage from one ecological and cultural condition to another. We move from a landscape dominated by Native American peoples to a landscape dominated by people who have migrated to the Chesapeake literally from all over the world. We move from an estuary with oysters innumerable to one where that shellfish has radically diminished. We move from skies darkened by flocks of passenger pigeons to skies so clear that those flocks have become difficult even to imagine. And so on and on.

There are any number of ways we could weave together these various stories to construct a more integrated narrative of ecological change in the Chesapeake as a whole. We could construct a catalog of different human activities—food production, trade, transportation, migration, manufacturing, government—and examine the ways each has related to changes in the larger environment. (If we wanted, we could then connect each in turn to our favorite One Big Cause to construct the One Big Story that we hope will explain all of them together.) To make this catalog of causes and effects less abstract, we could take one event—for instance, the extinction of the passenger pigeon—and trace out the many environmental and cultural relationships that brought it about: the reproductive and flocking behaviors that rendered this species so susceptible to mass hunting, the elimination of pigeon habitat with the spread of farming, the invention of firearms and nets suitable for large-scale hunting, the expansion of the railroad and telegraph systems which made it easy for market hunters to locate nesting areas and transport dead pigeons by the millions back to urban markets. Assembling such a catalog would not only display the many causes that eventually spelled doom for the passenger pigeon, but would suggest pretty compellingly that no one of those causes by itself can explain the extinction. One needs to see all of them operating together to understand just how complex, just how overdetermined, the pigeon's fate really was. Even if in the end all these causes are bundled together as One Big Cause called, say, "the market," to tell One Big Story of "market destruction of species," the fascination of this *particular* story will nonetheless reside in the very *particular* (and multiple) causes that converged to destroy this *particular* species while other species experienced no such disaster. As always, the devil is in the details, and we lose as much as we gain if for the sake of our story we suppress those details.

And remember: the extinction of the passenger pigeon is just one story about one animal. My task at the end of this book is to narrate the environmental history of the entire Chesapeake, a much more daunting assignment. Toward that goal, I can hardly help but resort to the historian's traditional technique of trying to impose order on the intricate and seamless web of the past by dividing it into periods. Sometimes these periods are chained together to make them fit the One Big Story the historian wants to tell; sometimes they simply serve as arbitrary dumping grounds for great gobs of disconnected facts and events that just happen to have occurred within the designated time slot. But when they really do the work we expect of them, a historian's periods serve to highlight chunks of time that really did differ from one another in important and intriguing ways. We may not be able to explain all the differences or say why events flowed in the directions they did, but the act of periodizing at least begins to suggest the questions we need to ask, the phenomena we need to explain, and so point to the stories we need to tell. If we approach the task of delimiting them in a reasonably heuristic and open-minded way, refusing to force events into the Procrustean bed of our prior theories or narratives, then we at least have a chance of discerning big causal patterns while not losing track of the devil's own details. Periods, simply put, are a tool for recognizing difference and parsing the flow of time as a first step toward narrating change. They divide the palimpsest of the landscape into layers and pose the difficult riddle of how to explain the transitions from each succeeding layer to the next.

Fortunately for the task of synthesis, the book's contributors are in rough agreement about the periods they discover in the Chesapeake past. Although a number of them pay close attention to the large natural cycles and processes to which I have already alluded—and it is clear that the Chesapeake has in the deep past undergone dramatic environmental change that has nothing to do with human influence—none relies on natural processes to periodize the events of the last half millennium. Instead, it is human actions and human social organization that structure the flow of narrative time. As is typically true of colonial and frontier histories, the great divide occurs sometime in the sixteenth or early seventeenth centuries with the arrival of European sailors and settlers. Before then is prehistory: Indian Country. Most of these authors seem to agree that with the exception of local burning, local harvesting of fish and game species, local clearing of land and planting of crops, the effects of native peoples on the Chesapeake landscape were limited in scale and relatively stable. Different Indian groups had developed elaborate techniques for exploiting the resources of the Bay and the surrounding lands. They had made of the landscape a complex cultural space into which their material and spiritual lives were more or less seamlessly integrated. They had made it their home.

But for the purposes of this book's narratives—and this too, for good or for ill, is characteristic of the way colonial and frontier stories usually get told—Indian Country occupies the time before the Big Change began. It is where the story starts, the moment at which "nature's" time gives way to "human" time, which is to say, "white people's" time. There are, of course, many reasons to be suspicious of such a formulation, because it so easily privileges the historical experience of one group over another. There is no reason to doubt that the native peoples of the Chesapeake were just as dynamic and changing in their relationships with one another and with the environment as were the European and African peoples who invaded their land. The Powhatan confederacy that greeted the colonists at Jamestown is only the most visible evidence of this dynamism. Indian effects on the landscape may now seem less significant, but that may be because we define "significance" in ways that the native peoples whose stories we are supposedly telling might not be inclined to recognize.

Part of the problem, inevitably, is the absence of written documents with which to trace the stories of these earlier times, which have become faded and flattened as they have been pressed into the archaeological strata from which they must now be extracted if they are to be read at all. In the buried records of the earth, even the largest changes (the extinctions of the great Pleistocene mammals, for instance, which are sometimes attributed to native hunting, and sometimes not) are rendered ambiguous. Cultural subtleties and the lived experience of people and other creatures can be read only crudely from such evidence. By the time we have the records we need to tell richer stories, Indian Country is contested terrain and the available documents, even when we work very hard to overcome their biases, tell mainly the stories of the invaders who were doing the contesting. On the evidence offered in this book, the Big Story of anthropogenic environmental change begins with the arrival of the Europeans and their African slaves—but it is important always to remember that this has as much to do with the way we have chosen to recognize change and tell this story as it does with the subtler and more complicated realities of the actual past.

Be that as it may, even the pollen at the bottom of the bay begins to record a change in Chesapeake ecology with the coming of the colonists. Ragweed pollen starts to creep up, very slowly at first, then much more quickly as pastures, fence rows, and abandoned fields—good weedy habitats all—spread across the landscape. The initial slowness of ragweed's rise is important, for it defines one period about which the volume's contributors agree: from the first arrival of colonists to the middle of the eighteenth century—say, from 1607 to 1750, if you like your periods sharp—early colonial impacts on the Chesapeake landscape were fairly minimal. The population remained small, and settlements were concentrated in narrow strips of land located not much more than a few hundred yards from the shore. Aside from limited grazing, interior and upland areas experienced little new pressure from the colonists.

Farming during this period was still heavily devoted to subsistence, with tobacco the main commercial crop. Contrary to the long-standing belief that tobacco exhausted the soil, on plantations of the colonial Chesapeake it was raised on a long-fallow rotation, which did a decent job of preserving soil fertility and preventing large-scale erosion. Critics from the far side of the Atlantic might complain about the unkempt appearance of the American landscape and the wastefully regressive ways Chesapeake farmers worked their fields, but this said less about actual damage to the soil than about the different land-labor ratios, and the different aesthetics, of Europe and America. Europeans had the cheap labor they needed to tidy up their fields and farm intensively; Americans did not. This did not necessarily mean that the land itself was worse off as a result.

To be sure, the ecology of the Chesapeake was beginning to undergo subtle changes. Species were introduced that would gradually work their way across the American landscape. Although some of the earliest important crops (most notably tobacco and maize) were of New World origin, a variety of Old World grains, garden vegetables, pasture grasses, herbs, and orchard trees were putting in their first appearances on colonial plantations. With them came the weeds: opportunistic species like dandelion, mullein, and a few natives such as ragweed whose reproductive strategies permitted them to move aggressively onto the disturbed soils that colonists were so fond of creating. A good part of this disturbance came from another set of introduced species: grazing animals like cattle, pigs, horses, and sheep. Because their need for food was great, these animals brought new pressure to lands well beyond the limits of cultivated fields. With them came fences, roads, and eventually plows and other machine technologies. All would gradually change the landscape, inscribing it with new cultural practices and meanings.

How one tells the story of this early period of settlement depends a lot on one's narrative frame of reference. If one is generally pointing the story toward the dramatic changes that would transform the Chesapeake after the middle of the eighteenth century, this can look like a pretty peaceful time. Ecological impacts were limited in geographical scale, and colonial relationships with the Chesapeake environment were reasonably benign. This is the perspective that Carville Earle, Ronald Hoffman, and Lorena Walsh offer. Earle and Hoffman go so far as to suggest that Chesapeake farmers during the first century and a half of settlement developed a traditional wisdom for stable, conservative use of the land which was handed down from generation to generation until a new class of scientific farmers, obsessed with the Enlightenment faith in progress, rejected that wisdom in the late eighteenth century and set off on a disastrous new course.

There is a good deal of truth to this way of narrating the period, for farming practices did change after 1750 and impacts on the land markedly increased. Earle and Hoffman make an important contribution by warning us not to accept at face value accounts that criticize early colonial farmers as being more ecologically destructive than the "progressive" farmers who succeeded them. But if, for instance, we use earlier Indian communities and land-use practices as our chief frame of reference, then the changes brought by the colonists perhaps look more significant. If we look forward from 1600 instead of backward from 1800, we will probably be struck by the colonists' more sedentary living patterns, their more bounded conceptions of property, the wanderings of their grazing animals, their forced labor systems, and their many introductions of alien species, all more or less radical breaks from the earlier cultural landscapes of the Chesapeake.

One additional set of introduced species is especially striking: the Old World diseases that brought disastrous epidemics to Indians throughout the Americas. Although Douglas Ubelaker and Philip Curtin are surely right to caution us against painting too utopian a portrait of the pre-colonial disease environment, there can be no question that smallpox, measles, and other diseases were devastating to Indian communities in the Chesapeake, as elsewhere. However numerous Chesapeake Indians may have been before 1600, the epidemics surely brought massive depopulation and complex cultural changes whose full implications we can only guess. The microscopic invasion had to have had macroscopic effects on the landscape.

There is another way the narrative frame will change the way we interpret this period. If we assume that early tobacco cultivation and its ecological effects were stable, that the agriculture and settlement system of the first century and a half could have persisted indefinitely in a self-sustaining way, we will be inclined to assign responsibility for subsequent ecological change to elements that appeared in the system only after 1750. This is essentially what Earle and Hoffman offer as their narrative: the stable and ecologically benign regime that Charles Carroll of Annapolis so carefully practiced on his lands collapsed when his son, Charles Carroll of Carrollton, introduced crops and techniques whose effects on the land were ecologically destructive and anything but self-sustaining. The important assumption in this narrative is not that the landscape of Carroll *père* differed from that of Carroll *fils;* it certainly did. Rather, the assumption is that the two are radically disconnected from each other, so that the one did not contain the seeds of the other. Nothing in the father's world pointed toward the son's.

If, on the other hand, one sees the initial colonial system as dynamic and unstable, already containing elements that would eventually move it in the direction of subsequent change, then one is likely to construct the narrative of this period with a different telos. Whatever the engines of change to which one appeals—rising populations of people and animals, shifting ratios of land and labor, the evolving demands of a market economy, the changing cultural values of the colonists, what have you—the break in the mid-1770s will probably seem at least a little less radical. My own bias in comparing early colonial settlements with their Indian predecessors is to emphasize the extent to which the Chesapeake plantations from the very beginning were embedded in a transatlantic economy unlike anything that had preceded it. Trade and markets had certainly existed in Indian Country, but their scale was quite different and the range of commodities which they drew into their orbit was not at all the same in its ecological and economic implications. To tie the Chesapeake to external markets was to change its ecological dynamics, even if the effects of those markets were initially small. Add to this the emerging imperial systems within which the Chesapeake colonies were embedded, to say nothing of the evolving power of the states which stood behind those systems, and one discovers all sorts of dynamic instabilities built into the colonial enterprise from the beginning. The implication of *this* narrative frame is not to deny that change accelerated sometime after the middle of the eighteenth century; it unquestionably did. Rather, the assumption is that the seeds of change were present much earlier. The worm was already in the bud.

Whatever the larger narrative frame, we are left with a landscape which until 1750 or so was changing only slowly and locally. Not until then did the curve of ragweed pollen begin to slope more steeply upward. As the pollen count rises, we can define a second period in this narrative of environmental change, running roughly from 1750 to 1820. This is a period of transition, in which the changes apparent earlier continue to proliferate in a gradual but accelerating way. Economic production and exchange for purposes other than subsistence still focused primarily on the export trade, but internal markets were becoming more important, especially in the vicinity of the small but growing city of Baltimore. As population rose, settlements moved farther and farther inland from the coast. Forests were cut at an increasing rate, and the uplands experienced their first major wave of clearing. Chesapeake farms began a general shift in emphasis away from tobacco toward grain, with wheat being the dominant crop despite a variety of new pests, especially the black stem rust and the Hessian fly, which also appeared during this time. Grain production was accompanied by more systematic and extensive disturbance of the soil, primarily by plowing but also by draining wetlands and planting steeper slopes. Greater rates of erosion followed, reflected in the increased sediment loads carried by the rivers. In the Bay itself, oysters began to be harvested in a more intensive way.

Almost none of these phenomena were entirely new; what makes them feel different is their increasing scale. It is typical of environmental history that we can rarely divide the flow of time with anything like the precision or certainty we feel when we say that 1776 was a watershed in the changing politics of this same period. But it does seem clear that by 1790 the Chesapeake was well on its way toward the revolutionary transformations that would characterize the next century. The economy was increasingly commercial, the United States had made its republican break from England, progressive forms of agriculture were being imported from Europe, and new techniques for making many forms of production more efficient were appearing with greater regularity. By the time we reach our next period, which we can locate more or less arbitrarily between 1820 and 1900, we have fully entered a new era in which commercial markets, technological innovation, and urban-industrial growth impose a new order on the Chesapeake landscape.

This is the great era of market expansion in the Chesapeake and the United States more generally. Whatever changes had characterized the years before 1820, they pale when compared with the rest of the nineteenth century—undoubtedly one reason some may be inclined to view the years before 1750 as a more "stable" time. Unlike earlier periods, internal markets claimed an enlarging share of economic production from exports, so that national and regional economies became more important as the engines of growth and environmental change. This process was aided by the symbiotic development of cities and the transportation networks which concentrated economic demand in urban markets. On the water, travel along the coast and across the Atlantic was speeded and made more regular by innovations in sail and steam technology and in the evolving institutional infrastructure of the shipping industry. On the land, improved roads were followed by canals, and they in turn by railroads—the effects of which on the landscape are almost impossible to exaggerate. All of these extended the hinterlands of the cities they fed. In the case of the Chesapeake their effect was to concentrate population growth and economic development on the city of Baltimore while changing the rural countryside by uniting urban and rural markets as never before. (Later, Washington, D.C., would emerge as a major center in its own right, its growth fueled not by railroads or hinterland markets but by the expansion of the national state and its bureaucracy.)

New markets meant new opportunities for profiting from the unexploited resources of land and sea. In the Chesapeake Bay, fishing and oystering reached unprecedented levels by midcentury, so much so that for the first time human harvesting began to threaten the resource stocks on which it depended. Although the Chesapeake would long remain the greatest single center for the American oyster industry, its production peaked during the 1880s and then began a slow but steady decline. Other fisheries followed their own unique patterns, but all would eventually face the problem of bringing harvests into line with the reproductive rates of the Bay. The same was true in the estuaries, where intensive new forms of market hunting threatened waterfowl populations at about the same time. The only species to go extinct during the nineteenth century were those whose behavioral and reproductive patterns rendered them uniquely vulnerable: the passenger pigeon's extraordinarily concentrated nesting strategy, the Carolina parakeet's habitat requirements (to say nothing of its highly marketable plumage). But the populations of many other migratory birds also plummeted in the face of the new market pressures to which they were subjected.

On the land, agricultural clearing reached its maximum extent as Baltimore emerged as a leading flour-milling center. Canals and railroads meant that Chesapeake farmers found themselves competing with grain raised on the much more fertile soils of the western prairies, making it harder to produce wheat cheaply enough to earn a profit. Fortunately, Baltimore and other cities at the same time encouraged shifts toward more diversified and profitable products (vegetables, orchards, and dairy herds) to satisfy the urban demand for food. To keep up with this demand, Chesapeake farmers used growing quantities of fertilizers (rock phosphates from the Carolinas, guano from the south Pacific, even buffalo bones from the Great Plains) in an effort to restore the declining fertility of their soils. And to fight off a growing array of pests, they turned to toxic compounds that relied on heavy metals, especially arsenic but also copper and lead, to kill the six-legged enemies of their crops. All these substances would soon show up in the nutrient cycles of the Bay.

Finally, this was the era in which industrial manufacturing exploded across the American landscape. Early factories in the United States relied on waterpower for their motive force, and the sites for such development were relatively limited in the immediate vicinity of the Chesapeake—though upstream sites like Harper's Ferry undoubtedly made their contributions to the waters of the Bay. With the coming of the steam engine and the development of the anthracite coal industry, however, factories could move into cities to be near their labor supplies, and Baltimore certainly participated in that process. Coal consumption meant injection of new particulates into the Chesapeake atmosphere and ash dumps, which produced locally toxic sites. Add to this the new metals used in construction and manufacturing—iron, lead, copper, zinc, mercury, cadmium, arsenic, and others—and one begins to understand the exotic new pollutants that start to show up in the Bay's sediments and food chains during this period.

None of these phenomena disappear as we move into our final period, which, again arbitrarily, we can trace from 1900 to the present. Urbanization and the drift away from the farm have continued, as has the increasing interconnectedness of the Chesapeake with the world economy. More and more of the regional landscape is occupied by people whose livelihoods do not derive in any direct way from working the land or water. Even those who live in the country are now fully integrated into an essentially urban economy. Linkages between city and country have proliferated in part because of the general shift of transportation away from railroads toward automobiles, diesel trucks, and highways. Whereas railroads encouraged nineteenth-century cities to expand along dense linear corridors, automobiles have brought promiscuous growth in all directions.

No less important to this outward sprawl of city and suburb has been the extension of various amenities powered by electricity (radios, televisions, telephones, and all manner of labor-saving appliances) which have significantly diminished both the isolation of rural areas and the hardships of rural life, thereby making it possible for urban people to reside farther and farther from the cities where they work. One consequence of the out-migration is that a growing proportion of those who live in rural areas regard the countryside as a residence and playground, not a workplace. Leisure-time amenities have thus had a growing role in defining the meaning and value of the rural landscape. Accompanying all these changes has been a dramatic rise in the consumption of fossil fuels, with coal burning increasingly concentrated in the enormous boilers of electrical generating stations and petroleum fueling the bulk of the land transportation system. The diffuse settlement patterns of the twentieth-century Chesapeake, like those of the United States more generally, are dependent as never before on intensive energy consumption. Much of what is most distinctive about the modern landscape is implicitly predicated on this fact, in many more ways than people typically realize.

One defining feature of this most recent period of Chesapeake environmental history, then, is the emergence of an ever more interconnected regional landscape, in which the boundaries between city and country have become ever harder to locate. No less important has been the emergence over the past one hundred years of concerns about human impact on the Bay's ecosystem, concerns that were almost entirely absent before. At the start of the twentieth century, they provided the political energy for what came to be called the conservation movement, in which activist governments sought to regulate private uses of public resources to make them more efficient and sustainable. In the Chesapeake, conservation principally took the form of regulating fisheries, protecting migratory waterfowl, controlling soil erosion, safeguarding municipal water supplies, and rationalizing sewage systems. Such activities depended for their success on growing cadres of scientists, engineers, and government bureaucrats, all of whom brought new managerial perspectives to the regional environment.

Viewing the Chesapeake as a management problem meant taking responsibility for things which had been treated as free gifts of nature requiring little or no conscious attention on the part of those who exploited them. Conservation, on the other hand, meant responsible management, which in turn encouraged a new tendency to view the Bay and its watershed as an integrated whole. The managerial impulse usually implied an assumption that if only nature were properly understood it could be appropriately controlled, a perspective that could have unexpected environmental consequences. Among the most famous of these was the widespread use of DDT after World War II to control mosquitoes and other insect pests. Although in many ways an attractive alternative to its predecessors (compared with arsenic, DDT looked remarkably benign), its chemical longevity and its concentration in fatty tissues as it moved up terrestrial and aquatic food chains had serious implications for wildlife and marine organisms. What had earlier seemed appropriate control of nature—good management, wise use—soon seemed ecologically destructive and dangerously shortsighted.

Controversy about DDT in the wake of Rachel Carson's *Silent Spring* of course helped spawn the post-1960 environmental movement. But for the purposes of narrating the long-term story of Chesapeake environmental change, perhaps it is equally important to note the paradox it revealed at the heart of twentieth-century attitudes toward nature. Management implied that nature could and ought to be controlled to prevent harm to people and to the parts of the environment which people most valued. At its worst, this assumption could lead to an arrogant belief that ecosystems were infinitely malleable in the service of human ends. At its best, the impulse toward management encouraged an ever more careful effort to understand the complex ways in which human actions affected ecosystems for good and for ill. It could produce an environmental problem like DDT, but it could also produce the response to that problem. In either case, it defined a fundamental tendency of the post-1900 period which set the twentieth century apart from its predecessors: an inclination to view the Chesapeake environment as a system and to respond to that environment in systematic ways. This book is itself an expression of that inclination.

As the network of human institutions and activities in the Chesapeake became ever more tightly integrated, both within the region and with other parts of the nation and the world, managers and the public at large gained a growing sense that changes in one part of the region tended to have repercussions for the entire system. The construction of a highway in one location meant complex changes for markets, human settlements, and natural habitats at other locations. Pollution in one medium at one place could turn up in a different medium at another place, sometimes surprisingly far away. A laundry detergent used in Baltimore, or an agricultural fertilizer used in the Shenandoah Valley, could reappear with unexpected effects on aquatic vegetation and bird life far down the estuary.

Such effects were systemic and could only be understood in systemic terms. Just as important, efforts to manage or control them became equally systemic. And so the twentieth century has been marked by far more coordinated and integrated responses to environmental change than were typical of earlier periods. The obvious example is the much expanded body of environmental laws and regulations with which federal, state, and local governments have reacted to perceived problems in the Chesapeake environment. When successfully enforced, such laws changed the behavior not of one individual or corporation, but of the entire system. In the 1980s and 1990s, government intervention on behalf of the environment came under increasing attack, so it is perhaps worth noting that state regulation is far from being the only example of systemic environmental manipulation in the modern world. The bureaucratic structure of the government regulators is matched by corporate managerial hierarchies that are no less systemic in their perspectives and effects on the Chesapeake and other ecosystems.

Even the seemingly uncontrolled behavior of "free" markets inevitably has far-reaching environmental consequences in the highly interconnected world of the twentieth century. Individual behavior quickly produces collective effects in the modern world. The appearance of a new product and its widespread adoption by millions of consumers, each freely exercising his or her independent choice in the marketplace, can quickly produce environmental change that is every bit as systemic, with consequences every bit as unexpected, as the most heavy-handed of government regulations. Like other twentieth-century landscapes, the Chesapeake is now a tightly coupled world in which change at one location all too easily produces change everywhere.

And so we return to the surface of the palimpsest, having worked our way up layer by layer through the past landscapes that lie beneath the surface of the present Chesapeake. Along the way, we have followed any number of narrative paths, each of which offers a different story of how the worlds that were before became the world that is today. What are the big patterns here? It depends on the tale you want to tell. An Indian world has become a multicultural world of immigrants. A native ecosystem has been invaded by growing numbers of alien species and exposed to growing numbers of exotic substances, all with unknown effects. At the same time, some things have vanished forever: skies once darkened by passenger pigeons will never see those birds again. Economies oriented mainly toward subsistence have been replaced by economies oriented mainly toward the market. Farming has given way first to industry and then to a growing service sector. A rural countryside has become a diffuse urban-suburban-exurban region whose residents make their livings by drawing paychecks instead of by fishing or raising crops. Paths have gone from roads to railroads to highways, and cars and trucks now carry the loads once borne by people and horses. A technological system dependent mainly on the flux of solar energy has given way to one dependent on fossil fuel. Loosely coupled human communities have been absorbed into far more tightly coupled political, economic, and environmental systems governed by the decisions of managers and professional experts. And as human effects on the Chesapeake ecosystem have grown, so have human concerns about protecting the ecosystem from the worst of those effects.

If your inclination is to turn these many patterns into One Big Story with One Big Cause pointing toward One Big Moral, you are more than welcome to do so. There are plenty of candidates ready at hand. For myself, though, I prefer a more ambiguous ending. In contemplating the dilemmas we now face in managing the Chesapeake—dilemmas that the essays in this book depict in striking detail—it seems to me that one option no longer open to us is to abandon management altogether. Returning to pristine nature is not an option, for the writing on the palimpsest cannot be scraped away. At least in the linear world we now inhabit, time's arrow points in only one direction and there is no obvious way we could return to the conditions of earlier centuries even if we wanted to do so. More important, it is crucial to remember that the history of the past five hundred years has not been unambiguously one long story of decline and environmental destruction. There have been gains as well as losses, and we would do well to remember both.

Among the most valuable of the gains have been the ironic lessons of management itself. The modern impulse toward management encouraged the belief that environmental problems could be solved, that nature could be controlled, if only people used their understanding of nature's laws to reshape the world around them. Despite many important successes, the failures and unexpected consequences of management have often revealed just how difficult the control of nature can really be. The potential for hubris is everywhere, for nature's laws apparently include the one apocryphally attributed to Murphy. Whether one considers the effects of smallpox on Chesapeake Indians, the self-defeating consequences of unrestricted harvests for oysters and birds, the unanticipated longevity of the DDT molecule, or the remarkable skill with which cowbirds and ragweeds and viruses have seized the opportunities people have given them, one is repeatedly struck by nature's unfailing ability to find new applications for Murphy's Law. Things always seem to go wrong.

And yet it would be spectacularly unwise to conclude from this that just because anything can go wrong, nothing can ever go right. The lesson of Murphy's Law is not that we should never act, but that we should act with care, thinking as hard as we can about the likely consequences of what we do, knowing that even our best efforts will in all likelihood have results that will surprise and trouble us. The same managerial attitude that sometimes leads to arrogant action can also lead to wise restraint. Another, friendlier, word for it is "stewardship." By taking responsibility for the extraordinarily complicated relationships that make up an ecosystem like the Chesapeake, we open ourselves to the knowledge and wisdom that can lead to an ever deepening appreciation for the complexity of the world in which we live. The best efforts to manage nature are the ones that have been most attentive to the intricacy of nature's systems and most respectful of the autonomy of nature's creatures. It is, after all, the managerial perspective of modernity—the impulse to see the world whole as an integrated system of mutually interacting creatures and relationships, all capable of being understood at least to some degree—that makes a book like this one possible. We cannot turn away from these insights. Instead, we must try as best we can to decipher the palimpsest: reading the landscape in time, puzzling out its many riddles, and using its stories as our only available guides for charting our way into the uncertain future.

Index

Addington site, 114
Addison, Joseph, 261
Africa, diseases from, 127, 128, 129–30
Africans, 139–40, 142–44, 243, 270–74, 312. *See also* slave labor
agrarian reform: and agricultural depression, 279–81, 287, 288, 297n1; and crops, 296, 299n11, 310; and drainage, 292, 293, 295–96, 297, 300n13; and environment, 279–81, 291–97; and landscape, 293–94, 296–97, 300n11, 367; and market economy, 279–82; and productivity, 286–87
agriculture, 220–44; biological thought in, 305–6; and birds, 326–28, 331–32, 341, 346, 360; and crop yields, 304–5; and deforestation, xix, 13, 40, 41; and environment, xix, 60, 123, 195, 237, 241–44, 363–64, 365, 372; fallow field system in, 222, 235, 238–40, 242, 280–81, 294, 297, 311, 312, 363; and field abandonment, xix, 41, 73, 74, 123, 239, 312; and forests, 57, 177; hoe cultivation, xix, 123, 235, 238, 240, 282–83, 290, 291, 294, 307; and market economy, 220, 241–44, 279–83, 366, 368; of Native Americans, 123, 250, 254, 266, 281; prehistoric, 100, 109, 115–16, 119; runoff from, 65, 78, 81, 193; and SAV, 337; and settlement patterns, 220–25, 244; slash-and-burn, xviii, xix, 119, 122–23, 155, 173–74, 235, 239, 326–28; and waterfowl, 334; and wildlife, xxi, 101. *See also* agrarian reform; crops; fertilizers; gardens; plowing practices; tobacco
Agriculture, U.S. Department of, 315, 318–19
Aikial Swamp (Md.), 111
algae, 45, 57, 58, 194, 212, 337
Algonquian tribe, 136, 137, 139
Allegheny Plateau, 7
Allied Signal, 10
American Agriculturist, 316
Amerindians, 98, 100
Anacostia River, 51–53
Anburey, Thomas, 173, 174
Andrews, Jay, 214
animals: changes in, xxi, 81, 99, 100, 103–4, 371; deer, 109, 111, 113, 114, 118–21, 161–62, 184, 237; disease in, 19, 101, 304–6, 306, 314–20, 316, 317, 318–19; and European settlement, 101–2, 160, 161; extinctions of, 86–87, 98, 102, 103; extirpation of, xxi, 98–99, 102, 103; and forest patterns, 153–54; fossil records of, 87, 88–89, 94–97; and human activity, 101; megafauna, 81, 87, 98, 99, 110–11; and Native Americans, xix; non-native, xxi, xxii, 101, 103–4, 372; prehistoric, 86–100, 113–14; range changes of, 84, 102; wild, 104, 153–54, 160, 163, 184, 185, 237. *See also* birds; domestic livestock; fish
Anne Arundel County (Md.), 225–28, 231, 232, 233
Appalachian cycle, 5–7, 10
Appalachian Mountains: coal in, 5, 11–12, 15–16, 33; forest patterns in, 151, 152; sediment deposition in, 4–7; and timber market, 13
Appalachian Plateau, 5–7, 42, 50, 51
Archaic period, 109, 110, 111–13, 121
Arundel Formation, 11
Atlantic Ocean, 15, 16, 33–34, 70

Atlas of the Breeding Birds of Maryland and the District of Columbia (Robbins), 325
Audubon, John James, 334, 336

Bacon's Castle (Surry County, Va.), 263, 264
Baird, Spencer, 205, 214
Baltimore (Md.): Fall Line site of, xxii, 7, 8; industry in, 10–11, 210, 307, 312–13, 368; and market economy, 366, 367; oyster shell foundation of, 75; as timber market, 157
Barrie, R., 212
Bartram, John, 172, 173, 176
Batiuk, Richard, 193, 215
Batts, Thomas, 174
Bayard, Ferdinand, 173, 175
Bay Commission, xxiii
Baylis, John, 180
Baylis, William, 180
Beverly, Robert, 196, 204, 212
Bigger, William, 270
Biggs, Robert, 202–3
Binford, Lewis, 120
birds, 322–50; conservation programs for, 336, 345–49; distribution of, 322–25, 328–29; extinctions of, xxi, 86, 98, 101, 323, 324, 333, 336, 347, 356, 360, 368; extirpation of, xxi, 101, 103, 323, 328, 329, 332–33, 336; and forests, 322, 326–34, 336, 340–41, 344, 347–48; future of, 344, 347–48, 348; and habitat changes, 98, 328–31, 368; history of, 81–104, 323–25, 326–31; and human activity, 322, 328, 331–33, 338–39, 344–45, 349, 371; and hunting, 335–36; and land use changes, 232, 340–41, 345–46, 348; migratory, 87, 325–26, 330, 333–34, 337–38, 344, 369–70; nonnative, 322, 337–40; passenger pigeon, 98, 153, 154, 323, 324, 329, 333, 335–36, 347, 356, 360–61, 368; in Pleistocene period, 86, 87, 88–89, 90–93; and pollution, 322, 341–42, 344–45, 347, 348; range changes of, 84, 100, 103, 346–47; and SAV, 322, 329–30; scientific names of, 349–50; waterfowl, 118, 329–31, 334–38, 341–43, 348, 368
Blanton, Wyndham, 141
Blue Ridge Province, 7
Bordley, John Beale, 287
Bordley, Stephen, 269–70
botanical history. *See* gardens
bound labor. *See* slave labor; tenancy system
Bowen, J. T., 214
Boynton, Walter, 193
Breisch, Linda, 209, 211
Brooks, William, 210–11
Brown, G. M., 316
Brush, Grace S., 40–58, 115, 122, 192, 194, 211, 238, 242, 355
Bureau of Animal Industry (USDA), 318
Burnaby, Andrew, 172, 173
Bush River, 65, 73
Byrd, William, 204, 212, 266, 268

Calvert, Cecilius, 256
canals, 12, 61, 367, 368
Carr, Lois, 305
Carroll, Charles (of Annapolis), 270, 279, 281–90, 291, 297, 365
Carroll, Charles (of Carrollton), 279, 281–83, 290–97, 365
Carson, Rachel, 370
Carter, Landon, 158–59, 273, 307
Catoctin Furnace, 11
Catesby, Mark, 261
Cenozoic period, 34
Charles County (Md.), 134, 135, 141, 225–28, 230, 231, 233
Charlottesville (Va.), 174–75
Chase, Joan W., 133
Chesapeake, translation of, 123
Chesapeake and Delaware Canal, 61
Chesapeake Bay: formation of, xviii, 1, 9, 44; interdisciplinary views of, xvi, xviii, xxii–xxiii, 356, 357–58; physical characteristics of, xviii, xx, 60–62, 80–81
Chesapeake Bay Critical Area legislation, 348

Chesapeake Bay Institute (CBI), 75–76
Chesapeake Bay Program (multijurisdictional; USEPA), xxiii, 194, 215
chiefdoms, 117, 118, 120–21
Choptank River, 62, 63
circulation patterns: effects of, 67–70, 359; estuarine, 68–72, 78–79; and fish species distribution, 203; physics of, 60–61, 64–65, 74; and pollution, 66, 78–80, 195; and shoreland erosion, 73, 80; and winds, 60–61, 67, 70–72
Civil War, 312, 313, 314–17
Clarksville site (Va.), 132–33
Clayton, John (of Williamsburg), 261
Clayton, Rev. John, 258
climate, 16–23; and agrarian reform, 297; and animal range changes, 86–87, 100; and Atlantic Ocean, 16, 33–34; and birds, 326, 344, 346; and crop diseases, 308, 309; and estuary, 40–41; and extinctions, 87; and extirpation, 99; and fish, 203; and forests, 35, 40, 50, 51, 53, 55, 58, 151–53, 154, 156; future of, 35–36; and gardens, 254–56; and glaciation, 15, 28–34; historical, 19–23, 25–28; and human activity, xxi, 110, 124, 306; and landscape changes, 184, 356; Little Ice Age, 23, 25, 35; and plant domestication, 250; and precipitation, 16, 18–19, 23, 28, 30, 33, 152, 184; prehistoric, 15, 33, 112, 115; and sea level changes, 61; statistics on, 16, 18–19. *See also* storms
coal deposits, 3, 5, 11–12, 15–16, 33
coal industry, 11–12, 368, 369
Coan River, 115
Coastal Plain: agriculture in, 155, 161; bird populations in, 328–29, 333, 336, 345; flooding in, 156; forests in, 41, 151–52, 153, 159; landscape of, 53; Native Americans on, 120, 133–34; open land patterns in, 183; sediment deposition in, 5, 7, 8, 10, 42, 43
Coast Guard, U.S., 211
Cole, Kenneth, 99
Collinson, Peter, 173, 266, 268
Columbus, Christopher, 251
Connor, Edward F., xxii, 167–86
Conococheague River, 168
conservation programs, 369–73; and animal populations, 86, 102, 104; and aquatic pollution, xxi, 214–16; and bird populations, 336, 345–49; and European settlers, 162; and fish, 204, 206, 214–16; for migratory birds, 325–26, 369–70; and oysters, 75, 201, 210, 211, 215; and plant species, xxii; for soil erosion, 305, 370; and water quality, 12
Cooper, Sherri, 192, 194, 211
Cory, Robert, 214
Cowles, Raymond, 213
Cresswell, Nicholas, 173, 174
Crisfield (Md.), 75
Cronin, Eugene, 205
Cronon, William, 355–73
crop rotation, 280–83, 287, 294, 296, 297n1, 311, 312
crops: and colonial agriculture, 220; disease in, 19, 304–20, 366; diversification of, 241–44, 280–83, 299n11; and environmental history, 363–64; and European settlers, 149, 150–51; exploitation of, 249; and forests, 154–55; grain, 241–44, 282, 291, 292, 294, 296, 304, 306–9, 318, 319, 366, 368; maize, 222, 234, 235, 237, 240, 241. *See also* tobacco
Cunningham, James, 261
Curtin, Philip D., xv–xxiii, 124, 127–44, 306, 365
Custer, Jay, 192, 203
Custis, John, 266, 268, 270

dams, xxi, 74, 176, 192, 205, 207, 208
Danckaerts, Jasper, 158
Davis, William, 286
De Bary, Anton, 318, 319
deforestation: and agriculture, 72, 238; and birds, 322, 332, 334, 336, 344; and climate, 27, 35; and ecosystem, 58; and European settlers, xix, xxii, 40–41, 51, 55, 57; and slash-and-burn practices, xviii, xix, 122; and wildlife, 98, 101

Delaware Bay, 112
Delaware tribe, 131, 136, 137, 140
Delmarva Peninsula, 8
Dent, Richard, 113
disease: from Africa, 127, 128, 129–30; and Africans, 142–44; in animals, 19, 101, 304–6, 314–20; cattle fever, 306, 314, 316, 317, 318–19; in crops, 19, 304–20, 366; and environment, 19, 128–29, 137–44, 365; and European settlement, 127–31, 132, 137, 138–39, 140–42; and germ theory, 317–20; and Native Americans, 127–34, 137–40, 143; in oysters, 209, 211; in trees, 41, 46, 55, 57, 306, 355
Dismal Swamp (Va.), 111
Dodge, J. R., 315, 318
domestic livestock: and agrarian reform, 296; and bird populations, 332; and colonial agricultural systems, 235, 237–38, 240, 242; disease in, 306, 314, 316, 317, 318–19; diseases of, 314–17; and environmental history, 363, 364; and forests, 160; and landscape changes, 355; and nutrient levels, 193
Doughoregan Manor, 270, 289–93, 297, 299nn10–11, 300n12
Doutt, J. K., 102
drainage: and agrarian reform, 292, 293, 295–96, 297, 300n13; and European settlers, 159–60, 163; and forest patterns, 152–53; and landscape, 5, 366; and wetlands, 41
dredging, 61, 195
Dresler, Paul, 214
drought, 334

Earle, Carville, 141, 279–97, 364, 365
Earll, Edward, 198
Eastern Shore, 10, 113
ecosystem: and human activity, xviii–xix, xxi–xxii, 13, 128, 168, 358–59, 369–73; and management practices, 370–73; and market economy, 282, 358, 363, 365–69, 371, 372; and pollution, 192, 195, 368, 371; and technology, 358, 367; and transportation systems, 40, 168, 364, 367, 368, 369, 371, 372
Eddis, William, 158
Endangered Species Act, 204
environmental history, 357, 362–65. *See also* pollen stratigraphy
Environmental Protection Agency, xxiii
environmental research, xxiii, 185–86
erosion: and agrarian reform, 279, 280, 281, 289, 293–96, 300n12; and agriculture, xix, 123, 220, 237, 238, 242, 305, 363; and bird populations, 332; conservation programs for, 305, 370; and crop diseases, 310; and ecosystem, 192; of islands, 72–74, 80; and land clearing, 13, 155, 156; and landscape, 73, 152, 366; shoreline, 73–74, 80
European settlement, 149–63, 167–86; and animal habitats, 101–2; and Bay physical characteristics, 80–81; and birds, 322, 323–24, 327–28, 329; and disease, 127–31, 132, 137, 138–39, 140–42; and ecosystem, 192, 355, 362, 363, 364; and fish, 204; and forests, 51, 53, 55, 57, 58, 154, 162–63; and gardens, 249–52; health of, 140–42, 156; land use practices of, xix, xxi–xxii, 167; and Native Americans, 117–21, 124, 137–40, 221, 234, 235, 237
extinction: of birds, xxi, 86, 98, 101, 323, 333, 336, 347, 356, 360, 368; defined, 84–86, 325; long-term trends in, 102–4; of mammals, xxi, 98, 160; prehistoric, 81, 86, 87, 99, 103, 362; of SAV, 57–58
extirpation: of birds, xxi, 101, 103, 323, 328, 329, 332–33, 336; defined, 85, 325; of mammals, xxi, 98–99, 102; of SAV, 337

Fairbanks, W. L., 214
Fallan, Robert, 174
Fall Line, xxii, 7, 8, 10, 116, 119–20
Farmer's Institute, 319
Ferris, William, 269, 270
fertilizers: and agrarian reform, 289, 297n1; and bird populations, 332; and crop diseases, 309–10, 313; and ecosystem, xix, 13, 57, 193, 195, 279,

282–83, 368; oyster shells in, 75; and soil quality, 41, 311
fish: and agriculture, 238, 242, 243; anadromous, xviii, xxi, 113, 115, 118, 120, 159, 191, 196, 204–8, 215; benthic species of, xxi, 40, 57, 58, 81, 192, 194, 211–12, 215, 242; catch records of, 204–6; colonial accounts of, 194, 204, 212; and conservation efforts, 204, 205, 214–16; depletion of, xxi, 13, 204, 237, 242; and human activity, xxi, 40–41, 57, 367–68; non-native, 196, 214; and oxygen levels, 194; and pollution, 206, 207; as prehistoric food resource, 111–15, 118–20, 124, 203; and SAV, 343, 348; and timber trade, 159. *See also* oysters; shellfish
Fisher, George W., xxi, 1–14
fishing industry: and anadromous fish, 196, 205–8; colonial, 237; crabbing, 205, 238; and distribution changes, 201–13; and ecosystem, 195–201; and legislation, 201, 369; and market economy, 367; moratoriums on, 205, 215; and species distribution, 203, 204–6; technological improvements in, 191; in York County (Va.), 196, 243
Fithian, Philip, 173, 175
flooding, 19, 156, 176, 184, 202
Fontaine, John, 171, 173
Force, P., 212
forest fires: evidence of, 53, 170, 173; and land clearing, 25, 51, 115, 122, 152–53, 155, 176–77; and open land patterns, 183, 184. *See also* agriculture: slash-and-burn
forests: and agriculture, 57, 177; and birds, 326–34, 340–41, 347–48; and climate, 35, 40, 55; and environmental factors, 151–52; and European settlers, 40–41, 43, 149, 150, 161–62, 366; history of, 40–58, 111, 112, 355–56; and nutrient levels, 57, 58, 193–94; and overgrazing, 161, 237; patterns of, 55, 57, 171–74; restoration of, 41, 74, 123, 333; surveys of, 179–83. *See also* deforestation; timber resources
fossil records, 86, 87, 88–97
Frederick County (Va.), 168, 172–73, 177–86
freshwater discharge: and climate, 23; effects of, 67–70; and fish, 202–3, 207; and human activity, 60, 61, 65; levels of, 62–65; and oxygen levels, 76–78
Fry, Joshua, 175
Furnace Bay, 53
fur trade, 102, 160, 249

Galloway, Beverly T., 319
gardens: and Africans, 270–74; colonial, 252–60; eighteenth-century, 263–68, 271; and environment, 195, 252, 254–56; nineteenth-century, 274–76; non-native plants in, 249, 255–56, 262, 266–68, 273; plant species in, 263–69, 271; post-revolutionary, 268–70; and settlement, 249–52; and wealth, 251, 252, 262–63, 267–68; weedy plants in, 258–60
Gardner, William, 112, 113
Gates, Sir Thomas, 253
geology, 1–14, 41, 42–43
girdling. *See* agriculture: slash-and-burn
glaciation, xviii, 1, 3–9, 15, 28–34, 50, 86–87, 360
global warming, xxi, 13–16, 27, 32–36, 72, 103
Gooch, William, 169
Goode, George, 198, 205
Grant, Ulysses S., 317
Great Wicomico River, 215
Gross, Grant, 192
Guilday, J. E., 102
Gunpowder River, 65, 73
Guy, William, 268

habitat changes: of birds, 322, 323–25, 328–31, 333, 344, 345, 347, 348–49; and extinctions, 87, 98–99, 368; and human activity, 81, 371; in Pleistocene period, 97–99; of shellfish, 112; of waterfowl, 336–38
Hamill, W. S., 214

Handbook of North American Indians, 136
Hariot, Thomas, 194, 204, 212
Haven, Dexter, 210
Havre de Grace (Md.), xx, 66, 73
Heinz, G. H., 341
Hennessey, Timothy, 215
Heppenstall, C. A., 102
Herrman, Augustine, map by, 220–21
Hoffman, Ronald, 279–97, 364, 365
Hofstra, Warren R., xxii, 167–86
hog cholera, 304, 306, 314, 315–16, 317, 319
Holocene period, 83, 84, 98, 99–102
Hopkins, Sewell, 214
horticulture. *See* gardens
hunting: and birds, 322, 323, 328, 332, 335–36, 342, 344, 345; effects of, xix, 101; by European settlers, 237; and extinctions, 98–99, 102, 360, 362; laws on, 162, 342; and market economy, 368; by Native Americans, 121; by Paleo-Indians, 100, 109, 110–11; and slash-and-burn practices, 122; and waterfowl, 334–35; of wolves, 160–61; in Woodland period, 113–14, 115, 117–20

Indian old fields, 171, 175
industry: and anadromous fish, 207; and ecosystem, 60, 101, 192, 195, 368; and metal deposits, 2, 10–13, 43; oyster shell use in, 75
Ingen-housz, Jan, 309
Ingersoll, Ernest, 209
Iroquoian tribe, 131, 136, 138

James River, 5, 62, 63, 75, 114, 141, 156–57, 208
Jamestown settlement, 140–41, 253, 362
Jefferson, Peter, 175
Jefferson, Thomas, 19–21, 25, 307
Johnson, Paula, 213
Johnston, Joseph, 315
Jones, Hugh, 171, 251, 255, 258
Joppa Town, 73, 192
Josselyn, John, 256–57
Juhle site (Charles County, Md.), 134, 135

Kennedy, Victor S., 191–216, 242
Kent, Bretton, 198
Kercheval, Samuel, 170, 182
King, John, 212
Koch, Robert, 318, 319
Kutzbach, John E., 15–36

labor systems: and agrarian reform, 280–81, 297; and agricultural practices, xix, 118, 220, 222, 234, 239–41, 243, 363; and environmental history, 364, 365; and land clearing, 155, 280–81, 312; and settlement patterns, 222, 226–28, 231–33, 236; and tobacco culture, xix, 284, 286–87. *See also* slave labor; tenancy system
Land, Aubrey, 282
land clearing: and agriculture, 155–56, 159, 177, 241–42, 289, 305; and birds, 326–29; and environment, 192, 368; by European settlers, 150–51, 154–55, 173–74; and fish, 201–2, 207; and labor systems, 155, 280–81, 312; by Native Americans, 119, 122, 123, 175; and settlement patterns, 366; and soil erosion, 13, 155, 156; and wildlife, 160. *See also* agriculture: slash-and-burn; deforestation; forest fires
land cover. *See* plants
land grants: and settlement patterns, 186, 224, 225, 229, 231; survey records of, 177, 179, 180, 183
Landsberg, H. E., 21
landscape: colonial accounts of, 170–77, 185–86; colonial survey records of, 168, 177–86; history of, 355–57, 364, 366–67. *See also* forests; gardens
land survey records, 168, 169, 177–86
land use: and aquatic life, 40–41, 57; and birds, 323, 340–41, 345–46, 348; and deforestation, 40–41; and environmental history, 364; and European settlers, xix, xxii, 167, 186; and market economy, 305–6; and stream-

flow, 23; and tenancy system, 283–86; and tobacco culture, 286–87; and tree species, 53, 55, 57–58. *See also* agrarian reform; agriculture
Langley, Batty, 268
Lawrence, Thomas, 258
Lawson, John, 255, 263, 265–66
Lederer, John, 171, 174
Lee, D. S., 102
Lee, Robert E., 315
Liebig, Justus von, 309
Little Gunpowder River, 73
Love Point, 73
Lubber, Jack, 273
Lynch, James F., xxiii, 322–50
Lynnhaven Bay, 114

Madison, James, 19
Magothy River, 53, 54
management practices: adaptive, 81; and animal populations, 104; and ecosystem, 370–73; and oxygen levels, 75–78, 194; and oyster harvesting, 209, 211. *See also* conservation programs
A Map of Virginia (Smith), 19
market economy: and agriculture, 220, 241–44, 279–83, 366, 368; and crop diseases, 304–5, 306, 309, 313; and ecosystem, 358, 363, 365–69, 371, 372; and forest resources, 151, 155, 156–58, 160, 162, 242; and marine resources, 204–5, 243–44; and natural resources, 10–14; and passenger pigeons, 335, 360–61; and plants, 273–76; and waterfowl, 334, 336; and wildlife populations, 237; in Woodland period, 115–16, 120
Martin, Paul, 87
Maryland: deer hunting laws in, 162; environmental studies in, xxiii; farms in, 304, 312; fishing laws in, 201; land policy in, 168–69; oyster harvesting in, 210–11; oyster legislation in, 75, 201, 210; settlement patterns in, 221, 224–34
Maryland Game and Fish Commission, 336
Maryland Historical Society, 268
Maryland State Archives, 273
Mather, Cotton, 143
Mauch Chunk (Jim Thorpe, Pa.), 12
Mauzy, John, 180
Maycock Point site (Va.), 114
McCannon, James, 270
McGuire, Diane, 263
Mecklenburg, C. W., 134
Medieval Warm Period, 25, 53, 115
Menard, Russell, 305
Mesozoic period, 4, 5, 34
metal deposits, 2, 10–13, 43, 53
Michaux, François André, 161
Michel, Louis, 171, 209, 212, 215
Migratory Bird Treaty Act, 334
Miller, Henry, 109–24, 192, 196, 203
Miller, Naomi F., 258
mining, 12–13, 40, 195
Mitchell, James, 212
Mitchell, Robert, xxii, 167–86
Mobjack Bay, 198
Monticello, 19, 25
Moss, James, 273
Mountford, Kent, 191–216, 242
Mount Vernon, 25
Mount Vernon (painting; Ropes), 269

Nanticoke River, 53, 55, 115
Nanticoke tribe, 131, 136, 137, 139
Native Americans: agriculture of, 123, 250, 254, 266, 281; in Archaic period, 111–13; and birds, 326–28; and Chesapeake Bay, 123–24; culture of, xix; diseases in, 129, 130, 137–40, 143; and ecosystem, 109, 110, 192, 360, 362–63, 365; and European settlement, 117–21, 124, 137–40, 221, 234, 235, 237; fishing by, 195–96, 197, 204; health of, 100, 124, 127, 129, 132–35; and open land patterns, 175–76, 184; oyster harvesting by, 74, 112, 134, 196; populations of, 121, 131–32, 136–37; settlements of, 203; in Woodland period, 109, 113–21, 132–33
Natural History of Carolina, Florida, and the Bahama Islands (Catesby), 261

navigation, 72–73, 74, 191–92
Newell, Roger, 194, 211
New-England Rarities Discovered (Josselyn), 256
Nichol, A. J., 210
non-native plants: and environment, xxi–xxii, 213–14, 255–56, 258, 363, 364, 372; in gardens, 249, 255–56, 262, 266–68, 273; impact of, 238; import of, 252–53, 262–63, 274; and sediment dating, 46
Norfolk (Va.), 157
Northern Neck Proprietary grant, 168, 169
nutrient levels: and agriculture, 41, 193, 242; and aquatic habitats, 192–95, 214, 215; and forests, 57, 58, 193–94; and freshwater discharge, 64, 65; and oxygen levels, 75, 77, 78, 194; in sediments, 44–45; and slash-and-burn practices, 123; and water quality, 193–94, 214, 215

Odum, William, 195
open land patterns, 174–77, 179, 182, 183–85, 186
Opperman, Tony, 114
oxygen levels: and forest changes, 57, 58; and freshwater discharge, 64; and human activity, 80–81; and management practices, 75–78, 194; and nutrient levels, 75, 77, 78, 194; and oysters, 61, 81, 215
oysters: catch records on, 210–11; changes in, 196, 198; conservation programs, 75, 201, 210, 211, 215; depletion of, 13, 209–13; disease in, 209, 211; distribution of, 203–4; and European settlement, 74–75, 238; and fishing technology, 199, 200; history of, 209–13; and market economy, 313, 366, 367; and Native Americans, 74, 112, 134, 196; and navigational hazards, 191–92; and oxygen levels, 61, 81, 215; and salinity, 74, 80–81, 203; statistics on, 205, 206; and water quality, xxi, 194, 211; and Woodland peoples, 113–18, 123

Paleo-Indian period, 109, 110–11
Paleozoic period, 3–7, 11, 15, 33
Paradiso, J. L., 102
Parkinson, John, 261
Parnell, J. F., 102
Pasteur, Louis, 317–18, 319
Patuxent River, 73, 113, 305
Pennsylvania, xxiii, 12, 168
Percy, George, 250
Perry, M. C., 341
Phillips, J. C., 334
Piankatank River, 215
Piedmont Indians, 116, 120
Piedmont Plateau: birds in, 328, 333, 336; forests in, 41, 151, 152, 153, 159; landscape of, 7–8, 10, 161, 169, 174, 183; Native Americans in, 133–34; sediment deposition in, 5, 42, 43; settlement patterns in, 156
Pierce, Joan, 253
Piscataway (Md.), 115, 192
Piscataway (Conoy) chiefdom, 120
plankton, xxi, 58, 66, 81
plants: and climate, 15–16, 28, 31, 35; at European settlement, 167, 168, 170, 171, 173; export of, 249, 260–61, 274–75; and forests, 51, 53, 57, 152–53, 154; import of, 262–63, 271, 274; markets for, 274–76; range changes in, 103; restoration of, xxii; and sediment stratigraphy, 46–50, 53; weedy, 258–60, 364; wildflower species, 250. *See also* non-native plants; submersed aquatic vegetation; tree species
plate tectonics, 3, 33–34
Pleistocene overkill theory, 87
Pleistocene period: animal habitats in, 86–99, 100; climate of, 83; extinctions in, 103, 362; sediment deposition in, 4, 8–9, 44
plowing practices: and agrarian reform, 287, 289–90, 292, 296–97, 299n11; and birds, 332; and crop diseases, 307,

310; and deforestation, xix, 41, 155–56; and environment, 241–42, 279, 282–83, 364; and landscape, 72, 294, 355; and market economy, 366
Pocomoke Sound, 74
pollen stratigraphy, 35, 75, 170, 254; and environmental history, 363, 364, 366; and forest history, 40, 45–57, 51, 155, 355; and glacial landscape, 28–30, 31, 33
pollution: from agricultural runoff, 65, 78, 81, 193; and aquatic life, 205, 207; and birds, 322, 341–42, 344–45, 347, 348; and chemical disposal, 44–45, 195; and circulation patterns, 66, 78–80, 195; and ecosystem, 192, 195, 368, 371; and refinery cleanup, 10–11; and water quality, xix, xxi, 12, 13, 58, 78, 305. *See also* conservation programs; global warming
Pope's Creek site (Md.), 113, 115
Port Tobacco, 73, 192
Pory, John, 153
Potomac River, 5, 50, 73, 343; aquatic life in, 113, 205, 208; freshwater discharge of, 62, 63; and mining, 13; prehistoric settlements at, 112, 113
Potter, Stephen, 115, 116, 120
Powhatan chiefdom, 120
Powhatan confederacy, 362
Prince, Robert, 268
Principio Furnace, 11
Pritchard, Donald W., 60–81, 194, 195

Radcliff, Lewis, 213
Raney, Edward, 198
Reveal, James L., xxii, 238, 249–76
Revelle, Roger, 35
Rhoads, S. N., 102
Richmond (Va.), xxii, 7, 8
Rigg, Robert, 180
Robbins, C. S., 323, 325
Robert Cole's World: Agriculture and Society in Early Maryland (Carr, Menard, and Walsh), 305
Robison, William, 170
Rountree, Helen, 117
Ruffin, Edmund, 170, 309
Ruffin, Frank, 315
Rutherford, Robert, 180
Rutherford, Thomas, 180

salinity, 19, 343; and estuarine circulation pattern, 68–70; and oxygen levels, 76–78; and oysters, 74, 80–81, 203
Salmon, Daniel, 319
Sandy Point, 73
Sarudy, B. W., 269
SAV. *See* submersed aquatic vegetation
Schubel, Jerry R., xxi, 1–14, 60–81, 194, 195
Schuylkill Canal (Pa.), 12
Scott, Joseph, 310
Scott, Upton, 274
sea level, 44, 50, 66, 69, 71, 203; and climate, 35–36; and glaciation, 9, 13–14, 28, 31–32, 61; and landscape, 8–9; and shoaling, 74, 80
sea nettle jellyfish, 191–92, 212–13
sediment deposition, 19, 332; and agrarian reform, 280, 289, 293–94, 295–97, 298n7; and ecosystem, 192–93, 195, 215; and former ports, 73, 192; and glaciation, 3–9, 11; and land use practices, 57–58; and tree distribution, 41–43. *See also* shoaling
sediment stratigraphy, 44–45, 48–50; and benthic species, 211–12; and chronologies, 46–47; and climate history, 25; and forests, 40, 50–53; and prehistoric settlement patterns, 115, 122; and shellfish habitat changes, 112
seed stratigraphy, 48–50, 50, 55, 58, 86, 98
settlement patterns, 7, 168–69, 174, 220–44, 366; and agriculture, 220–24; and birds, 331–32; and ecosystem, xix, 169–70, 185–86, 367, 368–69, 371; and environmental history, 363, 364, 365; of European settlers, 167–70, 186, 224–36; of prehistoric peoples, 111, 112–15, 117, 133, 331–32; and water transportation, 221–22

sewage systems, 78, 370
Shannon site (Montgomery County, Va.), 134
Sharrer, G. Terry, 84, 238, 304–20
Shawnee-Minisink site (upper Delaware River valley), 111
shellfish: catch records of, 205, 206; distribution of, 203–4, 268; and fishing technology, 199, 200; prehistoric harvesting of, 109, 112, 113, 117, 118, 123. *See also* oysters
Shenandoah River, 168, 169
Shenandoah Valley, 7, 167–86
shoaling, 73, 74, 80
Shriver, Andrew, 313
Silent Spring (Carson), 370
Silver, Timothy, xxii, 149–63, 201, 237
slave labor: and agrarian reform, 280–81, 291, 292, 293, 297, 299nn10–11, 364; and agriculture, 220, 225–29, 233, 235, 237, 239, 240, 243, 312; and gardens, 270–71; and land clearing, 155, 157, 158; and land use, 283; and settlement patterns, 234; and timber market, 158. *See also* Africans
Smith, Bruce, 114
Smith, Hugh, 205, 208
Smith, John, Captain, 19, 72, 117, 215, 256
Smyth, John, 173, 308
Snow, Dean R., 139
soil quality, 10, 120, 184, 368; and agrarian reform, 280, 281; and agriculture, xix, 123, 193, 238, 239–40, 242, 311, 319, 363; and fallow field system, 311, 312; and fertilizers, 41, 311; and settlement patterns, 186, 221–22, 226–29, 236
Soldier's Delight, 43
Sparrow Point, 11
spawning fish. *See* fish: anadromous
Spectator (London), 261
Spesutie Island, 73
Spotswood, Alexander, 168, 171
St. Mary's City (Md.), 196, 198, 256
St. Mary's County (Md.), 196, 222–23, 225–28, 231, 232–33, 243
Staver, Lori, 213
Steadman, David W., xviii, 81–104, 323
Stevens, Sanderson, 122
Stevenson, Charles, 198, 205, 207, 209, 210
Stevenson, Court, 193, 213
Stewart, Michael, 113
Stewart, R. E., 323
storms: cyclonic, 16, 23, 184; tropical, 19, 61, 62, 64, 79–80, 152, 192, 213
Strachey, William, 117, 121, 150, 154, 195
Strickland, William, 310
submersed aquatic vegetation (SAV): and birds, 322, 329–30; and fish, 343, 348; and human activity, 337, 371; and land use practices, 57–58; and sediment deposition, 192, 193, 215; and waterfowl, 334–35, 336–38, 342, 343, 348
Susquehanna Flats, 73
Susquehanna River: anadromous fish in, 207, 208; and coal industry, 11–12; dams on, 74; fish species distribution in, 204; and forests, 84, 151; freshwater discharge from, 62–65, 76, 202–3; mining effects on, 13; physical characteristics of, 5, 28, 61; and sediment deposition, 8–9, 10
Susquehannock tribe, 136, 137, 139
Swan Point, 73, 80

Tangier Sound, 74, 210, 211
tenancy system: and agrarian reform, 280–81, 291, 297, 298nn4–5, 299n11, 300n13, 364; and agricultural depression, 280, 297n1; and land use, 283–87
tidal marshes, 41, 42, 111
tides, 60, 61, 65–67, 71, 72
timber resources: and environment, 13, 40, 186; and market economy, 156–59, 242, 249; and tobacco culture, 286–87
tobacco: and agrarian reform, 279, 280–81, 291, 292, 297n1, 299n11; and colonial agricultural systems, 235, 237, 239–40; and deforestation, 41; and disease, 304, 306, 310–12; and environment, 241, 286–87, 363, 365, 366; and gardens, 253–54; and labor systems, xix, 284, 286–87; and market

economy, 243, 282; and settlement patterns, 220–21, 225–34
Tollifero site (Va.), 132–33
Tradescant, John the Elder, 260–61
Tradescant, John the Younger, 260–61
transportation systems, 8, 12; and bird populations, 332, 360; and ecosystem, 40, 168, 364, 367, 368, 369, 371, 372
tree species, 172, 329, 356; and climate, 23, 26; common names of, 180, 181–82; and disease, 41, 46, 55, 57, 306, 355; exportation of, 260–61; extinctions of, 55, 57; and sediments, 41–44, 46, 50–51; surveys of, 179–83; types of, 50–55, 57, 58, 84
tributaries, xx, 23, 60, 70, 71, 192, 203. *See also specific rivers*
Truitt, Reginald, 213
Turner, Randolph, 121
Tyron Palace (New Bern, N.C.), 267
Tyson, Isaac, 10

Ubelaker, Douglas, 124, 127–44, 306, 365
Upper Marlboro, 73

vegetation. *See* plants
Verrazzano, Giovanni da, 149, 162
vertical stratification, 64, 65, 76–78, 80–81, 194
Virginia, 28, 195; farms in, 304, 312; land policy in, 168–69; laws in, 161–62, 256; oysters in, 75, 210–11, 215; settlement patterns in, 221, 222, 234
volcanic activity, 25, 27, 34

Walsh, Lorena S., 193, 196, 220–44, 305, 364
war, 116, 120, 133, 304, 306, 307, 312–17
Warden, D. B., 323
Waselkov, Gregory, 112, 115, 116
Washington, D.C., xxii, xxiii, 7, 8, 19, 367
Washington, George, 19, 25
water quality: and birds, 329–30, 332, 349; and land use practices, 57, 123; and nutrient levels, 193–94, 214–15; and oysters, xxi, 81; and pollution, xix, xxi, 12, 13, 58, 78, 305; and SAV, 337
wealth: and agriculture, 237, 240, 241, 242–43, 281; and gardens, 251, 252, 262–63, 267–68, 270, 274; and settlement patterns, 224–34, 236, 244
Webb, Thompson, III, 15–36
Webster, W. D., 102
Weiss, Jacob, 12
Weld, Isaac, 173, 174, 175, 310
wetlands, 41, 234, 328–29, 341
Wharton, James, 195, 196, 204, 209
Whetstone Point (Md.), 11
White, Gilbert, xv
White Oak Point site, 112–16, 123
Whyte, Thomas, 114
Wilson, Alexander, 334
Winchester (Va.), 174, 175
witness markers, 168, 170, 177, 179–83
Woodland period, 109, 110, 113–23, 132–33
Wright, Harrison, 204
Wyoming (Pa.), 204

Yentsch, Anne E., xxii, 238, 249–76
York County (Va.), 196, 234, 235, 236, 243
Younger Dryas period, 31, 51

Library of Congress Cataloging-in-Publication Data

Discovering the Chesapeake : the history of an ecosystem / edited by
Philip D. Curtin, Grace S. Brush, and George W. Fisher.
p. cm.
Includes bibliographical references and index.
ISBN 0-8018-6468-2 (pbk. : alk. paper)
1. Human ecology—Chesapeake Bay Region (Md. and Va.)
2. Natural history—Chesapeake Bay Region (Md. and Va.) 3. Nature—
Effect of human beings on—Chesapeake Bay Region (Md. and Va.)
I. Curtin, Philip D. II. Brush, Grace Somers. III. Fisher, George
Wescott, 1937–
GF504.C54 D47 2001
975.5'18—dc21

00-042405